W0079490

MECHANICAL PROBLEMS IN MEASURING FORCE AND MASS

IMEKO TECHNICAL COMMITTEE
ITC Series No. 8

ITC Series include publication by the 15 Technical Committees of the International Measurement Confederation (IMEKO):

Higher Education (TC1)
Photon-Detectors (TC2)
Measurement of Force and Mass (TC3)
Measurement of Electrical Quantities (TC4)
Hardness Measurement (TC5)
Vocabulary Committee (TC6)
Measurement Theory (TC7)
Metrology (TC8)
Flow Measurement (TC9)
Technical Diagnostics (TC10)
Metrological Requirements for Developing Countries (TC11)
Temperature and Thermal Measurement (TC12)
Measurements in Biology and Medicine (TC13)
Measurement of Geometrical Quantities (TC14)
Experimental Mechanics (TC15)

Inquiries ITC-series:

IMEKO - Secretariat
H-1371 Budapest, P.O. Box 457
Hungary

Editor ITC 8:

H. Wieringa
c/o TNO-IWECO
P.O. Box 29
2600 AA Delft
The Netherlands
phone: +31/15 608608
telex: 38192 iweco nl

Mechanical Problems in Measuring Force and Mass

Proceedings of the XIth International Conference on Measurement of Force and Mass, Amsterdam, The Netherlands, May 12–16, 1986
Organized by: Netherlands Organization for Applied Scientific Research (TNO) on behalf of IMEKO – Technical Committee of Measurement of Force and Mass

edited by

H. WIERINGA

TNO-IWECO
Delft, The Netherlands

1986 **MARTINUS NIJHOFF PUBLISHERS**
a member of the KLUWER ACADEMIC PUBLISHERS GROUP
DORDRECHT / BOSTON / LANCASTER

Distributors

for the United States and Canada: Kluwer Academic Publishers, 190 Old Derby Street, Hingham, MA 02043, USA
for the UK and Ireland: Kluwer Academic Publishers, MTP Press Limited, Falcon House, Queen Square, Lancaster LA1 1RN, UK
for all other countries: Kluwer Academic Publishers Group, Distribution Center, P.O. Box 322, 3300 AH Dordrecht, The Netherlands

Library of Congress Cataloging in Publication Data

ISBN-13: 978-94-010-8464-2 e-ISBN-13:978-94-009-4414-5
DOI: 10.1007/978-94-009-4414-5

TABLE OF CONTENTS

Preface

Nowadays electrical force transducers, in which various electrical conversion principles are applied, are widely used. Transducers for forces from 1N till 10 MN are commercially available and used for industrial as well as research purposes. They not only serve to measure forces but also for weighing purposes. Directly converting a force into an electrical signal is not possible. This must be done step by step. For instance, in a strain gauge based transducer the conversion chain is: force - stress - strain - resistance change - bridge output. At every conversion point in this chain parasatic influences can interfere with the results and may cause a loss in accuracy.

To surmount the problems related to obtaining sufficient accuracy and reliability for these transducers, much research has been done all over the world in the past 35 years. As a result, new materials, new techniques, improved constructional designs and compensation circuits have been found to overcome the parasitic influences. The object of the IMEKO Conferences on behalf of the Technical Committee on Measurement of Force and Mass (TC-3) is to exchange experiences, to discuss problems and to obtain knowledge about practical applications.

In this book the papers have been collected that will be discussed at the 11th International Conference on Measurement of Force and Mass. The topic of this conference is "Mechanical Problems in Measuring Force and Mass". The TC-3 members express their thanks to the authors who took the trouble to forward papers comprising their knowledge and experiences. The TC-3 members hope that these contributions will help you to solve your problems as to Measurement of Force and Mass and to improve your knowledge about these measurements. Moreover, the TC-3 members expect the Conference to contribute to the development and deepening of personal contacts between the participants form all over the world.

Furthermore they like to express their thanks to the Paper Selection Committee members for their evaluation of the papers contributed and for their advice and support in drafting the programme. Acknowledgement of thanks is also expressed to the Netherlands Organization for Applied Scientific Research (TNO) and in particular to the fellow-workers of the TNO Corporate Communication Department and TNO-IWECO; Institute of Mechanical Engineering. Without their efforts this Conference would not have been possible.

May 1986
H. Wieringa
Chairman IMEKO Technical Committee
on Measurement of Force and Mass.

Message to the Readers

Allow me to greet you all, dear readers of this book, which is rich in information covering our field of common interests. We may have met personally at the 11th IMEKO Conference on Measurement of Force and Mass in Amsterdam and witnessed high-level technical sessions and vivid discussions from 12 to 16 May 1986. This will remain a memorable event for me. After 18 years of activity I have to resign from the fuctnion of scientific secretary of the IMEKO Technical Committee on Measurement of Force and Mass (TC3) on account of a higher position within the confederation.

Close co-operation with chairmen like dr. K. Hild (FRG) and then ir. H. Wieringa (NL) was a great scientific experience and I feel honoured by their friendship, too. The same goes for many colleagues from many countries who have regularly attended TC3 conferences and helped to disseminate up-to-date knowledge on our subjects. The proceedings have been published and distributed widely: the response from the industry, the user of weighers and the national offices of measures proves that our work was not in vain. The 11th Conference is another milestone in the history of the Committee. Thanks and appreciation are due to the Dutch colleagues, first of all to Mr. Wieringa, of TNO and to the Royal Institution of Engineers in the Netherlands, Division for Automatic Control, member organization of IMEKO.

In my office, as Secretary General of IMEKO, I am responsible for carrying out the resolutions of the General Council (comprising 30 member organizations from countries of the five continents), for the activities of 15 technical committees and for the success of the forthcoming 11th World Congress, scheduled for October 1988 in Houston/Texas. On the occiasion of the Congress we plan to celebrate the 50th anniversary of the strain gauge. I hope you will be able to offer your co-operation - please contribute, because otherwise none of our goals can be fully accomplished.

May 1986
Assoc. Prof.Dr. Tamás Kemény
Secretary General
International Measurement Confederation

IMEKO TC-3 ACTIVITY IN BRIEF

The IMEKO Technical Committee "Measurement of Force and Mass" has been proposed to be founded during the 4th IMEKO Congress in Warszawa, Poland, July 1967. The participants of the Round Table Discussion entitled "Strain Gauge Technique in Industrial Weighing", chaired by Prof. Peter Stein form Arizona, USA, and Mr. Tamás Keméņy from Budapest, Hungary, decided to continue an exchange of experience in the field of force and mass measurements. The proposal discussed by the IMEKO Secretariat was approved by the General Council of the International Measurement Confederation IMEKO.

Thus was established the Technical Committee TC-3 "Measurement of Force and Mass, which was appointed to the following tasks:

- the organization of symposia on still unpublished results of experiments and practical experience,
- the helding of round table discussions on different topics,
- the promotion of the exchange of experiences among IMEKO Member Organizations.

The following meetings have been organized up till now:

1. Braunschweig, Federal Republic of Germany, 1969: "Precise Measurement of Force and Weight with Strain Gauge Techniques"
2. The Hague, The Netherlands, 1971: "The Characteritic Properties of Force Measuring Devices and Electromechanical Weighing Machines"
3. Ostrava, Czechoslovakia, 1972: "The Measurement of Force and Mass in Controlled Systems"
4. Udine, Italy, 1974: "Recent Developments in Force Measuring Devices"
5. Szeged, Hungary, 1974: "Up-to-date Verifiable Weighing Machines"
6. Odessa, Union of Soviet Socialist Republics, 1977: "Industrial Weighing"
7. Braunschweig, Federal Republic of Germany, 1979: "Measurement of Force and Mass"
8. Kraków, Poland, 1980: "Weighing Technology"
9. London, United Kingdom, 1983: "Weighing and Force Measurement in Trade and Industry".
10. Kobe, Japan, 1984: "Recent Advances in Weighing Technology and Force Measurement"

During IMEKO Congresses, TC-3 also organized the following round table discussions:

IMEKO VI, Dresden, German Democratic Republik, 1973:
"Force Transfer to Force Measuring Devices"
IMEKO VII, London, United Kingdom, 1976:
"Measurement of Force and Mass"

IMEKO VIII, Moscow, Union of Soviet Socialist Republics, 1979:
"Practical Problems Concering Force Measuring Devices".
IMEKO IX, Berlin-W, 1982:
"Problems in Respect to the Production and Use of Resistance Strain Gauge Force Transducers as Transfer Standards"
IMEKO X, Prague, Czechoslovakia, 1985:
"Transportable Weighing Systems".

Working groups created by IMEKO-TC3 are:

- Terminology and Calibration Procedure of Load Cells (A. Bray)
- Force Sensors of Robots (R.A. Mitchell)
- Long Term Stability of Strain Gauge Load Cells (R.F. Jenkins)

The first TC-3 chairman was Dr. Kurt Hild, now retired. At present experts in force and mass measurements from 19 countires act as technical committee members:

H. Wieringa (The Netherlands)-Chairman
T. T. Kemény (Hungary)-Secretary
A. Bray (Italy)
M. Dubois (France)
A. Gizmajer (Poland)
R.F. Jenkins (United Kingdom)
K. Hasche (German Democratic Republic)
G. Iordacheschu (Roumania)
A. Larsson (Sweden)
R.A. Mitchell (United States of America)
J. Lukas (Czechoslovakia)
T. Ono (Japan)
M. Peters (Federal Republic of Germany)
D. Prokić (Yugoslavia)
A. Pusa (Finland)
V.M. Sitnichenko (Union of Soviet Socialist Republics)
P.K. Stein (United States of America)
J. Thomas (Denmark)
Tsai Cheng-Ping (China)
J.C. Vanderschueren (Belgium)
W. Weiler (Federal Republic of Germany)

The papers for the 11th Conference at Amsterdam were evaluated by J.A.J. Basten (The Netherlands), R.F. Jenkins (United Kingdom), T. Kemény (Hungary), T. Ono (Japan) and H. Wieringa (The Netherlands).

DEVELOPMENT OF MULTICOMPONENT FORCE TRANSFER STANDARDS BY ONERA FOR FRENCH BNM

Daniel GIRARD
Office National d'Etudes et de Recherches Aérospatiales (ONERA)
92320 CHATILLON (FRANCE)

SUMMARY

This paper describes three force transfer standards built for the comparison of the various force standard machines developed in France by ONERA for BNM (National Bureau of Metrologie).

The first one measures the compression axial force up to 300 kN and three moments. The second one, of same capacity in traction, measures the six components, and the third measures only the axial force up to 50 kN in traction but is unsensitive to the other components.

Their sensitivity makes it possible to appreciate variations of axial force of a few 10^{-5} of their maximum value and for the two 300 kN dynamometers, force, off-centering of about 0,3 micron for the first one and 0,02 micron for the second.

1. INTRODUCTION

Since the recommendations given by the Technical Committee IMEKO "Measurement of force and mass" in The Hague in september 1971, one component strain gauge dynamometers have been generally used as force transfer standard in many countries, to qualify and compare various force standard machines and material testing machines [1 to 6]. But some defects in construction, machine assembly or connection between bench and dynamometer (off-centering, inclinations, faulty planeity, insufficient stiffness, backlash...) can generate, under the action of the applied force, parasitic components and erroneous measurements. To quantify errors and bring corrections on axial force, it is necessary to measure simultaneously the main component and the eventual parasitic components (fig. 1).

DETERMINING AND ELIMINATING OF THE INTERACTIONS

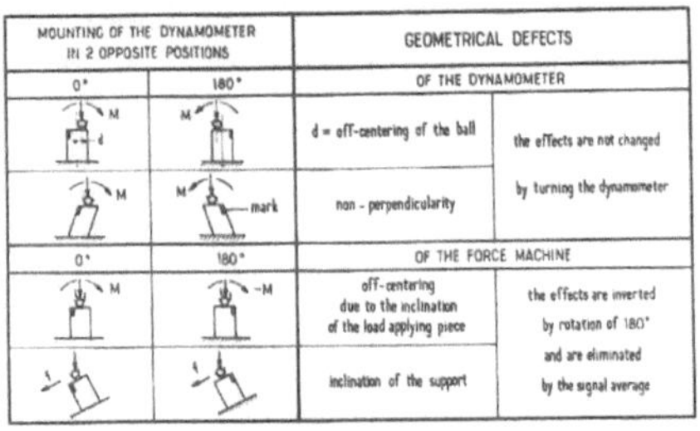

MOUNTING OF THE DYNAMOMETER IN 2 OPPOSITE POSITIONS		GEOMETRICAL DEFECTS	
0°	180°	OF THE DYNAMOMETER	
M, d	M	d = off-centering of the ball	the effects are not changed by turning the dynamometer
M	M, mark	non - perpendicularity	
0°	180°	OF THE FORCE MACHINE	
M	-M	off-centering due to the inclination of the load applying piece	the effects are inverted by rotation of 180° and are eliminated by the signal average
		inclination of the support	

FIGURE 1. Determining and eliminating

Wieringa, H. (ed), Mechanical Problems in Measuring Force and Mass.

That is the reason why, as early as 1975 the French National Bureau of Metrology (BNM) decided to develop multicomponent force transfer standards to qualify and compare the French force standard machines in order to increase the national level of quality. Their qualities (linearity, repeatability, stability v,s temperature) have to be such as the accuracy of the axial force measurement is better than 1.10^{-4} of the capacity. The experience acquired by ONERA in the field of dynamometry and the qualities of the various force machines operating at the Modane center [7 to 9], have led the BNM to entrust it with the study, equipment and qualification of a first then two others force transfer standards of different shapes and capacities.

The first dynamometer, studied and manufactured in 1976-1977 works in compression up to 300kN and measures four components : the main axial force and the three moments. Il was presented at the IMEKO.TC 3 conference at ODESSA in 1977 [10] [11]. It has been used in 1979 to compare the three French force standard machines, of capacities 250-300 kN approved by the BNM, belonging to the Laboratoire National d'Essai (LNE), Etablissement Technique Central de l'Armement (ETCA) and ONERA Modane. The results of this comparison were given at the 8th IMEKO.TC 3 at KRAKOW in 1980 [13]. Among the very interesting results, it was demonstrated that it would be better to use a dynamometer measuring six components in order to determine, completely, all parasitic efforts acting on the transfert standard and to obtain the best precision in the main force measurement.

Taking advantage of this experience, a second 300 kN transfer standard was studied in 1981. It works in traction and can measure the six components of efforts.

The third force transfer standard is defined by BNM to be used by manufacturers to calibrate their material testing machines. It is more simple and cheaper. It measures only the axial force up to 50 kN in traction and includes elastic flexures so that parasitic components do not affect the measurement. It was realised in 1980.

These three standard dynamometers are characterized by a structure designed by ONERA PARIS by means of the finite elements calculation. The architecture is realised by electro-erosion in a single block of metal. It is rigid along the longitudinal axis for the main component and well uncoupled in regard to the other components. The central part measuring the main force, works in shear that give the advantage of high rigidity, better sensitivity and linearity in regard to the other means of working (tension, flexion).

Experimental studies undertaken by ONERA several years ago [7, 8] had shown that the Z.85 WD V6 fast steel was the best material on creep and hysteresis point of view. But this alloy is very expensive and difficult to elaborate for great size dynamometers. That is the reason why the three force transfer standards were manufactured in the V.300 carbon steel (45 S.CD 6) well known by ONERA. According to the fact that transfer standards should be used only in one sense (traction or compression) the elastic defects are so tolerably reduced. The V.300 steel used was, vacuum cast, forged, then thermally treated so that ultimate strength be 1 600 M.Pa.

2. GENERAL CONDITIONS FOR CALIBRATIONS

The ONERA center in Modane has different force calibration machines able to one to six components. Forces and moments applied on dynamometers, thanks the machines described hereafter, are producted by 50 kg discs the masses of which are known within 0,2 g (4.10^{-6}), subjected to the earth gravity ($g = 9{,}8024\ m/s^2$ in Modane).

2.1. One component calibration machine B.1.2

This bench, of capacity 250 kN, of overall precision 3.10^{-5}, approved by the BNM, is sketched on figure 2. It is of mixed type (dead weight and lever) [11], [12]. The vertical forces can be transmitted to the dynamometer in two opposite directions : upward, through the lever (ratio 5) and downward, directly. This makes it possible to apply positive and negative forces (tension and compression) and to cross the zero without dismantling the dynamometer. Moreover, these forces can be exerted in several ways : either continuously, by increasing and decreasing values, or discontinuously, coming back to zero, or backwards (pilgrim step) after each loading step. These various possibilities allow a very fine study of the dynamometer hysteresis.

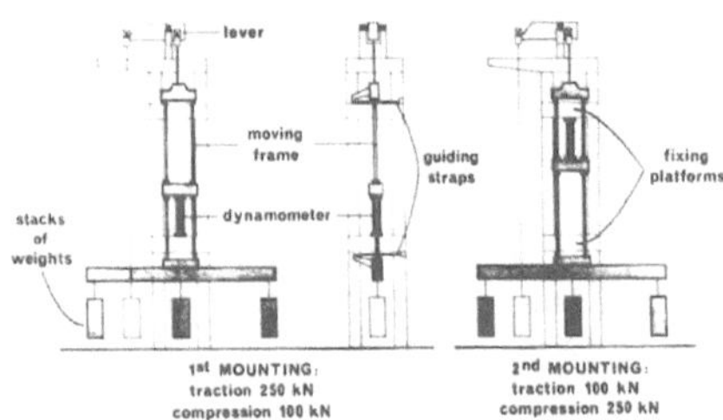

FIGURE 2. ONERA B1.d dead weight machine of 100-250 kN.

FIGURE 3. Mounting of 300 kN dynamometer on ONERA B1.d six component calibration machine.

2.2. Six-component calibration machine B.1.4

The three orthogonal forces and three moments are applied by the mean of this bench which allows the application, whether separately or simultaneously, of the six force components. These are applied perfectly independently from one another, with a 10^{-4} precision in both magnitude and direction. The maximum capacities of this bench are 100 kN for forces and 35 KNm for moments [9]. Figure 3 shows the 300 kN n° 2 dynamometer on B.1.4 bench.

Each of the six components and their fifteen two by two combinations had been applied on the three dynamometers.

2.3. Electric equipment and measuring instruments

As shown on the figure 4 the strain-gauge bridges of the dynamometer include different resistors in order to make thermal compensations (r_S, R_T) (zero, sensitivity) between 0 and 60°C and, for the main bridges, linearization by mean of resistors and semi conductor gauges R_L) and sensitivity adjustment (R_S).

During the calibration, all dynamometer strain gauge bridges are supplied with a continuous regulated voltage Ua of 8 Volts $\pm$ 0,04 mV ($\pm$ $5 \cdot 10^{-6}$). The two lead wires are supplemented by two other wires which permit the initial adjustement and the control, at any time, of the true value. The bridges output voltages are read with a resolution of 0,1 V (1 digit). For the three force transfer standards the sensitivity of the main bridge is adjusted in order to obtain around 2 mV/V for the nominal capacity, as many other dynamometers. With ONERA measuring instruments 2 mV/V represent 160 000 digits.

Sensitivity compensation is done by means of nickel resistors (R_T) connected into the bridge supply circuit and which values change with temperature. The measurement of bridge supply Up makes it possible to know the force standard temperature and to control its thermal stability during calibrations.

Two supplementary wires allow, by means of external standard resistances (Ro) an electric calibration of bridges and instrumentation and to determine the correspondence factor between different measurement devices of different laboratories.

3. FOUR COMPONENT FORCE TRANSFER STANDARD OF 300 KN

3.1. General description

The four component force transfer standard n° 1 measures the main axial force in compression up to 300 kN, the two bending moments and the torque. It is made on one cylindrical block (Ø 150, h 300 mm) (fig. 5).

The measuring central part includes four parallelipipedic bars, intermixed between four massive pillars, bound two by two with the two ends. The application of an axial force makes the bars work in shear. The four bars are marked N.S.E.W. as the four cardinal points. Bars are equiped with strain gauges electrically wired in wheatstone bridges, noted P,

FIGURE 4. Sketch of strain-gauge bridges. ►

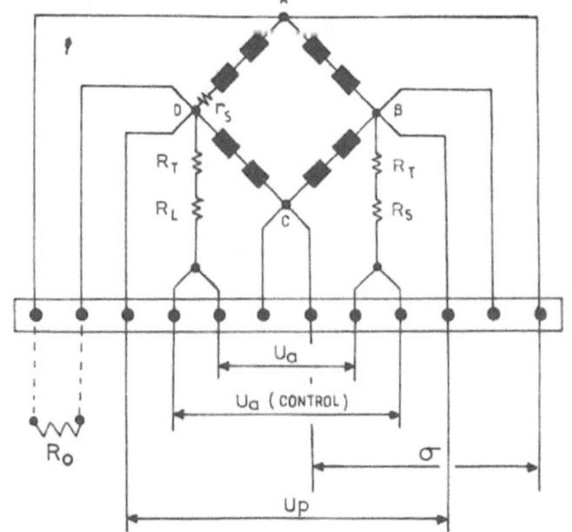

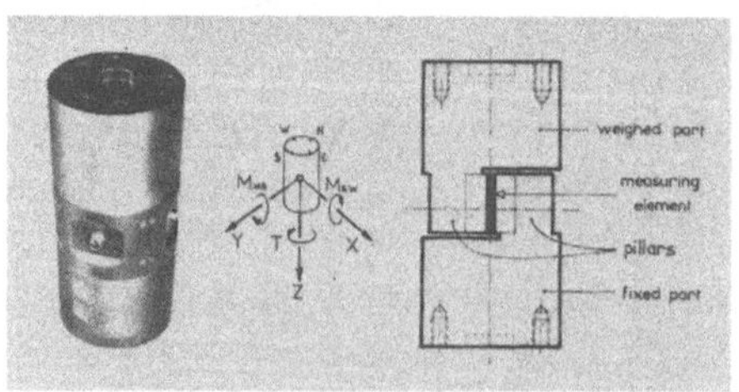

FIGURE 5. Four component force transfer standard: 300 kN n° 1.

M_{NS}, M_{EW}, T, which measure respectively the main axial force, the two bending moments around mutually perpendicular axes N.S and E.W and the torque around the vertical axis.

3.2. Preliminary calibrations

Many qualification tests have been carried out during the strain-gauging in order to adjust the dynamometric qualities of this force transfer standard.

The choice of gauges appropriated for the main bridge was made in order to minimize the signal creep and hysteresis. The residual creep, at full load, and relaxation, at the return to zero, is lower than 0.5 10^{-4} of the measurement range, for time intervals of two hours.

The main bridge has been linearised (2.3) ; its sensitivity coefficient of 460 digits per kN permits to detect axial force variation of 3 N (2 or 3 digits) i.e : 1 10^{-5} of the maximum value of 300 kN.

The sensitivity of the bending bridges permits one, to appreciate a moment variation of 0,1 Nm, which corresponds to an offcentring of the axial force at 300 kN of 0.3 micron.

The six component calibration has given the sensitivity coefficients of each of the four bridges as a function of the six force components. The linear interactions are relatively weak, the main bridge, in particular, is practically not interacted by the other components. The quadratic interactions are almost null.

3.3. Results obtained between 1977 and 1985

This force transfer standard has been calibrate in 1976-1977 in Modane. Then it has been used in 1979 to compare the three french force standard machines of 250.300 kN approved by the BNM [12].

The calibration has been checked again in june 1985 at ONERA Modane, before an international intercomparison at the European level, carried out by the community Bureau of Reference (BCR) of the Commission of the European Economic Communities, between France (LNE), Netherlands (TNO), Great Britain (NPL), and Italy (IMGC).

The axial sensitivity found is the same with an uncertainty of $0.2.10^{-4}$ as the value of the 1979 calibration. This testify to a very good fidelity of this force standard, calibrated on the same machine.

4. SIX COMPONENT FORCE TRANSFER STANDARD OF 300 KN

The second force transfer standard works in traction up to 300 kN and measures the six components (three forces and three moments).

4.1. General description

As the first one, this dynamometer is made of one cylindrical block (Ø 150 mm, h 300 mm). On each end two special pieces are added for the fixation on calibration benchs. They have two tapped holes (M.52 pitch 2) as many dynamometers of same capacity. It is presented on figure 6.

In the central part, four parallelipipedic bars have been machined, intermixed between a rigid rectangular frame and a central pillar. The application of an axial force makes the bars work in shear.

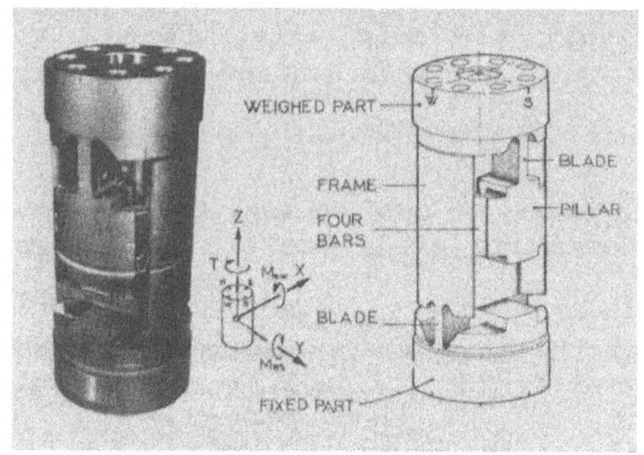

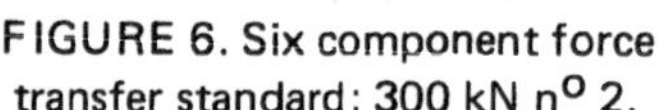
FIGURE 6. Six component force transfer standard: 300 kN n° 2.

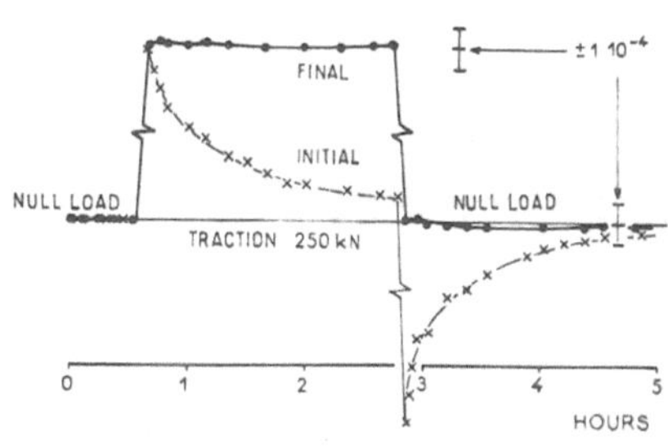

FIGURE 7. 300 kN n° 2 : creep of the axial bridge.

Four symmetrical vertical blades, two up and two down, perpendicular two by two are machined forming elastic pivots. The blades, stiff in the axial direction and flexible in the others, protect the central part from parasitic components and measure them by the mean of special extensometric equipment.

Measured stiffness are 5 550 kN/mm and 1 111 kN/mm respectively, for the central part and for the whole dynamometer. The central part is 1.2 stiffer than these of the first standard. The angular stiffness, in bending, is of 73.4 Nm per milliradian. It is 38 time more flexible than the first standard.

Each component is measured by strain gages (nominal resistance 350 ohms) bonded on the structure, and electrically wired as Wheatstone bridges.

The main bridge P , measuring the axial force Z is composed of suitable gauges, to reduce as much as possible creep and hysteresis. They are bonded on the four bars of the central part. The five other components are measured by five bridges located and wired on the elastic blades. On the higher one for X force M_{NS} bending moment and T torsion. On the lower ones for Y force and $M_{E.W}$ bending moment.

4.2. Adjustment of main bridge qualities

The elastic qualities of a force transfer standard are very important : creep and hysteresis of the dynamometer (metal and gauges bonded on it) fix the measuring precision [7].

Creep

Many experiments have been done for the first force transfer standard, to choose the most appropriate gauges for the main bridge in order to reduce, as much as possible, the signal creep at full load, and relaxation after unloading.

This second dynamometer is manufactured in the same metal thus, at first, the same gauges have been chosen (manufacturer, reference, alloy, support, sizes, resistance, ...) but they was not of the same fabrication batch. Results had not been as good as we could hope (fig. 7). Negative creep is rather important (4 to 5.10^{-4} in half hour). It has been necessary to look for other gauges and to carry out again many experiments to reduce as much as possible creep and to get results presented on figure 7 : $0.7.10^{-4}$ in two hours. Most part of those variations, which could be thermal effect, due to quick stress variations (thermo elastic effect) appear during the 5 minutes following loading or disloading and come back to only $0.3.10^{-4}$ 15 minutes after.

Hysteresis and linearity

Hysteresis as creep depends in sign and magnitude on metal and gauges [7].

Figure 8 presents residual hysteresis after the last equipment of the main bridge. Its maximum value is lower than 1.10^{-4} of the full range. On the same figure we can see that the measurement points for unloading are naturally linear. Those results are obtained without linearizing circuit as mentioned ($R_L = 0$) (4.2) as it was for the first standard. The sensitivity of the main bridge is exactly 2 mV/V at 300 kN. The difference

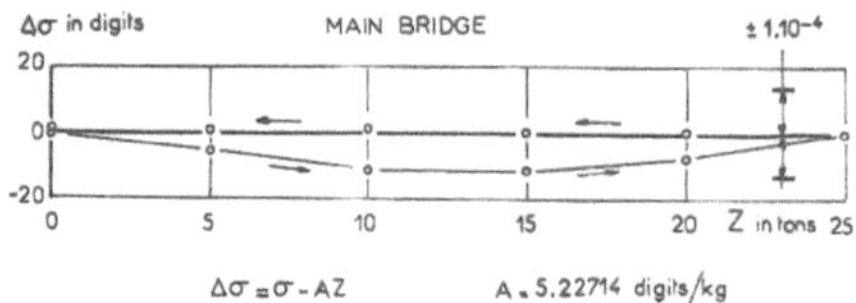

FIGURE 8. 300 kN n° 2 : hysteresis and linearity of the axial bridge.

SUPPLY VOLTAGE 8.0000 V.		FORCES X	FORCES Y	FORCES Z	MOMENTS M_{NS}	MOMENTS M_{EW}	MOMENTS T
BRIDGES	X	+ 4.5679	+ 0.0408	+ 0.00372	− 1.7855	+ 0.0765	+ 0.6377
	Y	− 0.0245	+ 3.9047	+ 0.00048	+ 0.0306	+ 0.4081	− 0.2908
	P	− 0.00245	+ 0.00122	+ 0.53333	+ 0.04081	− 0.03265	− 0.20916
	M_{NS}	− 17.1005	− 0.0755	0	+ 245.6917	− 1.1325	+ 0.2245
	M_{EW}	+ 0.1061	− 25.8190	− 0.00449	+ 1.0713	+ 246.0998	+ 0.1122
	T	+ 0.1449	+ 0.0173	+ 0.00051	− 0.9387	− 0.0408	+ 89.8897
		DIGITS / N			DIGITS / N.m		

MAIN COEFFICIENTS — INTERACTION COEFFICIENTS

FIGURE 9. 300 kN n° 2 : sensitivity coefficients of the six bridges.

of the signals, for a same axial unloading, repeated several times, does not exceed 2 or 3 digits for a signal at 250 kN of more than 130 000 digits i.e. : $0.2.10^{-4}$.

4.3. Six components calibration

The complete calibration (21 loading configurations : 6 simple and 15 combinations two by two) had been carried out on B1.4 bench [9]. The reduction center for the forces, where is placed the reference trihedrom has been chosen conventionnaly on the center of the dynamometer. Linear sensitivity coefficients of the six bridges versus the six components are given in the table of figure 9.

Quadratic coefficients (square and rectangular) are almost null. Interactions of one component on the others are relatively weak. This dynamometer n° 2 is better uncoupled than the first one.

To create 1.10^{-4} interaction of full scale on the main bridge, parasitic components have to reach independently :

X	Y	M_{NS}	$N_{E.W.}$	T
6 530 N	13 115 N	392 Nn	490 Nm	76,5 Nm

With the measuring conditions specified in paragraph 2.3 it is possible to detect an inclination of 1.10^{-5} radian, or an off centering of the axial force, at its maximum value, of 0.016 micron.

Inverse coefficient "matrix" of that presented above has been calculated. This matrix includes 36 (6 x 6) coefficients while n° 1 dynamometer's includes only 16 (4 x 4). It allows to obtain, from the bridge signals the exact value of the three forces and the three moments applied to the dynamometer.

The graph of figure 10 presents the results of axial calibrations up to 250 kN from the average of the four positions. The differences Z determined are smaller than 2.10^{-5} of the measuring range.

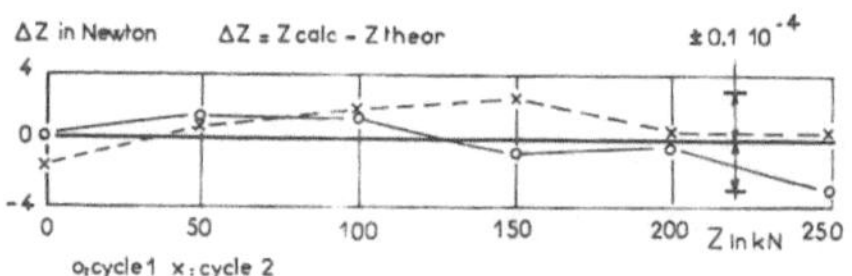

FIGURE 10. 300 kN n° 2 : axial calibration final results.

5. ONE COMPONENT FORCE TRANSFER STANDARD UP TO 50 KN

The 50 kN transfer standard measures only the axial force. But this one is insensible to the other components.

5.1. General description

It is made in one cylindar block (Ø 95 h 220 mm) like the others (fig. 11).

Its central part, measuring the axial force, is constituted by two bars, working in shear, intermixed between a rectangular frame and a central pillar. Two perpendicular blades, one on each side of the central part are used to protect the measuring bars from parasitic components. These uncoupling blades are thin and fragile so their movements are limited by slots and safety dogs manufactured between the central part and external ends. Maximum moments allowable and their corresponding deformations are

Moment (Nm)	Angle (10^{-3} rad)
M_{NS} or M_{EW} : ± 3.5	± 16
T : ± 3.2	± 19

Its fixation on calibration bench is done by mean of two pieces with tapped holes (M.26 Pitch 2) as similar dynamometers. Over all dimensions of this dynamometer are : diameter 95 mm length 340 mm.

The two bars of the dynamometric section are equiped with eight gauges, of 350 ohms, wired in one Wheatstone bridge as described above (4.2, fig. 7).

5.2. Experimental results

Creep and hysteresis

As mentionned before (4.3) previous studies are not sufficient to choose the best gauges without experiments if the metal and the gauges are not exactly the same. For this dynamometer too, special tests have been necessary to reduce creep as much as possible.

Final results are represented on figure 12. It's possible, as for 300 kN dynamometer n° 2, to see on this figure a relative variation up to $0.3.10^{-4}$ during 5 minutes due to a thermoelastic effect. After that, signal stability is very good during 1 hour and a half.

There is a relationship between creep and hysteresis. They increase and decrease at the same time, but it is not possible to minimize them together. Figure 13 gives the relationship between creep and hysteresis obtained with different gauges for successive experiments. It was the same for the two 300 kN standards.

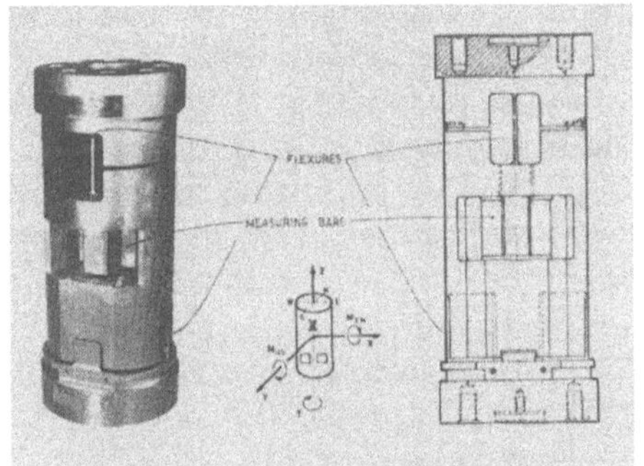

FIGURE 11. 50 kN one component dynamometer.

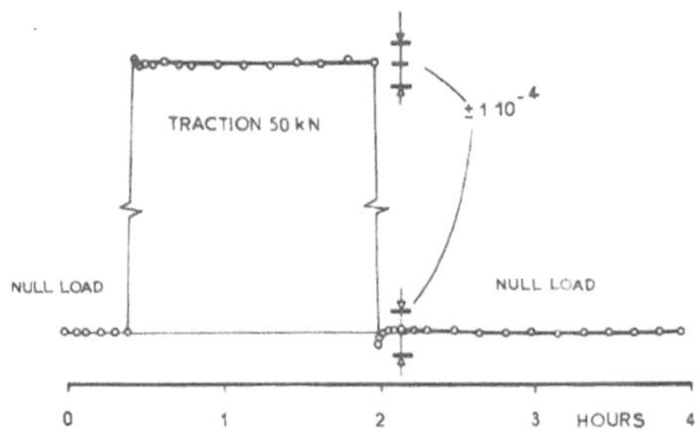

FIGURE 12. 50 kN dynamometer creep.

With minimum creep, calibration gives a positive hysteresis and non linearity shown on first graph on figure 14.

Linearity and repeatability

The initial non linearity has been cancelled by the mean of overmentioned linearizing circuit (4.2).

Calibration results, after linearization, are represented on the second graph of figure 14. Measurements corresponding to increasing loads are placed on a straight line with plus or minus 1 digit dispersion for 156 812 digits at full range ($0.12\ 10^{-4}$). Hysteresis for unloading points is $1.2\ 10^{-4}$ of the maximum value.

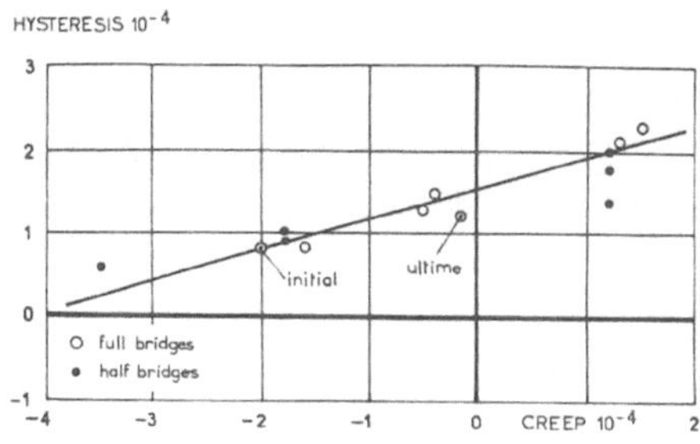

FIGURE 13. Relationship between hysteresis and creep.

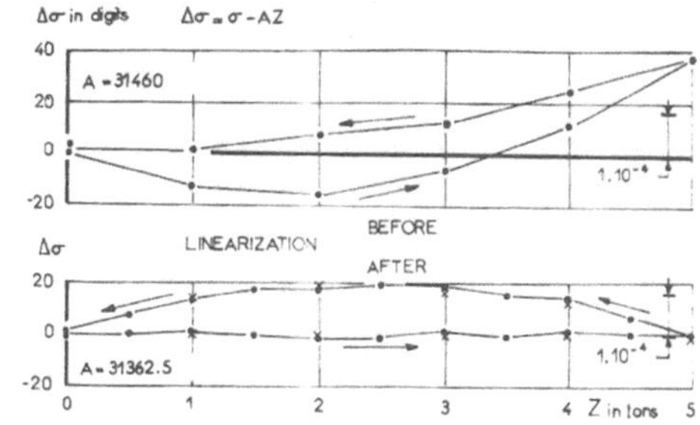

FIGURE 14. 50 kN dynamometer : hysteresis and linearity.

Multicomponent calibration

The six components calibration has given the sensitivity and angular coefficients represented in tabs figure 15.

Interactions are very small, for exemple parasitic efforts which could create separatively 1.10^{-5} uncertainty on the full range of the mean axial force are :

X	Y	M	T
35 N	26 N	1.7 Nm	6.4 Nm

BRIDGE SENSITIVITY		ANGULAR DEFORMATIONS		
efforts	coefficients	θ	φ	ψ
X	$+0.045\,X - 5.10^{-5}X^2$			
Y	+0.06			
Z	+ 3.2	0	0	0
M_{NS}	0	+4.57	0	0
M_{EW}	+0.9	0	+5.65	0
T	- 0.25	0	0	+1.78
in digits/N or Nm		in 10^{-3} rd/Nm		

FIGURE 15. 50 kN dynamometer : sensitivity coefficients.

6. CONCLUSIONS

This document points out the high qualities of the multicomponent force transfer standards designed and manufactured by ONERA for the French Bureau of Metrologie (BNM). It describes the different tests, qualifications and calibrations carried out at the Modane Test Center.

The first 300 kN standard was used, with a great success, in 1979, in the first national intercomparison between the three French 250-300 kN force machines approved by the BNM, belonging to LNE-ETCA and ONERA Modane. After checking on the ONERA machine, in june 1985, it is used again, at the present time, for an international intercomparison, organized by the European BCR (Community Bureau of References), between the forces standard machines of TNO (Netherlands) - NPL (Great Britain) - LNE (France) and IMGC (Italy).

The six component transfer standards 300 kN n° 2 makes it possible the complete knowledge of the parasitic components, transverse forces and moments, generated by the force machine and permits, in that way, a better precision in the axial force determination by taking in account all interactions. Working in traction, it completes the first standard, working in compression.

The 50 kN force standard, of another scale, is, thanks its flexures, completely insensitive to the parasitic components.

These two latter dynamometers, belonging to the BNM, are available since 1981 for national or international utilizations.

7. REFERENCES

[1] DEBNAM R.C. and WIERINGA H. - An intercomparison of force standard machines VDI Berichte nr 212, 1974.

[2] DEBNAM R.C. and JENKINS R.F. - Sources of measurement error during an intercomparison of force standard machines. Proceedings of the round-table discussion on "Measurement of Force and Mass" - 7th IMEKO Congress, London, may 1978.

[3] PETIK F. - Interactions between force measuring devices and standard of Force. Proceedings of the round-table discussion on "Force transfer to force measuring devices" 6th IMEKO, Dresden, june 1976.

[4] BRAY A. - Interaction dead-weight machine-load cell - Proceedings of the round-table discussion on "Measurement of Force and Mass" - 7th IMEKO Congress, London, may 1976.

[5] SAWLA A., WEILER W., PETERS M., BRAY A., LEVI R. and VATASSO M. - An comparison of force standard between the IMGC (Torino) and the PTB (Braunschweig) Proceedings of the 6th IMEKO TC 3 meeting, Odessa, september 1978.

[6] BRAY A., FERRERO C., LEVI R. and MARINARI C. - An investigation on parasitic effects on force standard machines - VDI Berichte 312, 1978.

[7] DUBOIS M. - Experimental study of strain gauge high precision dynamometers at the ONERA Modane Test Centre (IMEKO TC 3 meeting, The Hague, september 1971) - TP ONERA 995 ; 1971 - VDI Berichte 176, 1972. In French : revue "Mesures" - Mars 1972 - Bulletin d'informa

tion du BNM n° 9, juillet 1972.
[8] DUBOIS M. - Design and manufacture of high precision strain-gauge dynamometers and balances at the ONERA Modane Test Centre (NPL conference, Teddington, february 1973) - STRAIN, october 1974 - In French : TP ONERA n° 1 196 - Revue "Mesures", juin 1974.
[9] DUBOIS M. - Calibration of six-component dynamometric balances - Proceeding of the 7th IMEKO congress, London, may 1976 - TP ONERA n° 1976-79.
[10] DUBOIS M. - Multicomponent force transfer standard of 300 kN - IMEKO TC 3 meeting, Odessa, september 1977 - TP ONERA 1977-58.
[11] DUBOIS M. - Description du banc d'étalonnage dynamométrique de l'ONERA à Modane et de l'étalon de force, multicomposantes, destiné aux intercomparaisons de bancs - TP ONERA n° 1978-96 - Bulletin d'information du BNM n° 34, octobre 1978.
[12] DUBOIS M. - Progress in the precision of the 250 kN ONERA force standard machine (IMEKO TC 3 meeting, Braunschweig, september 1978 - TP ONERA n° 1978-100 - VDI Berichte 312, 1978.
[13] DUBOIS M., BOURATEU J.P., GOSSET A and PRIEL M. - Intercomparison of the three french 250-300 kN standard machines - 18th IMEKO TC 3 meeting, Krakow, september 1980 - TP ONERA 1980-105 - In french : bulletin d'information du BNM n° 41 - juillet 1980.

DEVELOPMENT, TESTING AND SPECIFICATIONS OF SUPER PRECISION FORCE TRANSDUCERS FOR INTERNATIONAL COMPARISON MEASUREMENTS

R.H.Hellwig
Dr.-Ing.
HOTTINGER BALDWIN MESSTECHNIK GMBH
Darmstadt

1.INTRODUCTION

The international interlacing of business and trade makes it necessary to compare the force standard measuring machines of the different countries. As this cannot be done in a direct way-especially for high forces up to 1 MN-special force transducers have to be used. Due to the high demands on repeatability, which must be less than $2 \cdot 10^{-5}$ for the whole measuring chain, only straingage transducers in combination with high precision electronic equipment are suitable. The different types of force standard measuring machines interact in different ways with the applied force transducers. The result may be the occurrence of undesired force components or asymmetric load. On one hand the amount of this error depends on the system and on the other hand on the design of the force transducer used.

For this reason it was the aim of this project to create a new type of force transducer which will compensate for all kinds of undesired force components and creep so that only the real vertical force is measured with a maximum deviation of $2 \cdot 10^{-5}$ which is 0.002% or 20 parts per million.

2. TRANSDUCER-DESIGN AND COMPENSATION METHODS

2.1 Design

With the help of a special designing method and the Finite Element Analysis calculation a force transducer was created, which you can see in figure 1 and 2. A 500 kN-type and a 1 MN-type were produced.

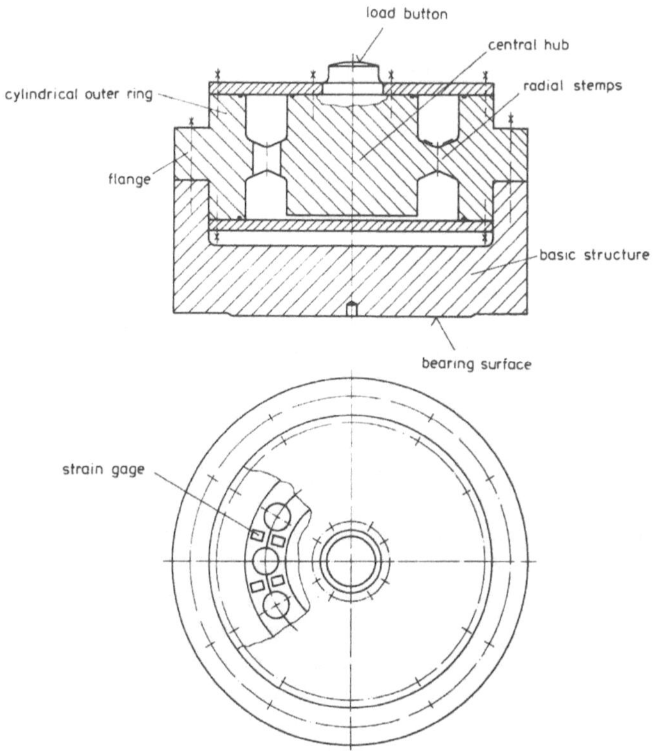

Figure 1: High precision force transducer C12

The force is introduced into the central hub by the ball shaped load head. This surface is ground with high precision in a special grinding machine.

When load is applied to the transducer, the lines of forces are running from the central hub across 12 profiled radial stems to a cylindrical outer ring and from here to the flange, which is connected with a potlike basic structure. Then the lines of forces are led down to the cylindrical bearing surface where they are concentrated. Because of the 12 radial stems, the application of 48 strain gages is possible.

During the development of the transducer different effective principles for reduction and compensation of undesired force components were worked out.

Besides, it was possible, to correct the creep error even when the transducer was finished.

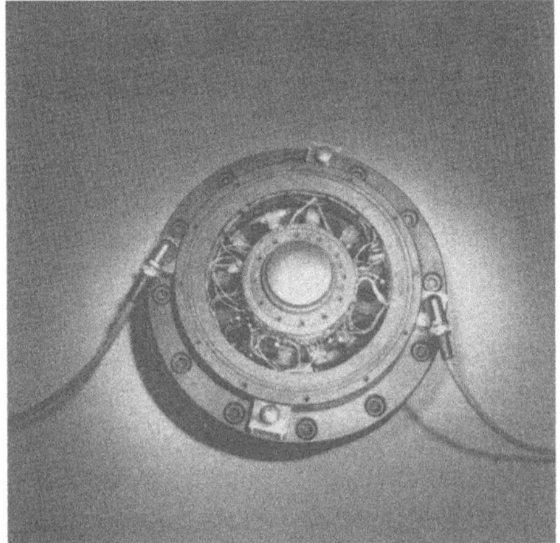

Figure 2: Picture of C12-transducer

To meet the high demands of single error specifications the following undesired deviations were compensated for in many single steps:

2.2 Output deviation produced by different positions

Every single force transducer has small deviations in its mechanical dimensions and in its strain gage applications. This is true even when the force element is designed symmetrically. The result is that the elastic displacement of the force element, when load is applied, will not move in exactly the same direction as the force vector, but tend to bend to one side. How much the force standard measuring machine responds to that bending movement depends on its design. To recognize the amount of the output deviation it is therefore necessary to test the force transducer in various axialsymmetrical mounting positions.
Usually this is done by turning the transducer clockwise by 60° after every single measurement.

If the deviation of the nominal sensitivity is put into a diagram, a sine-shaped curve will occur. (figure 3).

So a well-designed and precise force transducer shows the characteristics of a sine-shaped curve with a very small amplitude.

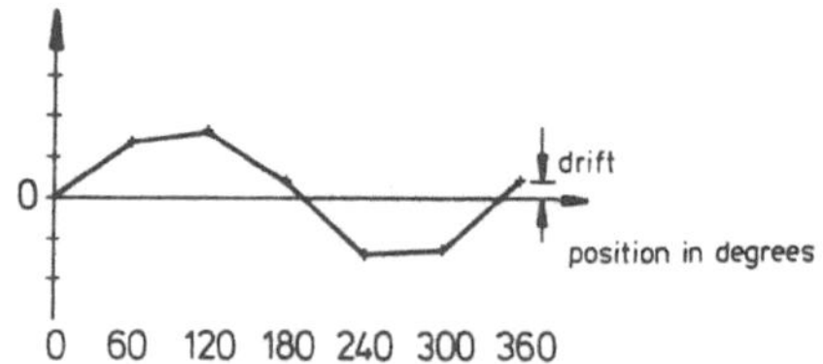

Figure 3: Output deviation of a force transducer in different positions

That is why one of the conditions of this project was to work out a special new method to compensate for the rotation effect of fig. 3 so that the sensitivity-deviation is less than $\pm 2 \cdot 10^{-5}$.

The measurement design for this condition is shown in figure 4.

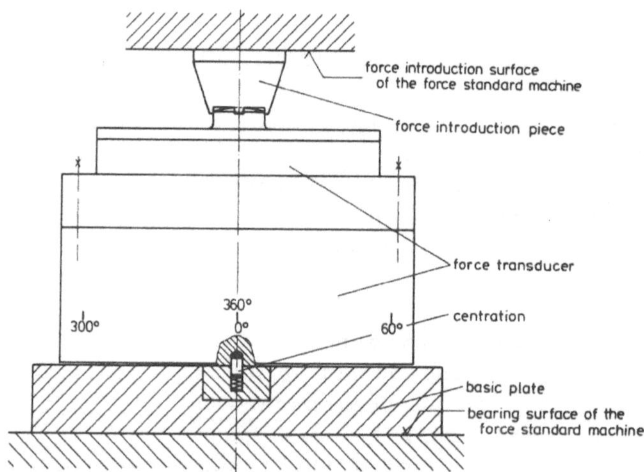

Figure 4: Measurement design for the investigation of output deviation produced by different positions

2.3 Output deviation produced by angular deflection of the force introduction surface of 0.2 degrees

In some cases the mentioned occurrence of bending moments or asymmetric load may be originally produced in the force standard measuring machine itself and has nothing to do with the transducer. This leads to an eccentric contact between the machine surface and the rounded head of the transducer. The force vector runs parallel but with the distance "a" to the measuring axis of the transducer. This is shown in fig.5.

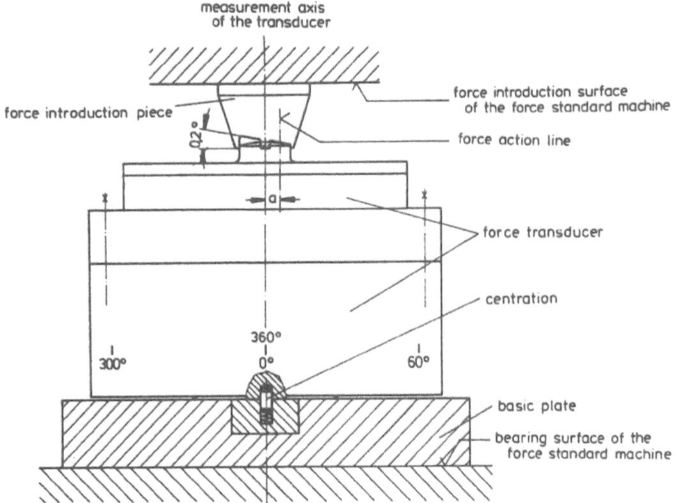

Figure 5: Measurement construction for the investigation of output deviation produced by angular deflection of the force introducing surface by 0.2°

The result is an output deviation which depends on the position of the force transducer as well. For this reason it is necessary to measure again at 7 different positions as described in 2.2.

The measurement construction is similar to that of fig.4, but the force introduction piece has an inclined surface of 0.2°.

2.4 Output deviation produced by angular deflection of the bearing surface of 0.2 degrees

Finally tests on the influence of an angular deflection of the surface bearing by 0.2° had to be made. In order to to that the transducer was put on a key with an angle of 0.2° degrees. This is demonstrated in figure 6. The measurements were carried out in the same way as described above.

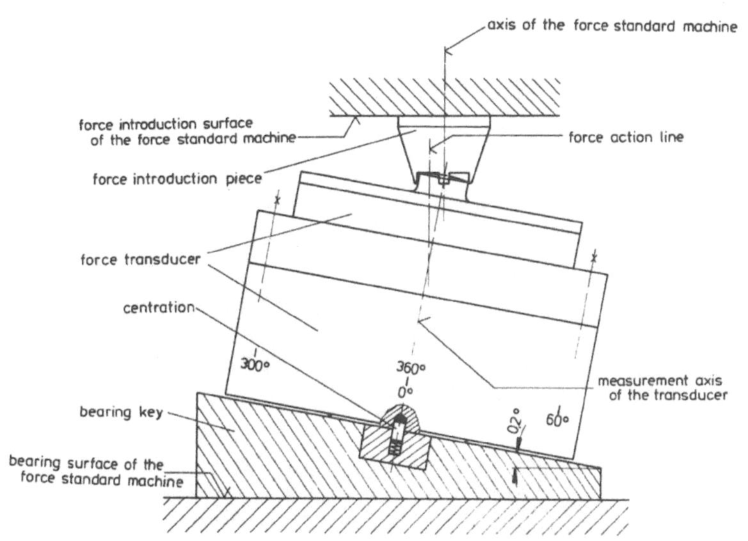

Figure 6: Measurement construction for the test of output deviation produced by angular deflection of the bearing surface by 0.2°

3. MEASUREMENT RESULTS

At the beginning, a number of measurements were made with two 500 kN-force transducers to study the theory of compensation principles in practice.

The experience with and results of the two models were used for the design of a 1 MN-type. Owing to this experience it was possible to meet the specifications of the first 1 MN-model at once.
The results of the various measurements are shown in the figures 7 and 8.

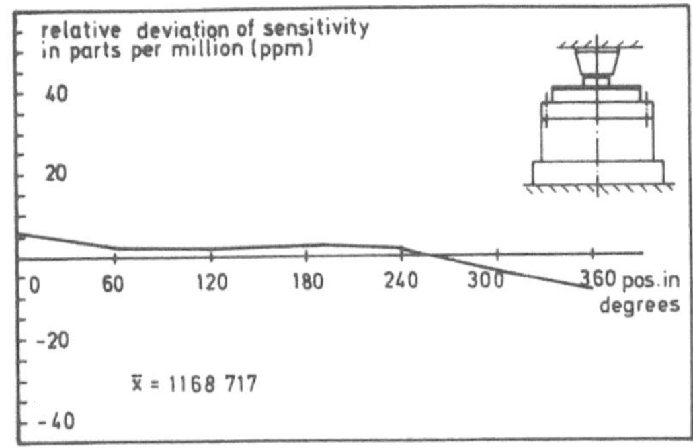

7a: Output deviation produced by different positions

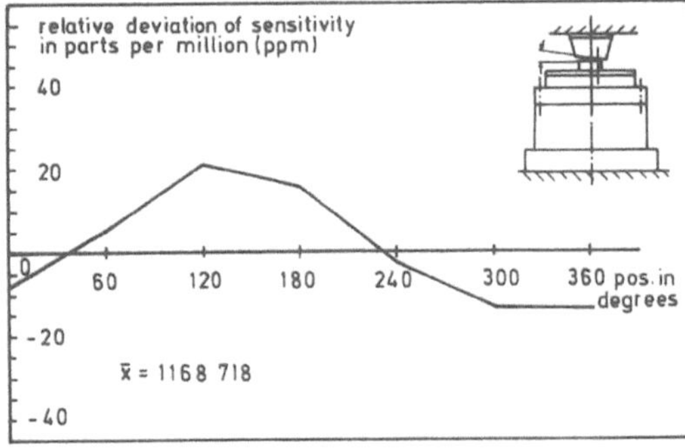

7b: Output deviation produced by angular deflection of force introduction surface by 0.2°

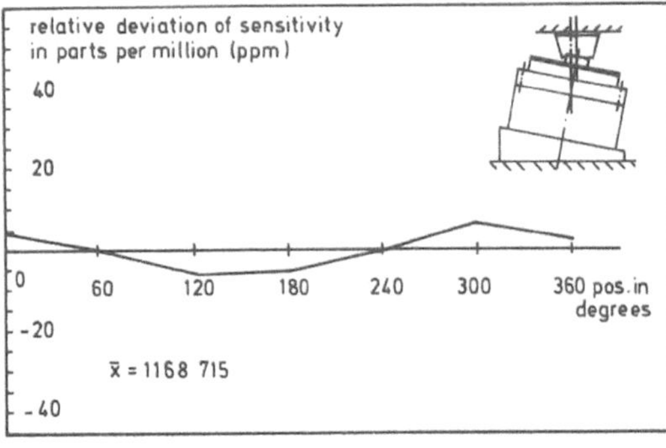

7c: Output deviation produced by angular deflection of the bearing surface by 0.2°

Figure 7: Results of measurements of the 500 kN-model in different positions

Figure 7b and 7c show a sine-shaped curve.
It was not possible to reduce the amplitude, because the measurement performance of 7b is ad verse to 7c. The deviation in 7b has its maximum value at about 120°, whereas for

this position a maximum can be seen in figure 7c. This was not the same for with the 1 MN-type in the figures 8b and 8c. In some diagrams a thermal shift can be discovered, which was caused by a temperature drift of several degrees during the ten hours' time of measurement.

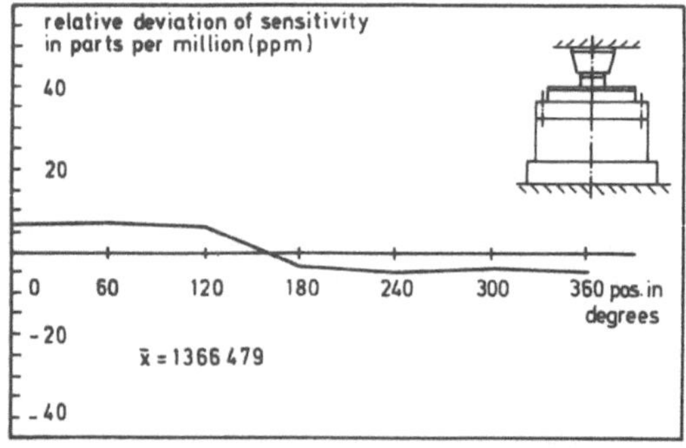

8a: Output deviation produced by different positions

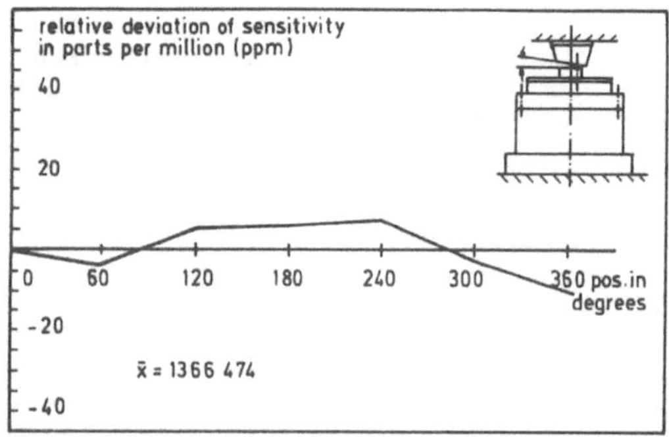

8b: Output deviation produced by angular deflection of force introduction surface by 0.2°

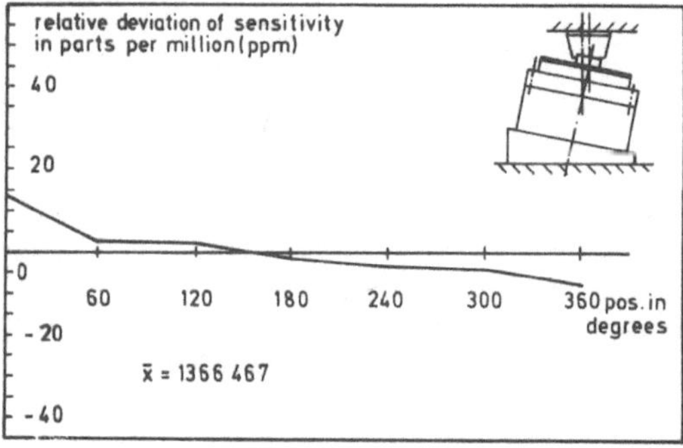

8c: Output deviation produced by angular deflection of the bearing surface by 0.2°

Figure 8: Results of measurements of the 1 MN-model in different positions

It is important to compare the average output values of the diagrams in figure 7 and 8.

output - sensitivity

	500 kN-model	1 MN-model
without wedge:	1.168.717	1.366.479
wedge on top :	1.168.718	1.366.474
wedge underneath:	1.168.715	1.366.467

Obviously the numbers are close to each other though the measurements were made on different days. Thus an important demand could be met.

It was mentioned in 2.1 that creep had to be compensated for to a minimum as well. There are many possibilities to influence the creep curve or to compensate for the error. But all these possibilities are disadvantageous because they must be realized during the manufacturing of the transducer. It is usually impossible to correct the creep curve when the transducer is completed.
In this case a new principle was worked out, which allows to correct the creep error on the completed transducer after the creep has been measured. This demand was met by using a special electric network.

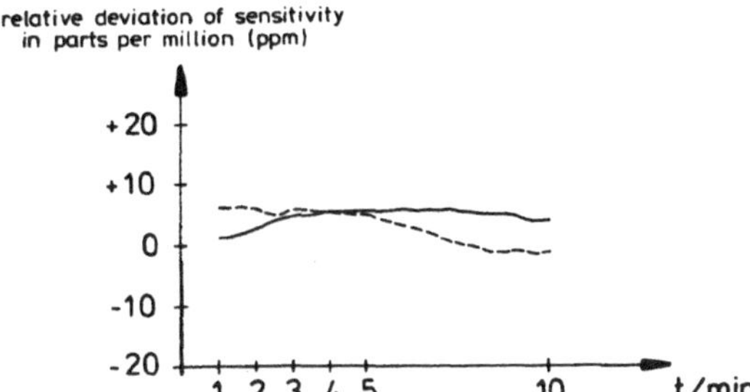

Figure 9a:Creep curve from 1 minute to 10 minutes after load is applied (loading creep)

To be able to study the creep behavior of the transducer better the so called "reverse creep" was also measured. This is the change in output signal immediately after unloading the transducer. Usually the values will creep slowly back to zero, but this can be quite different when the creep error is very small. In Figures 9a und 9b both curves are demonstrated.

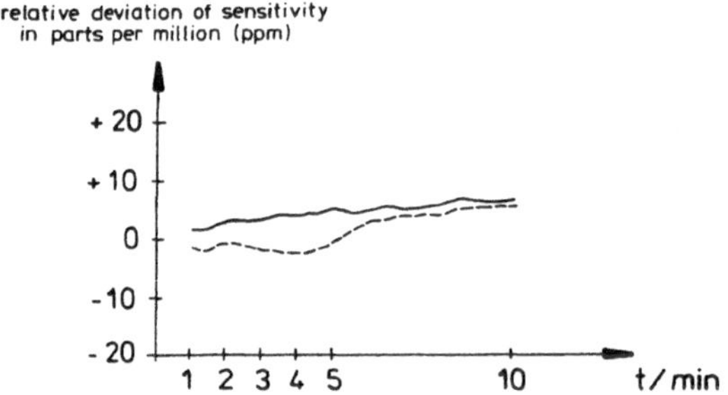

Figure 9b: Creep curve from 1 minute to 10 minutes after unloading (reverse creep)
---------- 500 kN
———— 1 MN

4. SUMMARY

By using modern calculation methods, manufacturing processes, and new technologies for compensating for force and creep errors it was possible to create force transducers of high precision.

The following undesired force components and creep had to be compensated for in many single steps down to a maximum deviation of 0.002%:

- Output deviation produced by different positions
- Output deviation produced by angular deflection of the force introduction surface of 0.2 degrees
- Output deviation produced by angular deflection of the bearing surface of 0.2 degrees
- Creep from 1 minute to 10 minutes after load has been applied (Creep error had to be compensated for after the transducer was completed).

But all these efforts would not have been successful without the existence of the programmable precision measuring instrument DMP 39, developed and built by HBM, and the high precision force standard measuring machine at the German PTB in Braunschweig.

MEASUREMENT OF FORCE BY MEANS OF FIBER-OPTIC SENSORS.

R. Kist, S. Ramakrishnan, H. Wölfelschneider

Institut für Physikalische Meßtechnik, Heidenhofstr. 8, D 7800 Freiburg

ABSTRACT

Various principles of measuring force by means of fiber-optic sensors such as fiber to fiber transmission modulation, microbending, elastooptic modification of the state of polarization (induced birefringence) are shortly reviewed. A novel method of measuring force by means of bending a Fiber Fabry-Perot-(FFP-) resonator is described. This interferometric FFP-sensor is easily applicable to AC force measurements, but makes special temperature compensation schemes necessary if DC forces are to be measured. The FFP force sensor can be designed to meet a wide dynamic range at high sensitivity.

1. INTRODUCTION

Force (F) is a physical quantity that can be derived from various measuring parameters such as displacement Δx, with $F = D \cdot \Delta x$ (D = spring constant), acceleration $\ddot{x}$, with $F = m \cdot \ddot{x}$ (m = mass), or pressure p, with $F = A \cdot p$ (A = area).

Hence the measurement of force is directly linked to these parameters, so that force sensors are based upon the same operation principles that characterize displacement, acceleration, and pressure sensors.

The new technology of fiber-optic sensors (1-5) has led to the development of a large variety of displacement sensors, to which in turn most acceleration and pressure sensors can be reduced. Thus fiber-optic displacement sensors are basic for a large set of fiber-optic sensors that measure various physical parameters (6).

Specific advantages of fiber optic sensors are their small weight and size, low optical power consumption, electrical isolation, immunity against electromagnetic contamination as well as hazardous and corrosive environments, and achievable high sensitivity.

Concepts of displacement sensors that can be used for the measurement of force, range from simple fiber to fiber coupling and light shutter configurations as well as external reflection and microbending devices to polarization- optic and interferometric approaches. A few examples will be described shortly,

Wieringa, H. (ed), Mechanical Problems in Measuring Force and Mass.

and the concept of a force sensor based on bending a Fiber Fabry-Perot (FFP) will be dealt with in more dtail.

2. INTENSITY-MODULATED FORCE SENSORS

In order to measure crack widening in various materials as function of the crack producing force a fiber to fiber coupling sensor is used at the Materials Testing Institute (Materialprüfungsanstalt) Stuttgart, Germany (7). The configuration is shown in Fig. 1a. The light flux coupled from the transmitting to the receiving fiber is a measure of the crack widening force as shown in the diagram (Fig. 1b) for tests on various steel samples. This method allows for dynamic measurements with frequency response up to 1 MHz, and in addition is suitable for operation at temperatures up to about 400 °C.

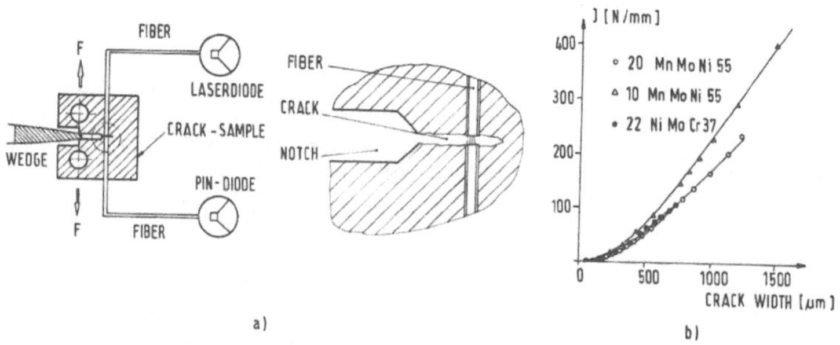

Fig 1a) Configuration of crack widening sensor,
b) Crack width as function of force acting on the material sample of three different pressure vessel steels.

Fig. 2 shows a pressure sensor that has been developed into an industrial prototype at the University of Manchester (8). It relies on the relative displacement of two grids within the collimated light beam that couples the receiving to the transmitting fiber via two grin lenses. One of the lenses carries a fixed holographic grid, the other grid is mounted onto a membrane and displaced vertically to the optical axis according to the pressure applied. This configuration allows to resolve pressure changes of the order of 0.1 mPa. With a membrane area of A m^2 this corresponds to a force resolution of $10^{-4} \cdot A$ N .

Fig. 3 shows an example of a reflection sensor developed at the IPM Freiburg in cooperation with Daimler-Benz (9) for measuring dynamic pressure (engine knocking) in the combustion chamber of an automobile. The sensor element containing the transmitting and the receiving fiber is configured in a spark plug size with a small membrane the motion of which modulates the light intensity reflected into the receiving fiber. Quasistatic pressure measurements showed the sensor to be responsive up to about $2 \cdot 10^6$ Pa which corresponds to a force measuring range (membrane diameter approximately 8 mm) up to about 100 N.

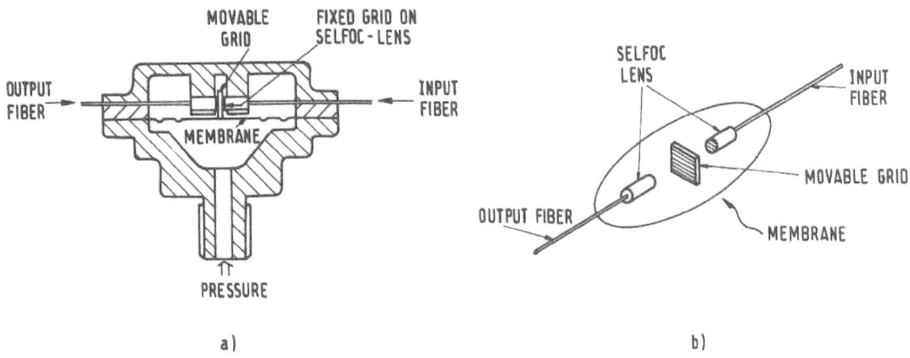

Fig 2) Industrial prototype of a pressure sensor based on a light shutter configuration with two grids displaced with respect to each other.

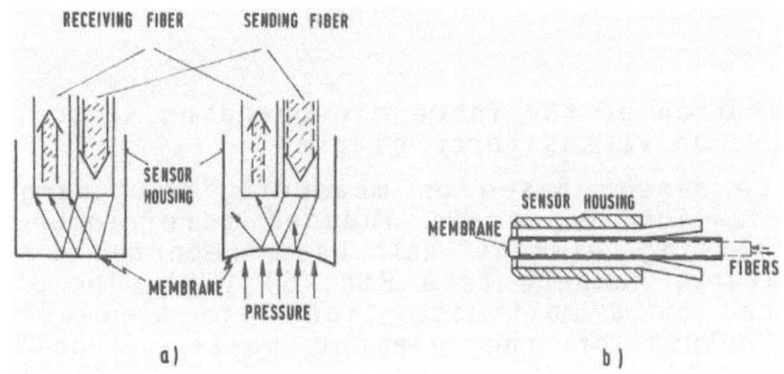

Fig 3) Reflection pressure sensor for automotive applications a) principle of operation, b) spark plug size sensor head.

The microbending effect has been used by several investigators to construct fiber-optic displacement sensors that can be applied to the measurement of pressure and force (10, 11). In these cases attenuation due to conversion of energy from guided into radiated modes at fiber bends is measured as function of the pressure or force that squeezes the fiber between two periodically structured plates. Typical sensitivities of such sensors with a spatial bending period of about 2 mm are of the order of a few percent transmission loss per 1 µm relative normal displacement of the two plates.

A different approach (12) for a microbending force sensor uses a fiber that is passed through small eyelets which can be displaced with respect to each other applying stress to the fiber thus causing microbending. The configuration consisting of a plate within a cylinder, both carrying the eyelets, is shown in Fig. 4a. Fig. 4b displays the transmission versus force diagram with a dynamic range of four orders of magnitude and a force resolution of about 10^{-4} N (corresponding to 0,1 µm resolution in displacement).

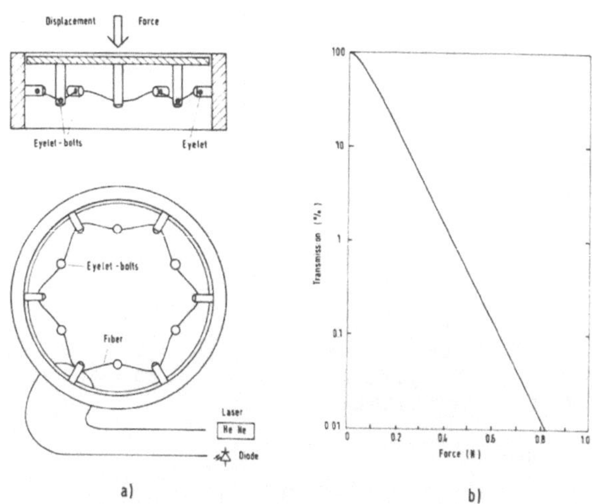

Fig 4a) Configuration of the force microbending sensor,
b) transmission versus force diagram.

Recently a force sensor based on measuring the change of the state of polarization by strain induced birefringence in an elastooptical sensing element has been reported by Philips Research Laboratory, Hamburg (see Fig. 5) (13). The element is optically powered by a multimode fiber via a polarizer. The light at the output of the element passes a polarization analyzer and is guided by a second fiber (or the feeding fiber in case of reflection mode operation) to the detector. The sensor is used to monitor respiration motion during NMR-tomography. Gating the NMR-measurements to the respiration rest phase leads to considerable quality improvement of the NMR-tomograms.

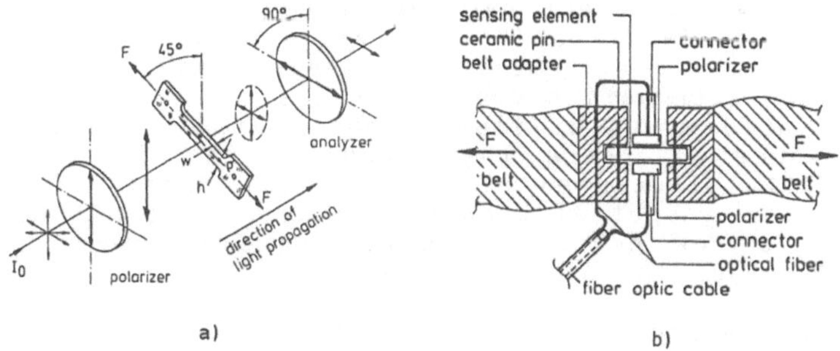

Fig 5) Elastooptic strain sensor a) measuring principle, b) sensor integrated into a belt for detection of respiration motion.

There have also been reports on hybrid fiber-optical sensors for force measurements, that contain an electromechanical sensing element with optoelectronic converters that is powered by an optical fiber with a second fiber carrying the optical signal to the detection unit. In one case (14) a quartz crystal attached to a cantilever is excited by light pulses from a fiber that are converted into voltage pulses. The resonant frequency of the quartz element depends on the force acting on the cantilever. Another approach (15) makes use of a taut wire which oscillates at a resonance frequency that is proportional to the square root of the strain force applied to the wire. The oscillation of the wire is detected by two optical fibers that are configured as a reflective sensor. The force sensor thus represents an optically powered oscillator whose resonant frequency is a function of the tension present in the wire.

3. PHASE MODULATED FORCE SENSOR

During the past few years the fiber-optic Fabry-Perot (FFP) resonator has developed into a versatile element for sensing a variety of physical parameters (16, 17, 18, 19). The FFP element is fabricated simply by cleaving or polishing the end faces of a piece of fiber and providing mirrors in form of external reflectors or metallic or dielectric reflective thin films at these ends.

The FFP length L may range from mm up to several m. A FFP with core refractive index n and powered by light with wavelength λ has a round trip phase

$$\phi = 4\pi nL/\lambda .$$

Any change of the optical length n·L and/or the wavelength will produce a phase change

$$d\phi = \frac{4\pi}{\lambda}(ndL + Ldn) - \frac{4\pi}{\lambda^2} nLd\lambda .$$

Variation of λ corresponds to application of the FFP as an optical spectrum analyser, whereas variation of n·L relates to FFP operation as a sensor element. Any parameter P that affects the optical length produces a phase change

$$d\phi = \frac{4\pi L}{\lambda}\left(n\frac{1}{L}\frac{dL}{dP} + \frac{dn}{dP}\right) .$$

In the case of P being temperature, dL (L·dT) is the thermal expansion coefficient (for silica: $5\cdot 10^{-7}/^{\circ}C$), and dn/dT the thermooptical coefficient (for silica: $1\cdot 10^{-5}/^{\circ}C$). In case of force measurement (P = F) dL/(L·dF) describes strain per unit axial stress applied, and dn/dF the elastooptical variation of the index of refraction under the influence of the external force.

In (19) we reported on a fiber-optic Fabry-Perot strain gauge consisting of a U-shaped FFP bonded to a cantilever (see Fig. 6).

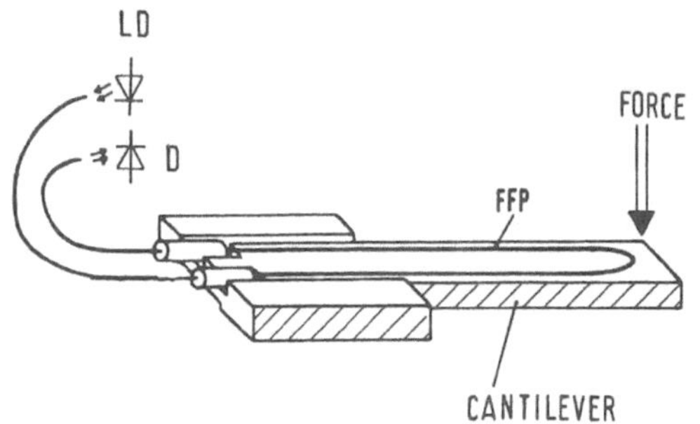

Fig 6)

Configuration of the U-shaped FFP bonded to a cantilever.

Bending of the cantilever under the action of a force produces strain to the FFP thus providing a change d(n·L) of the optical length and hence a phase change dØ. Theory of the FFP (17) tells that the optical intensity transmitted through the resonator reaches maximum (interference fringe) whenever the round trip phase is N·π, where N denotes the order of FFP-interference. Thus a monotonous strain variation produces a sequence of intensity maxima at the FFP output. Signal evaluation can be done either by fringe counting (with incremental resolution corresponding to the FFP free spectral range c/(2nL), c being the velocity of light) or compensating the effect of Δ (n·L) via a suitable wavelength change Δλ of the laser diode source by controlling the laser injection current (see 18). A third signal evaluation method, described in (19) applies to low finesse FFP resonators which exhibit a cosine shape of the transmitted intensity as function of the measuring parameter P that affects optical length and hence round trip phase. As in conventional signal evaluation for incremental sensors the cosine signal (which in general contains also an intensity bias) is differentiated twice thus providing two periodic signals which are phase shifted by π/2 and free of the intensity bias. The zero crossings of the sine and cosine signals are used to determine the sign of the parameter change ΔP as well as its value by incrementing or decrementing a counter. This provides a two bit per free spectral range resolution. An additional three bit resolution is achieved by electronicallly interpolating the tan(P)-function (derived as the ratio sin(P)/cos(P)) by means of a weighted AD converter. Thus this FFP strain gauge provides a total five bit per free spectral range resolution (19).

Since the position of the zero crossings is only dependent on the optical length of the FFP, this method is largely independent of intensity variations that may arise from bends in the linking fibers, coupling losses, source and detector degradation etc. This is an important feature of a fiber-optic sensor as to its suitability for industrial applications. Another favourable feature of the FFP strain gauge is the use of a suitable graded index fiber (instead of monomode fibers used in previous work) which considerably facilitates coupling to the feed and signal fibers.

A basic problem with most of the sensors in general is their susceptibility to temperature changes. The FFP strain gauge, too, is sensitive to temperature. The transmission signal of a silica fiber FFP of length L = 0,1 m for example would change by one free spectral range if temperature changes by about 0.37 °C. In order to use the FFP as a sensor for a quasistatic parameter such as DC displacement, force, and pressure, the influence of the temperature has to be compensated.

The concept to realize this involves an arrangement of two equal FFP resonators that are bonded symmetrically to a cantilever (or membrane in case of pressure measurement) as shown in Fig. 7. The two FFP resonators are powered by the same laser diode via a 3dB directional coupler. Bending the cantilever by the action of the weight force F = m·g (m = mass, g = graviational acceleration) provides a displacement Δx, that is monitored by an inductive position sensor. The (upper) FFP1 is strained whereas the (lower) FFP2 is compressed. The detectors D1 and D2 receive the periodic signal of FFP1 and FFP2, respectively. These signals move in opposite directions if the force changes by ΔF, i.e. towards higher orders of interference (from A_1 to B_1) for FF1, and lower orders of interference (from A_2 to B_2) for FF2, as shown in Fig. 8. This provides double resolution of the force measurement if a suitable (differential) signal evaluation scheme is used.

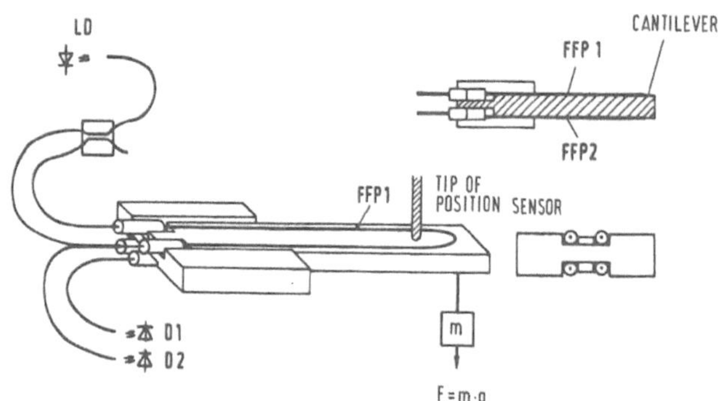

Fig 7) Double FFP force sensor with two FFP resonators bonded symmetrically to the cantilever.

On the other hand, a temperature change ΔT acts in the same direction for both the FFP resonators, i.e. the signals change from A_1 to C_1, and A_2 to C_2, respectively. Thus in a differential signal evaluation scheme the temperature effect is cancelled.

Fig. 9 shows the principle of the double FFP sensor with the block diagram of its signal evaluation electronics. The laser diode (Hitachi HL 7801E) is modulated in wavelength at f = 10 kHz by modulating the injection current with an amplitude of less than 10 mA. The modulated component of the two

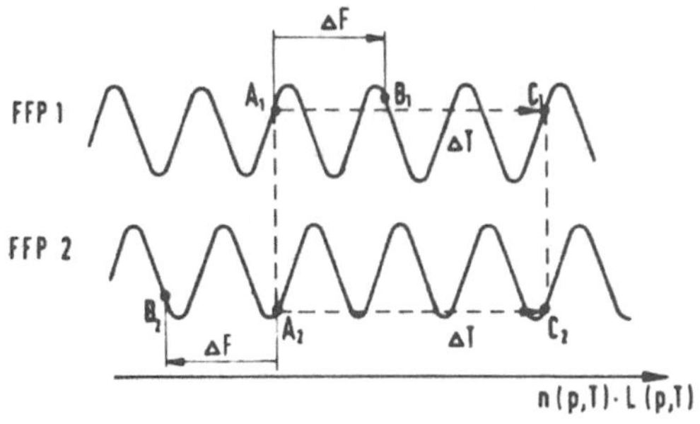

Fig 8) Effect of a force change ΔF and a temperature change ΔT on the double FFP sensor.

optical FFP signals is detected, amplified and used to provide the first and second derivative ($dI/d\lambda$ and $d^2I/d\lambda^2$) by synchronous demodulation at the frequencies f, and 2f, respectively. These derivatives are of sine and cosine shape, respectively, and do not contain any more the optical intensity bias present at the output of the two FFP resonators. The differentiated signals are used to identify the sign of the optical signal change and to correspondingly increment or decrement counters in order to provide digital signals A and B that are a measure of the change d(n·L) of the optical length in the resonators FFP1 and FFP2, respectively.

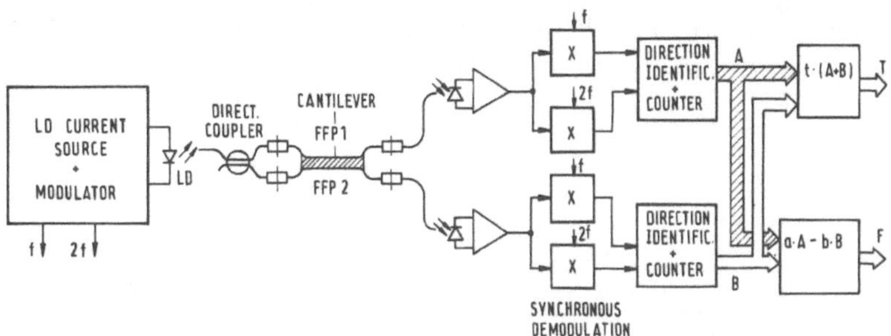

Fig 9) Double FFP sensor scheme with signal evaluation electronics.

The difference signal $S^- = aA-bB$, with a and b being suitable scaling factors to compensate for different lengths of FFP1 and FFP2. This provides a voltage proportional to the force F applied to the cantilever. The sum signal $S^+ = t\cdot(A+B)$, with t being another scaling factor, provides a voltage proportional to the temperature at the sensor.

Fig. 10a) shows the sensor signals U(FFP1) and U(FFP2) as a force F of about 2 N is periodically applied to the cantilever for a time interval of 10 seconds. Both signals are superimposed to a drift that simulates a fluctuating temperature, i.e. optical length n·L of the FFP. This simulation has been achieved by a drift component in the laser diode injection current which produces a drift in wavelength and hence a net increase in the interference order of both FFP resonators. The difference signal in the upper part of Fig. 10b is free from the drift, whereas the sum signal in the lower part displays the drift only, which demonstrates the proper operation of the signal evaluation scheme of Fig. 9.

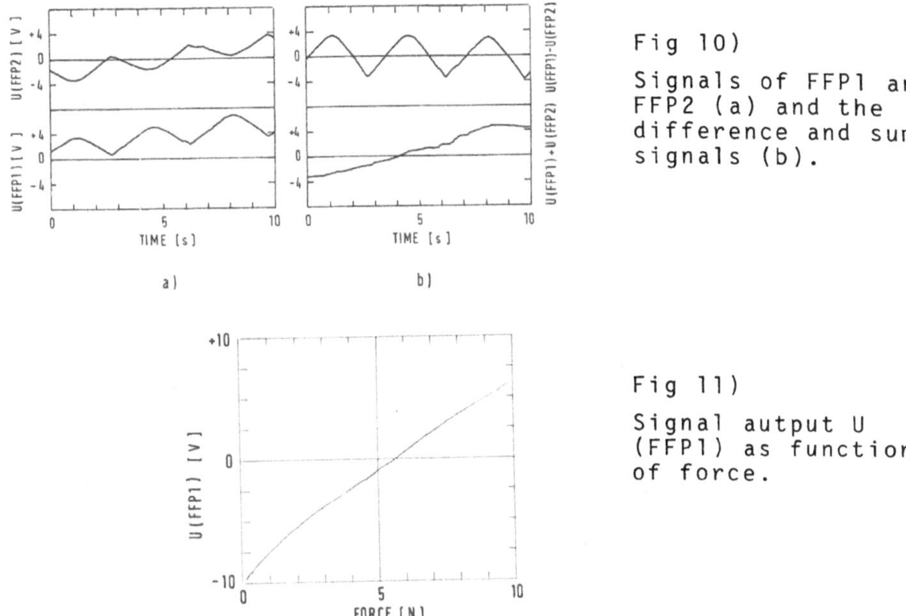

Fig 10)

Signals of FFP1 and FFP2 (a) and the difference and sum signals (b).

Fig 11)

Signal autput U (FFP1) as function of force.

In Fig. 11 the signal of FFP1 is plotted against Force in the range 0 to 10 N.
Fig. 12 displays U(FFP1) oscillating in time at about 3 Hz as the cantilever is periodically bent by a 5 N weight acting via a spring loaded balance. During the free oscillations of the corresponding 0,5 kg mass the sensor electronics continuously evaluates the change in the interference order of the FFP1 resonator into a corresponding output voltage.

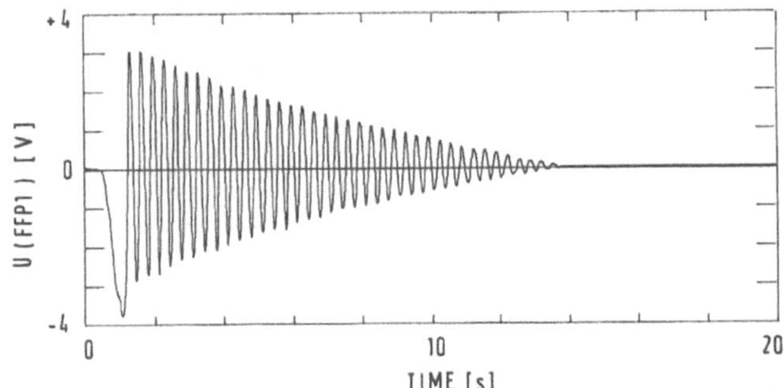

Fig. 12) Output voltage U(FFP1) oscillating as the cantilever is periodically bent by a 0,5 kg mass acting via a spring loaded balance.

REFERENCES

1. Giallorenzi, T.G. et al.: Optical Fiber Sensor Technology, IEEE J. Quant. Electron. QE-18 (1982), 626-665.
2. Kist, R.: Meßwerterfassung mit faseroptischen Sensoren, Technisches Messen 6 (1984), 205-212.
3. Proceedings of the 1st Internat. Conf. on Optical Fiber Sensors (OFS), London (Apr. 1983).
4. Proceedings of the 2nd Internat. Conf. on Optical Fiber Sensors (OFS), Stuttgart (Sept. 1984).
5. Proceedings of the 3rd Internat. Conf. on Optical Fiber Sensors (OFS), San Diego (Febr. 1985).
6. Ulrich, R.: Faseroptische Wegaufnehmer als Grundelemente für Sensoren, Automatisierungstechnische Praxis, Heft 3, S. 117-123, und Heft 4, 178-183 (1985).
7. Zimmermann, C.: Use of fiber optical components as active sensors in dynamical material testing, 2nd Internat. Technical Symposium on Optical and Electro-Optical Applied Science and Engineering, Cannes (Nov. 1985).
8. Jones, B.E., Spooncer, R.C.: Optical fiber pressure sensor using a shutter modulator and two-wavelength intensity referencing, 2nd Internat. Conference on Opt. Fiber Sensors, Stuttgart (1984).
9. Wiesmeier, A., Schwerdt, P.: Anwendung faseroptischer Sensoren im Kraftfahrzeug, VDI-Bericht Nr. 515 (1984), 31-37.
10. Fields, J.N., Asawa, C.K., Ramer, O.G., Barnoski, M.K.: Fiber optic pressure sensor, J. Acoust. Soc. Am. 67 (1980), 816-818.
11. Lagakos, N., Litovitz, T., Mohr, R., Meister, R.: Multimode optical fiber displacement sensor, Appl. Optics, 20 (1981), 167-168.
12. Spenner, K.: Microbending pressure and displacement sensor, 3rd Internat. Conf. on Optical Fiber Sensors, San Diego (1985).
13. Martens, G., Helzel, T., Kordts, J.: Optical detection of respiration motion and heartbeats for NMR imaging applications, 2nd Internat. Technical Sympos. on Optical and Electro-Optical Applied Science and Engineering, Cannes (Nov. 1985).
14. McGlade, S.M., Jones, G.R.: An optically addressed vibrating quartz force sensor, Colloqu. on Optical Point Sensors for Process Control, London (Jan. 1984), IEE Coll. Digest No 1984/7.
15. Jones, B.E., Philp, G.S.: A vibrating wire sensor with optical fiber links for force measurement, Proc. Conf. Sensors and their Applications, Manchester (Sept. 1983).
16. Yoshino, T., Kurosawa, K., Itoh, K., Ose, T.: Fiber-optic Fabry-Perot-Interferometer and its sensor applications, IEE J. Of Quantum Electronics, Vol. QE-18 (1982), 1624- 1633.
17. Kist, R., Sohler, W.: Fiber-optic spectrum analyzer, J. of Lightwave Technology, Vol. LT-1, (1983), 105-110.
18. Kist, R., Drope, S., Wölfelschneider, H.: Fiber-Fabry-Perot (FFP) Thermometer for medical applications, 2nd Internat. Conf. on Optical Fiber Sensors, Stuttgart (1984), 165-179.
19. Kist, R., Ramakrishnan, S., Wölfelschneider, H.: The Fiber-Fabry-Perot and its applications as a fiber-optic sensor element, 2nd. Internat. Technical Sympos. on Optical and Electro-Optical Applied Science and Engineering, Cannes (Nov. 1985).

FIBER OPTIC LOAD SENSOR AS A MICROCOMPUTER PERIPHERAL

D.G. SPOREA, A. DUMITRICA and N. MIRON

CENTRAL INSTITUTE OF PHYSICS, LASERS DEPARTMENT, MAGURELE, R-76900, ROMANIA

1. INTRODUCTION

The last years have been marked by the spread of optical fibers use for remote sensing and actuating. Optical fiber sensors can be applied to sense a wide range of physical perturbations. Sensing and control systems based on the fiber optic conversion of physical or chemical input variables into modulated light signals have been developed in various laboratories and industries.

As compared with other technologies these systems offer numerous advantages, which include high sensitivity, light weight and reduced dimensions. The signal transmission over optical fibers is unaffected by external interference (i.e. electromagnetic interference) and safe in hazardous environments.

There are two basic types of fiber optic sensor:

- one in which the parameter to be measured changes the transmission properties of the fiber itself;
- and another in which the optical fiber is only used as a signal path, carrying the light to and from the transducer.

Single-mode and multimode optical fibers are used for fiber optic sensors.

The working principle is based on amplitude or phase modulation, induced on the light transmitted along the fiber, by the investigated parameter. Amplitude modulation is employed mainly in multimode fiber systems. The phase modulation requires the use of single-mode optical fibers (1,2).

Multimode fiber sensors are easier to handle and the components associated to these fibers are easily built and connected to the system.

Single-mode optical fibers do not suffer the noise problems caused by the modal interference effects, and therefore allow the development of high sensitivity coherent sensors. The highest sensitivity is achieved in fiber optic interferometers such as Mach-Zehnder, Michelson and Fabry-Perot (3).

Fiber optic sensors have already been built for indicating linear or angular position, measuring temperature, pressure, strain, acceleration, rotation rate, magnetic field intensity, electric field intensity, radiometric dosage.

In multimode fiber sensors the amplitude modulation is obtained by increasing light attenuation in the sensing element. Sensors with amplitude modulation have greater design flexibility, lower electric complexity and lower component cost, but require some form of compensation of the optical losses not connected to the measured phenomenon.

A fiber optic load sensor based on the "bending loss" principle has been developed. The transmitted light intensity is modulated by fiber optic bending. This principle along with the fiber optic load cell are described in the paper. Some methods employed for the compensation of the radiation source fluctuation are outlined. The connection of such a fiber optic sensor to a microcomputer is presented.

2. FIBER OPTIC LOAD SENSOR

2.1. Losses caused by fiber curvature

In principle, optical fibers whose axis is bent into a circle lose power by radiation and their transmission loss increases. The loss coefficient increases exponentially with the decrease of curvature radius. At a certain critical radius, the losses become noticeable and the rays guidance is lost (see Fig. 1).

The rays of the light launched into one end of a straight fiber are guided if they are totally internally reflected at the core-cladding interface. The cone of the light rays which can be guided is described by the numerical aperture of the fiber:

$$NA = \sin \Theta_c = (n_{core}^2 - n_{clad}^2)^{\frac{1}{2}}, \qquad (1)$$

where Θ_c is the acceptance angle, n_{core} and n_{clad} are the refractive indexes of the core and cladding respectively.

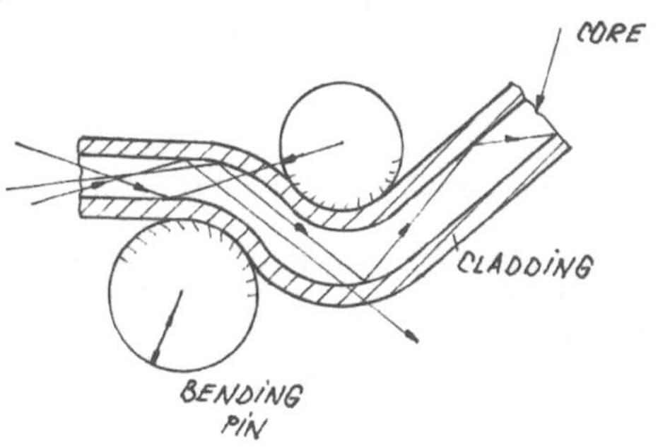

FIGURE 1. The "bending loss" phenomenon in optical fibers.

An explanation for the occurrence of the curvature losses is based on the WKB method. The analysis shows that there are two forms of modes:

- guided modes, which are waves propagating along the fiber core;
- unguided waves or radiation modes, which are waves radiating out of the fiber core into the cladding.

There are also evanescent radiation modes which are radiation modes with little loss.

Guided modes from the fiber core suffer from radiation loss when the fiber is curved.

The total number of modes that can be guided by a curved fiber is smaller compared with the same number in a straight fiber. The effective number of modes supported by a curved fiber can be computed as follows (4):

$$N_{eff} = N_{\infty}\left\{1 - \frac{g+2}{g+\Delta}\left[\frac{2a}{R} + \left(\frac{3}{2n_{clad}kR}\right)^{2/3}\right]\right\}, \tag{2}$$

where N_{∞} is the total number of modes for the straight fiber, g represents the exponent of the power law of the refractive index and Δ is the relative index difference.

It results that the permissible radius of the curvature increases with the fiber optic radius; thinner fibers can be bent more than the thicker ones. Also, large values of Δ permit more abrupt bends.

If the optical fiber is subjected to a spatial periodic bending, the mode conversion effects give very high radiation losses.

2.2. The fiber optic load sensor

The principle of the developed load sensor is depicted in Figure 3. When a load is applied on the upper plate the two rows of pins come closer and bend the optical fiber more or less.

The curvatures produced increase the transmission loss. The upper plate has eight pins with a diameter of 2.5 mm each, placed at 10 mm distance from one another. The lower plate has nine pins identically separated, but shifted 5 mm with respect to the upper pins. The optical fiber used was a graded index one (50/125 μm) having a numerical aperture of 0.2. The fiber is coated with 15 μm of acrylate. The measured optical loss for this fiber was 2.8 dB/km at 850 nm.

An example of the load-attenuation characteristics is given in Figure 2.

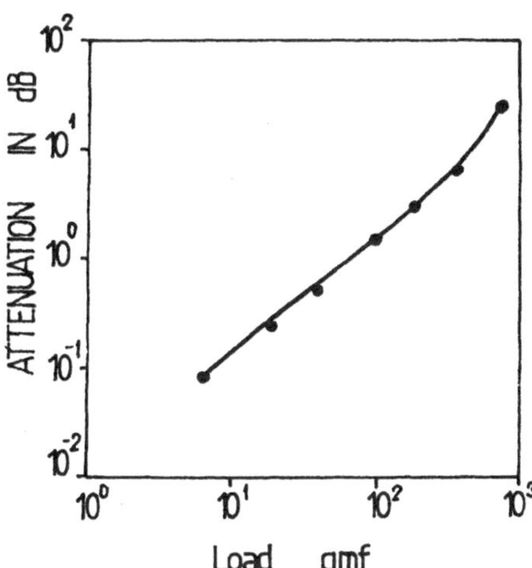

FIGURE 2. The load-attenuation characteristic for the optical fiber sensor.

3. THE EXPERIMENTAL SET-UP

The block diagram of the experimental set-up is shown in Figure 3. The radiation source employed was a light emitting

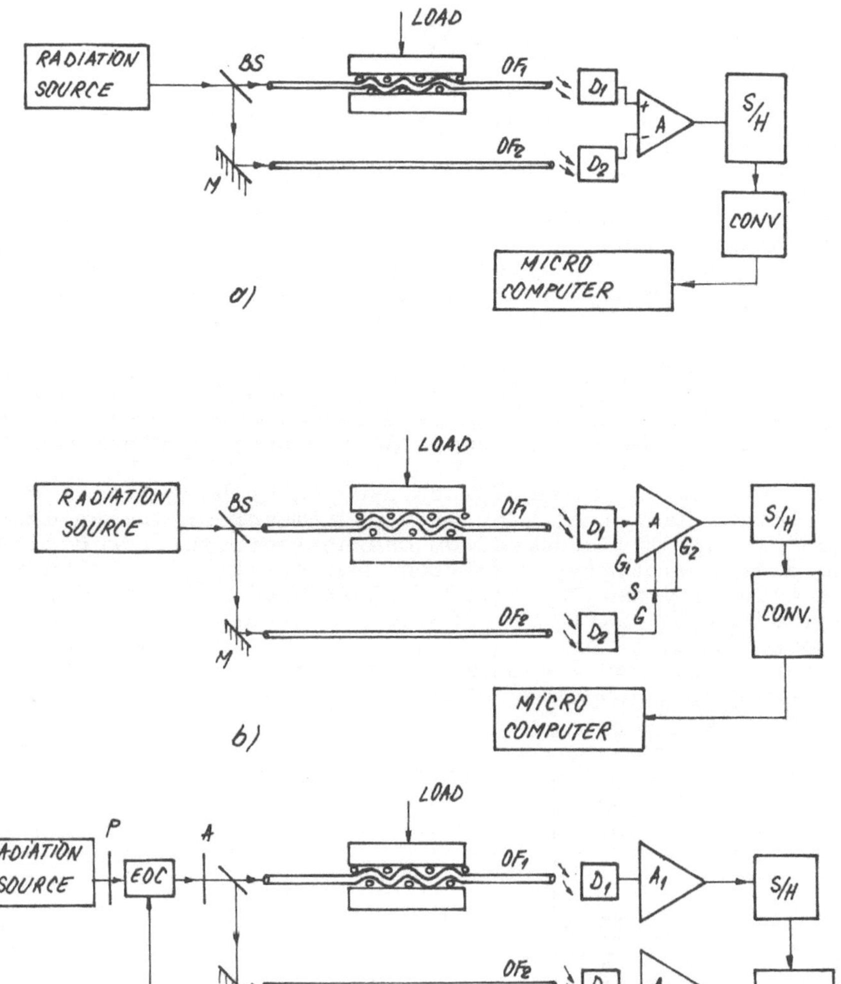

FIGURE 3. The block diagram of the experimental set-up for the fiber optic load cell. Various compensating systems: a) The differential amplification with common mode rejection. b) The optoelectronic automatic gain control. c) The electrooptic phase shifting method.

diode (LED) having a launch of -20 dBm at 850 nm or a He-Ne laser emitting at 6.32 nm with a power of 2 mW.

As the radiation source output signal amplitude is subjected to time variation and the transmission losses induced by the cell loading appear as amplitude changes, a second optical fiber (OF_2) was used to accompany the load cell optical fiber (OF_1), for compensation.

The radiation emitted by the source passes through a beam splitter and the resulting beams are coupled to the inputs of the two optical fibers OF_1 and OF_2. The two optical fibers are running parallel with the only difference that OF_2 does not pass through the load cell.

The signals transmitted by the optical fibers are detected by two photodetectors (D_1, and D_2); after detection they are amplified.

The signal provided by detector D_2 can be used to compensate for the intensity variation of the radiation source and the attenuation caused by the optical fiber bendings, other than those produced by the load cell. For example, when the sensor is used for remote measurements, its positioning can require the flexing of the optical fiber, which will give additional transmission losses. These losses correspond to a noise in the overall detected signal so that the S/N ratio decreases.

In the differential compensation scheme (see Fig. 3a), the common mode rejection capabilities of the operational amplifier cancel the influence of the unwanted perturbation appeared in the detected signal.

In the second solution (see Fig. 3b), the signal obtained from the reference optical fiber (OF_2), detected by D_2, controls the gain of a gain-controlled amplifier. The control signal is applied on the gate of a field effect transistor (FET). The FET working as a voltage control resistor is connected to the gain control inputs of an operational amplifier µA 733. The phase of the control signal is properly chosen so that the signal at the amplifier output increases when the radiation source intensity decreases and vice-versa. In this way the µA 733 amplification compensates for the time variation of the radiant source output power (5). By the same method - the optoelectronic automatic gain control - parasitic losses other than those introduced by the load sensor can be compensated for if the beam bifurcation through OF_1 and OF_2 is performed just in front of the load cell.

In Figure 3c, an electro-optic phase shift (6) was used to produce the same effect as the system in Figure 3b. Before the beam splitter, the radiation from the source passes through a polarizer, an electro-optical cell and an analyzer. When an intensity change of the radiation beam, not subjected to the load is detected by D_2, the voltage on the electrooptic cell is modified in such a way that the intensity at the beam splitter input is independent of the variations of the radiation source. The principle of operation can be used also to compensate for the transmission losses caused by other bendings than those generated in the load cell.

All these methods result in a significant improvement of the S/N ratio. They relax the conditions imposed to radiation source output power stabilization and extend the distance at which the sensor can be used.

The signal obtained at the D_1 output is coupled to the input of a sample/hold circuit.

The analog to digital conversion was performed using an A/D converter (ADC85-10) or a voltage/frequency (V/F) converter (VFC 12) working with a programmable timer 8253 for frequency measurement. One timer channel functions as a programmable one-shot and controls the gate input of a second channel, programmed for the interrupt on terminal count mode.

The end of the pulse provided by the first channel generates an interrupt request for the microcomputer. From the content of the second timer at this moment can be computed the digital value of the analog signal corresponding to the bending loss in the cell.

The microcomputer employed was a Romanian-made minicomputer TPD, using the CP/M operating system.

4. CONCLUSIONS

The study of a fiber optic load sensor based on the "bending loss" principle was investigated. Several solutions to compensate for unwanted transmission losses in the system were suggested. The sensor was coupled through an A/D converter or a V/F converter to a minicomputer for data processing.

REFERENCES

1. Giallorenzi TG, Bucaro JA, Dandridge A, Sigel GHJr., Cole YH, Rashleigh SC and Priest RG: Optical Fiber Sensor Technology, IEEE J. Quantum Elect. QE-18, p. 626-664, 1982
2. Culshaw B, Davies DE, Kingsley SA: Multimode Optical Fiber Sensors, Advances in Ceramics, vol. 2, edited by Bendow B and Mitra SS, Ed. New York, 1981
3. Munia Q and Weber MP: Fiber Optic Sensor in a Resonant Optoacoustic Cell. Optics Communications, 52, No 4, p. 269-273, 1984
4. Gloge D: Bending Loss in Multimode Fibers with Graded and Ungraded Core Index, Appl. Opt. 11, p. 2506-2512, 1972
5. Sporea D and Miron N: Optical AGC Minimizes Video Measurement Errors, Electronics Designer Casebook No 5, New York: McGraw-Hill Publications Co., 1982
6. Miron N: Signal Processing Methods in Laser Interferometry and Anemometry, Ph.D. Thesis, 1983

A NEW DESIGN FOR 6-COMPONENT FORCE/TORQUE SENSORS

KOZO ONO *, YOTARO HATAMURA **

* Hitachi Construction Machinery Co. Ltd., 650, Kandatsu-machi, Tsuchiura, Ibaraki, 300 Japan

**University of Tokyo, 7-3-1, Hongo, Bunkyo-ku, Tokyo, 113 Japan

1. INTRODUCTION

Recently, 6-component force/torque sensors have been drawing the attention of reserchers and engineers in various engineering fields especially in robotics. Most of the present industrial robots are based on position control. Therefore they are not applicable for inserting mating parts in precise assembly work or deburring and grinding in metal working. For those applications, force control is indispensable and 6-component force/torque sensors play a main role therein. Such force/torque sensors as expected in robotics must satisfy the following conditions.

(1) High accuracy and resolution to be useful for the above-mentioned applications.

(2) Compact size and light weight to save more handling capacity for a robot, which is always suffering from shortage of it.

(3) High rigidity to detect accurate data. The detected data of a force/torque sensor of low rigidity tend to contain parasitic effects from loads.

In order to satisfy the condition (1), a sensing block of a force/torque sensor must be constructed only with rigid or purely elastically deformed parts, and desirebly be of a mono-block construction. Previous work in this field (Charles Stark Draper Laboratory[1], Stanford Reserch Institute etc.) have skillfully solved this problem. Their sensors are based on a principle of, so to speak, a uniformly deformed beam as shown in Fig.1. According to this principle, local strains of the beam are the same order of those at its strain detecting parts, and, in consequence, its total deformation δ_1 becomes comparatively large, as illustrated in Fig.1. In order to reduce this deformation while keeping the same sensitivity, every dimension of a beam must be enlarged. Thus, this principle does not satisfy conditions (2) or (3). For solving this trade-off problem, semiconductive strain gauges are adopted for most of those sensors instead of metallic

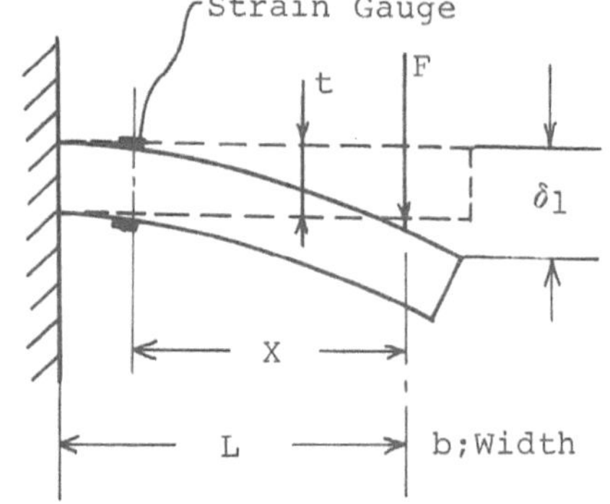

Fig.1 A Uniformly Deformed Beam

Wieringa, H. (ed), Mechanical Problems in Measuring Force and Mass.

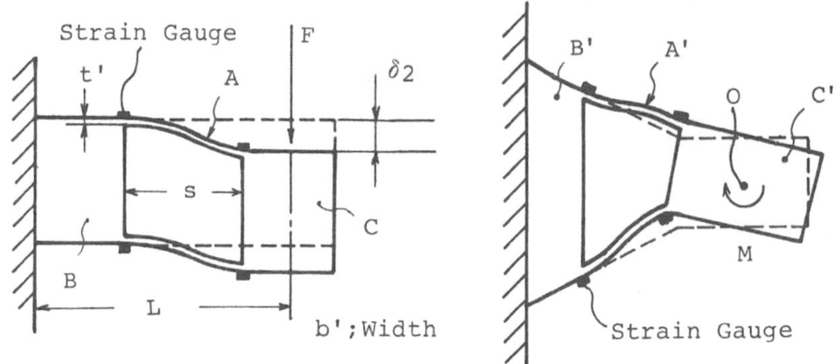

Fig.2 A Paralle Plate Structure Fig.3 A Radial Plate Structure

ones, which makes them very sensitive to temperature changes.

2. PARALLEL/RADIAL PLATE STRUCTURE

2.1 Fundamental structure

A Parallel Plate Structure (Fig.2), hereinafter called a PPS for brevity, is composed of a plurality of flexible plate-like beams A, arranged in parallel to one another connecting a fixed rigid body B and a movable body C. A PPS is suitable for a detecting device of a force F acting perpendicularly to its parallel plate-like beams A, becasuse a PPS is mostly rigid to any load component except to the force F shown in Fig.2.

A Radial Plate Structure (Fig.3), hereinafter called an RPS for brevity, is composed of a plurality of flexible plate-like beams A', arranged radially from a fixed point 0, connecting a fixed rigid body B' and a movable body C'. An RPS is suitable for a detecting device of a moment M acting about the central axis of radially arranged plate-like beams A', because an RPS is mostly rigid to any load component except to the moment M shown in Fig.3.

2.2 Comparison with a conventional technology

Now, let us make a quantitative comparison of a PPS and a uniformly deformed beam. The strain output $\varepsilon 1$ and the total displacement $\delta 1$ of a uniformly deformed beam shown in Fig.1 are calculated as follows.

$$\varepsilon 1=(6Fx)/(Ebt^2) \qquad (1)$$

$$\delta 1=(4FL^3)/(Ebt^3) \qquad (2)$$

The equivalent factors $\varepsilon 2$ and $\delta 2$ of a PPS shown in Fig.2 are calculated as follows.

$$\varepsilon 2=(3Fs)/(2Eb't'^2) \qquad (3)$$

$$\delta 2=(Fs^3)/(2Eb't'^3) \qquad (4)$$

Then, the ratios of the total beam displacement to the output strain $\delta 1/\varepsilon 1$, $\delta 2/\varepsilon 2$ are induced as follows.

$$\delta 1/\varepsilon 1=(2L^3)/(3xt) \quad \cdots\cdots (5)$$

$$\delta 2/\varepsilon 2=s^2/(3t') \quad \cdots\cdots (6)$$

Let r be $(\delta 1/\varepsilon 1)/(\delta 2/\varepsilon 2)$, then,

$$r=2(L/s)^2(L/x)(t'/t) \quad \cdots\cdots (7)$$

The value of L/s is on the order of 10 (s is actually very small but shown enlarged in Fig.2); L/x, 2 to 3; t'/t, 1/30 to 1/50. So r may be estimated to be about 10 to 20.

The above estimation shows that a force sensor based on a PPS has 10 to 20 times higher rigidity than one based on a uniformly deformed beam, assuming one uses the same strain detecting devices. The same tendancy exists for moment detection by an RPS. It also means that the PPS/RPS principle enables a force/torque sensor to be reduced in size and weight. This is one of the most important features for robotic applications.

2.3 Application for 6-component force/torque sensors

In Fig.2 and 3, every PPS and RPS is shown as a block with a rectangular or a trapezoidal hole therein. But these kinds of holes are not easy to machine. Therefore, such PPS/RPS as made of a block with two circular holes connected by a slit as

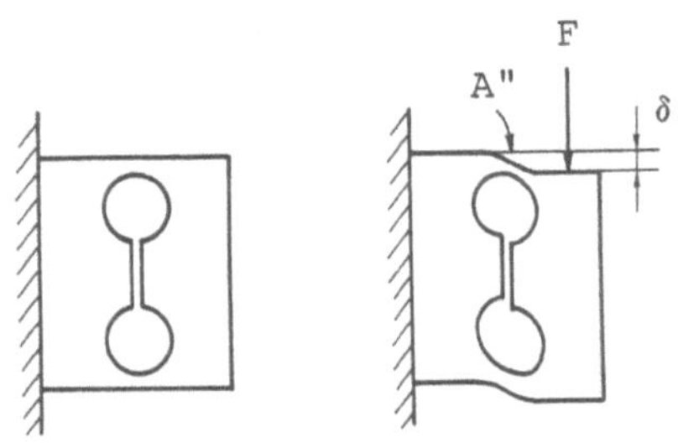

Fig.4 A Circular Holled PPS

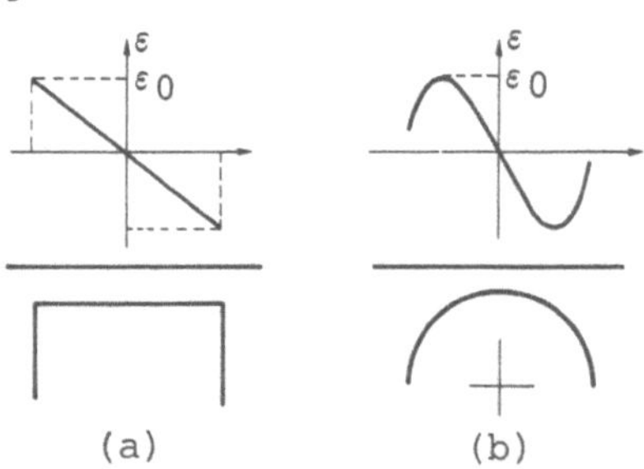

Fig.5 Comparison of Strain Distribution

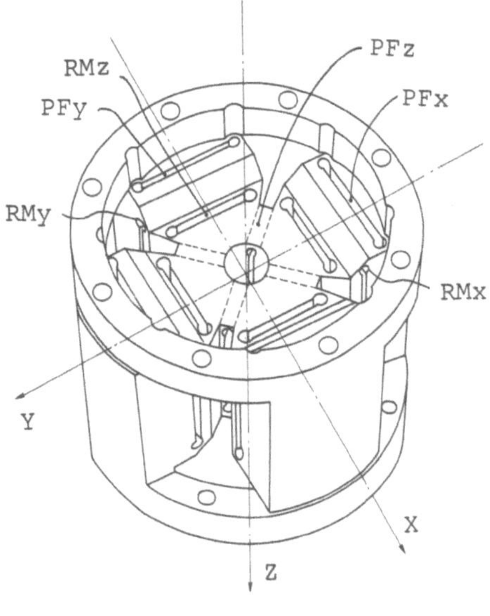

Fig.6 Sensing Block of the Proto-Type Model

shown in Fig.4 is proposed for better machinability. Such a circular holed PPS/RPS has additional favorable features. As shown in Fig.5, it is easier to mount strain gauges at the peak point of strain distribution along flexible plate-like beams A" owing to the difference of strain distribution between normal PPS/RPS (a) and a circular holed one (b).

Fig.6 shows a sensing block of a 6-component force/torque sensor cmmposed of 3 PPS's and 3 RPS's with each standard axis coincident with each coordinate axis. Each PPS of PFx and PFy is composed of two symmetrically arranged PPS's and each RPS of RMx, RMy and RMz is composed of two symmetrically arranged RPS's. PFx and PFy are detecting parts of force components of Fx and Fy respectively. RMx, RMy and RMz are detecting parts of moment components Mx, My and Mz respectively. A PPS designated as PFz is called a cruciform PPS which is a set of 4 PPS's arranged radially in the center of the sensing block, and functions as a detecting device of force Fz.

For this sensing block, circular holed PPS's and RPS's are used. This sensing block based on the PPS/RPS principle was proved to be effective for practical use after trial manufacturing and testing [2].

3. PROTO-TYPE MODEL

Based on the sensing block whose fundamental construction is shown in Fig.6, a proto-type model was developed for a commercial use. Several modifications were made to satisfy the demands of robotic engineers to reduce every dimensions as much as possible and to increase the rated moment as much as possible. Table 1 shows the calibration matrix of the proto-type model and Table 2 shows its total accuracy. The calibra-

Table 1 Calibration Matrix of the Proto-Type
(Unit of Coefficients; mV/ Rated load)

$$\{C_{ij}\} = \begin{pmatrix} 2239 & -33 & -54 & -187 & -664 & -248 \\ 6 & 2018 & 11 & -384 & -54 & -202 \\ -380 & 402 & 1515 & 59 & -239 & 1783^{*} \\ 23 & 56 & -94 & 4207 & -141 & -21 \\ -35 & 16 & 51 & 104 & 4373 & -73 \\ -28 & 7 & -7 & 40 & -44 & 1643 \end{pmatrix}$$

Table 2 Total Accuracy of the Proto-Type

Input Load	Maximum Errors (% of Rated Output) After Interaction Singal Compensation					
	Fx	Fy	Fz	Mx	My	Mz
Fx 10kgf	0.3	0.1	0.2	0.0	0.1	0.1
Fy 10kgf	0.1	0.4	0.6	0.1	0.0	0.1
Fz 10kgf	0.5	0.2	0.8	0.1	0.1	0.2
Mx 1.0kgf-m	0.4	0.3	1.3	1.1	0.4	0.3
My 1.0kgf-m	0.3	0.1	1.0	0.1	0.7	0.1
Mz 1.0kgf-m	0.5	0.6	2.1	0.2	0.3	0.9

tion matrix $\{c_{ij}\}$ is obtained by calibration procedures and is defined by the following matrix equation;

$$\{S_i\} = \{C_{ij}\} * \{L_j\} \quad (i,j=1,2,\cdots 6) \quad \cdots\cdots \quad (9)$$

where S_i is an output of each strain gauge bridge, and L_j is each rated load; Fx, Fy, Fz, Mx, My, Mz.

From the calibration matrix $\{C_{ij}\}$ and the output of strain gauge bridges $\{S_i\}$, actual load components $\{L_k\}$ are calculated. Let the inverse matrix of $\{C_{ij}\}$ be $\{D_{ki}\}$, then,

$$\{L_k\} = \{D_{ki}\} * \{S_i\} \quad \cdots\cdots \quad (10)$$

The above procedures are well known and disclosed in [3] for example.

If $\{L_k\}$ and $\{S_i\}$ are normalized by rated loads and rated strain output respectively, the matrix $\{D_{ki}\}$ would ideally be so composed that each diagonal component is equal to 1 and that all the other components are equal to 0. The total accuracy shown in Table 2 means the differences in percentage from the ideal value of 1 or 0.

Judging from the above data, the proto-type model was proved to have an accuracy of about 2% and the maximum error mainly came from the unusually large coefficient of C_{36}, which is designated with * in Table 1. The coefficient C_{36} shows the infuluence of the input moment Mz to the output S_3. Measured data of the calibration also showed the calibration curve of C_{36} had far bigger hysteresis (4%) than the others'. After several experimental stress analyses on the sensing block, it was found that the cruciform PPS (shown as PFz in Fig.6) deformed too much by a moment Mz, and that it was because its dimensions had been too much reduced to bear a large rated moment.

4. THEORETICAL AND EXPERIMENTAL ANALYSES OF PPS

Therefore, it is found to be necessary to know precise deformations and strain distributions of a PPS and an RPS. They were analized theoretically by an FEM method and experimentally by strain gauge technique for a simple PPS. A simple PPS was selected because it seemed simpler for analysis than an RPS and a cruciform PPS, and we believed its tendancy would represent the others'.

4.1 FEM analysis

A simple PPS whose dimensions are indicated in Fig.7(a) was analyzed by FEM. The model was divided into 1500 elements, and the number of grids was 2416. Typical results are shown in Fig.7(b) to (d). The applied load and the maximum deformation are shown by a thick arrow and a thin arrow respectively. For this analysis, the FEM program developed by Hitachi Laboratory of Hitachi Ltd. was used. This program was furnished with a special function to display minute deformations in enlarged form together with a total model with a nonlinear scaling. According to its rule, the maximum deformation of each calculated model is shown as the same dimension in each figure. So, the readers must realize that the scale factor for the maximum deformation in each figure in Fig.7 is different.

As shown in Fig.7, the maximum deformations of the PPS model by moments Mx and My are the same order as the one by a force Fz when the force is 10kgf and the moment is 1kgf-m. This ratio of a rated moment to a rated force was chosen according to the demands of robotics and is comparatively larger than those of conventional force/torque sensors.

From the results of the FEM analysis, the strain value of each strain detecting part of the PPS model was calculated. Those values showed that the strain under the moment My was comparative to the one under the force Fz.

4.2 Experimental analysis

The main aim of this FEM analysis was to know macroscopic deformations of a PPS. Therefore the division of elements was not so precise as would be used for calculating local strains. So it was suspected that the strains calculated above would not be accurate. Therefore, the strains were measured by a strain gauge method using an equivalent model to the one shown in Fig.7(a). One example of measured data is shown in Table 3. The symbols in Table 3 designate the positions of strain gauges following the same way as shown in Fig.8. The strains in Table 3 are shown as magnifications of a designed strain εd, which is a strain to be generated at a strain detecting part of ideally manufactured PPS/RPS under its rated load. The strain gauges Ga2 and Gb2 are put at the corresponding positions to the strain detecting parts of the proto-type model shown in Fig.6, and the other gauges are put at the adjacent positions for reference. It must be noted that the strains of the adjacent parts are also comparable to the designed strain εd even in the cases of loads other than a force Fz and a moment My.

Table 3 Measured Strains of a PPS

Applied Load	Measured Strain / Designed Strain (εd)					
	Ga1	Gb1	Ga2	Gb2	Ga3	Gb3
Fz	1.15	-1.36	1.20	-1.31	1.31	-1.31
Mx	-1.41	0.84	-0.10	0.10	0.94	-1.20
My	1.73	-2.88	1.78	-2.98	0.99	-3.25

← Strain Gauge Position (See Fig.8)

4.3 Considerations on the proto-type model

The above analysis showed that under large moments Mx and My, a PPS is deformed and generates strains in comparable amount to those under rated force of Fz. It is reasonable to presume that an RPS and a cruciform PPS have the same tendancy as a simple PPS. An experimental analysis was made for a cruciform PPS and it was found that abnormally high strain was generated under a moment Mz. This was the main reason for the

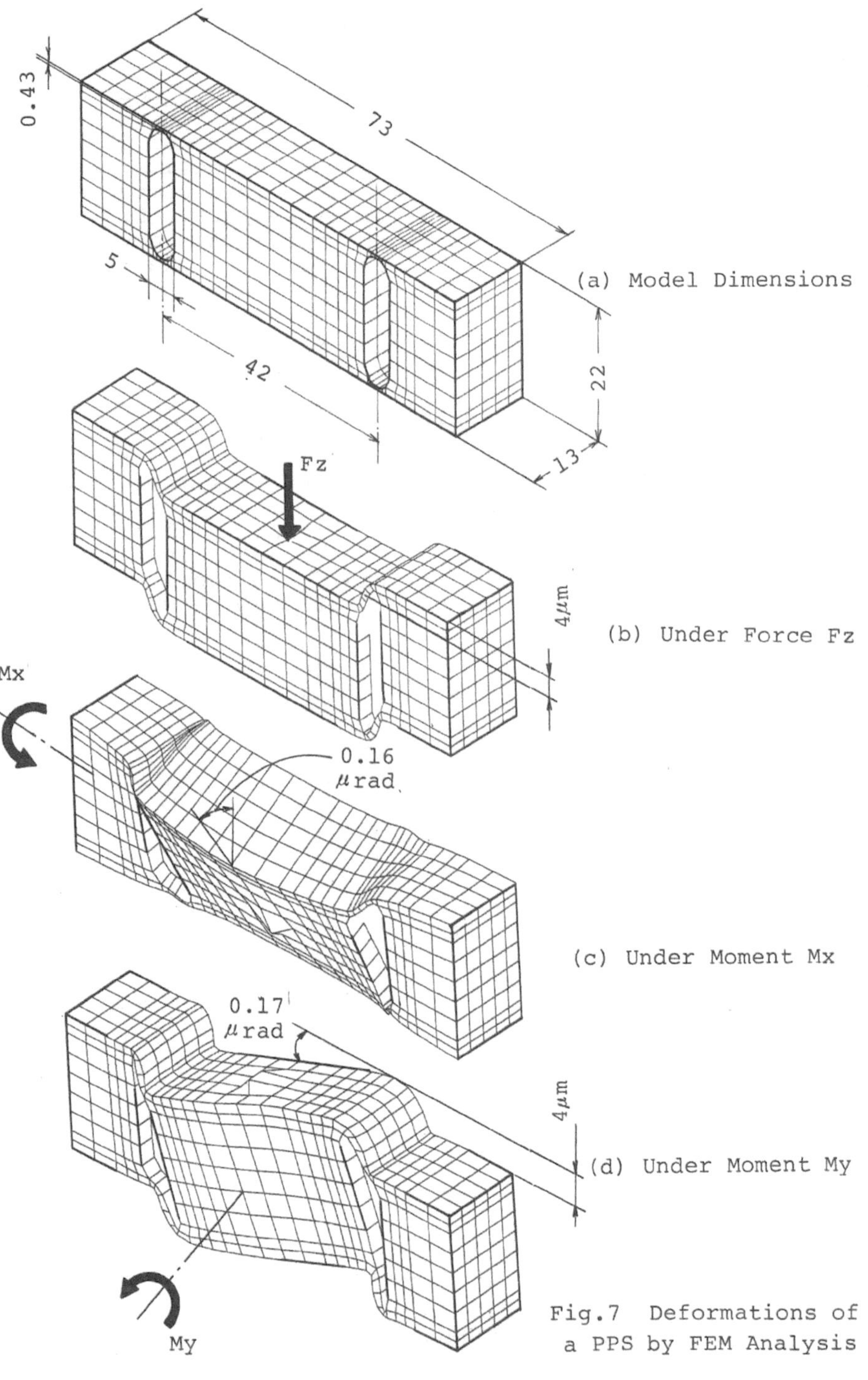

Fig.7 Deformations of a PPS by FEM Analysis

big calibration coefficient C_{36} mentioned in Chapter 3.

5. FINAL MODEL

5.1 A new design concept for load detecting devices

During the above mentioned analyses and considerations, a new design concept for load detecting devices was discovered. The main features of the new load detecting device is that one detecting unit alone can detect multiple load components while fully utilizing the characteristics of PPS/RPS. Fig.8 is an example of the new load detecting device. This device detects moments Mx and My as well as a force Fz. One example of strain gauge arrangements is shown in the attached table of Fig.8. The readers should notice that, even in this embodiment, the selectivity of rigidity of a PPS is effectively utilized.

5.2 A sensing block of the final model

Based on the above-mentioned new concept, a final model was designed. Fig.9 is the fundamental construction of its sensing block, composed of only 3 PPS's. In this embodiment, PPS's designated as PFx and PFy in the figure detect not only forces Fx and Fy respectively but a moment Mz also, and a cruciform PPS designated as PFz detects not only a force Fz but moments Mx and My also. In this final model, its rated moment was reduced from that of the proto-type model. The

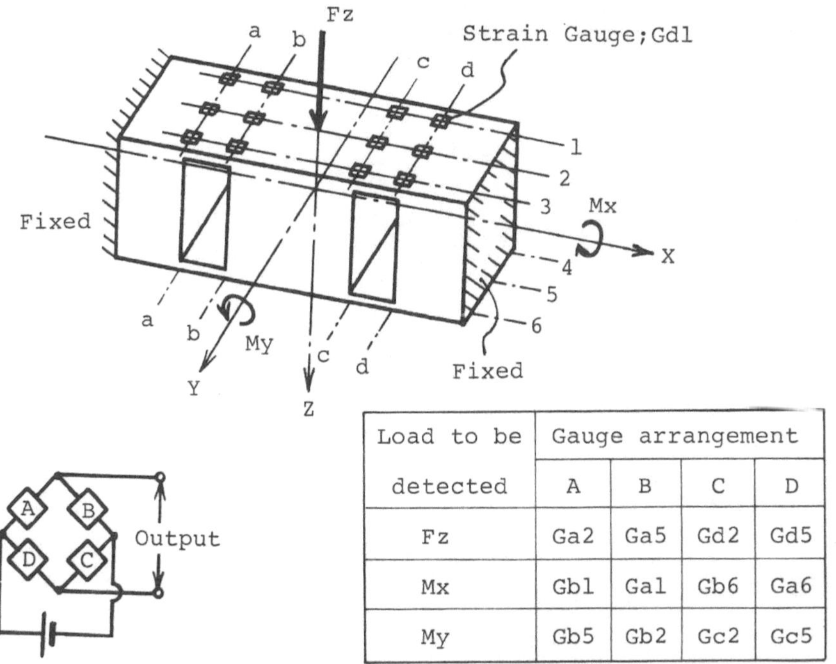

Load to be detected	Gauge arrangement			
	A	B	C	D
Fz	Ga2	Ga5	Gd2	Gd5
Mx	Gb1	Ga1	Gb6	Ga6
My	Gb5	Gb2	Gc2	Gc5

Fig.8 A New Load Detecting Device Using a PPS

robotic engineers would like force/torque sensors of even larger moment capacity. But there is a trade-off relation between performance and moment pacacity of force/torque sensors.

From this new concept of utilizing PPS/RPS for a detecting device of multiple load components, emerge two important merits. First, fabrication of a sensing block becomes far easier. Second, every PPS (or RPS) can take enough space to maintain necessary dimensions to keep good characteristics within the limitations of the total dimensions of a force/torque sensor. Because of the above merit, the performance of the final model was improved in great deal. Examples of its calibration matrix and total accuracy are shown in Table 4 and Table 5 respectively. Table 5 shows that the final model has an accuracy of 0.2%.

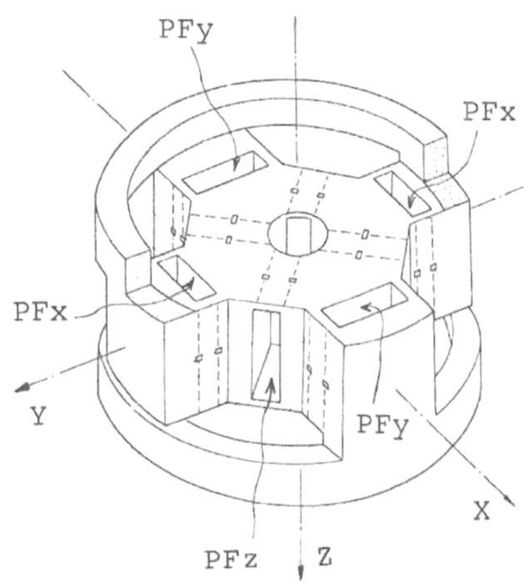

Fig.9 Sensing Block of the Final Model

Table 4 Calibration Matrix of the Final Model (Unit of Coefficients; mV/ Rated load)

$$\{C_{ij}\} = \begin{pmatrix} 2331 & -20 & 8 & -84 & -547 & -49 \\ 2 & 2139 & -2 & -801 & -25 & -39 \\ 141 & -110 & 3011 & -353 & 168 & 673 \\ -31 & 631 & -64 & 3231 & -198 & -799 \\ 722 & -30 & -25 & -45 & 3386 & 763 \\ -6 & -41 & 7 & 65 & 27 & 2993 \end{pmatrix}$$

Table 5 Total Accuracy of the Final Model

Input Load	Maximum Errors (% of Rated Output) After Interaction Signal Compensation					
	Fx	Fy	Fz	Mx	My	Mz
Fx 30kgf	0.1	0.0	0.0	0.0	0.1	0.0
Fy 30kgf	0.1	0.0	0.0	0.1	0.0	0.0
Fz 30kgf	0.0	0.0	0.1	0.1	0.0	0.0
Mx 1.5kgf-m	0.1	0.2	0.1	0.1	0.2	0.1
My 1.5kgf-m	0.2	0.1	0.1	0.1	0.1	0.1
Mz 1.5kgf-m	0.0	0.0	0.1	0.1	0.1	0.1

5.3 Commercial model

At the present time, 6-compoment force/torque sensors are commercially available as LSA6000 series models (those force capacities are from 10 to 30 kgf) from Hitachi Construction Machinery Co.Ltd.. For the time being, they are sold only inside Japan, but we intend to supply them abroad in the near future.

Each model is composed of a transducer unit and a data prcessing unit as shown in Fig.10. The transducer unit is constructed with a sensing block and analogue electronic circuit boards, and designed so compact (81mm diameter, 56mm height, 380gram weight) as to be applied not only for robotics but for any kind of applications. The data processing unit is a compact 16-bit micro-computer, having functions of interaction signal compensation, software variable filtering, over load checking, signal transmission control etc.

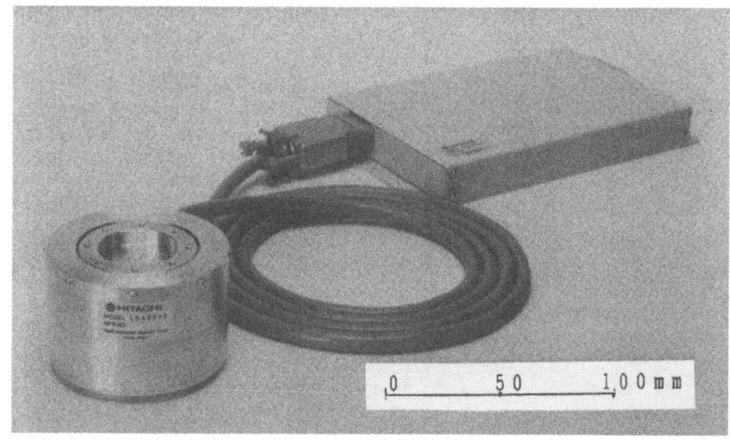

Fig.10 Hitachi 6-Component Force/Torque Sensor LSA6000 Series Model

6. CONCLISIONS

University of Tokyo has developed the PPS/RPS principle and a fundamental model of a 6-component force/torque sensor based on the principle. Hitachi Construction Machinery Co.Ltd. has improved it and developed commercial models of 6-Component force/torque sensors LSA6000 under a tight cooperation with University of Tokyo. Those force/torque sensors are ranged from 10 to 30kgf force capacity and have an accuracy of 0.2%.

REFERENCES

[1] P. C. Watson et al: Method and Appratus for Six Degrees of Freedom Force Sensing, U.S. Patent No.4,094,192, June 13, '78

[2] Y. Hatamura et al: Development of Six-Axis Force Sensor, Preprint of the Japan Society of Mechanical Engineers, Sept. 28, '84, pp.94-96 (In Japanese)

[3] M. Dubois: Étalonnage de Dynamometres et de Balances au Centre d'Essais de Modane-Avrieux, ONERA T.P. No.452, '67, pp.1-20

A Directly Weighing Suspension Balance With Frequency Variant Output

Th. Gast,G. Luce

Institut für Meß- und Regelungstechnik,Technische Universität Berlin

1. INTRODUCTION

This discourse will try to discuss the following question: What are the possibilities of measuring the force acting on a free magnetic suspended solid part (Fig.1)?

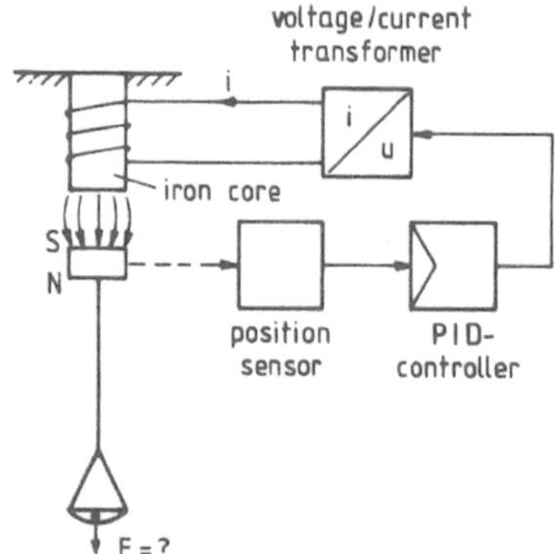

Fig.1 **Control Circuit of a Free Magnetic Suspension**

The position of the suspended part is detected with the aid of capacitive, inductive or optoelectrical arrangements and is fed to PID-control as actual value.The control loop is closed by a voltage to current transformer supplying an electromagnet as final control element.A well known and simple method of force determination is the measurement of the exciting current i, while the distance of the attracting magnetic poles is kept constant. However,no high accuracy will be achieved on this basis because,as a result of the non-constant permeability and the hysteresis of the ferromagnetic core and return path,the magnetic flux is not sufficiently correlated with the exciting field strength generated by the coil current i.Besides,the coil current causes additional errors by heating the electromagnet and the adjacent position sensor.Therefore,direct measurement of the magnetic flux or the flux density in the air gap between the magnetic poles is preferable.

2. PRINCIPLES FOR MEASUREMENT OF THE FLUX DENSITY

Direct measurement of the flux density can easily be achieved by inserting a Hall-probe into the air gap,but because of the temperature

Wieringa, H. (ed), Mechanical Problems in Measuring Force and Mass.

dependence of the Hall signal,the achievable accuracy will not be much better than by application of the first method.

Several principles for precision measurement of the magnetic flux density are known,f.e. Faraday effect,nuclear magnetic resonance,Zeeman effect a.o. These methods require a considerable test probe volume and field homogeneity better than 10^{-4} within a certain volume which can not be realized with the given assembly,where the air gap dimensions are 8X8X5 cmm approximately.Because of its comparative simplicity the Faraday effect,for example, should be taken into consideration (Fig.2):

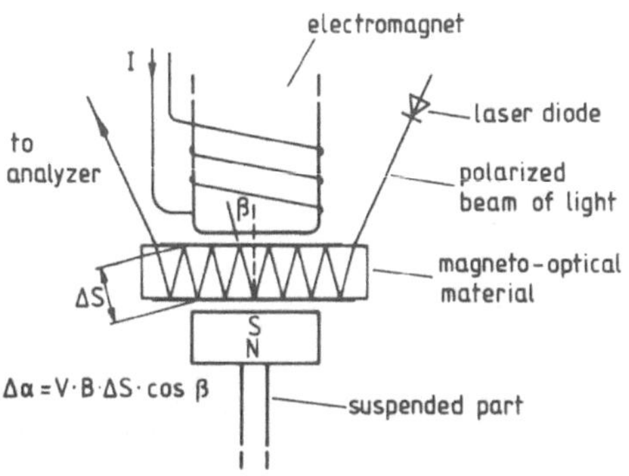

Fig.2 **Measurement of Flux Density with the Aid of the Faraday Effect**

The plane of polarization of a polarized light beam passing along the lines of a magnetic field rotates by an angle α which is proportional to the field strength and to the length of the light path.The proportionality factor is Verdets' constant v.For most materials v is very small.Thus a light path length of several meters within the air gap would be necessary to enable a sufficient sensitive field strength measurement.Japanese measuring devices based on this method have been reported recently.They use special crystalline magneto-optical elements with high Verdets' constants.But the sensor probes are still too large and the accuracy is still too low.The application of fibre optics,suggested for current measurement via magnetic field strength in high voltage appliances,is not possible because the necessary fibre length can not be installed within the air gap.

Another possibility consists in the series connection of a operative- and a measuring air gap (Fig.3).In the upper gap,a sufficiently large volume of homogeneous magnetic field can be achieved and the very exact measurement by nuclear magnetic resonance could be used.However,the errors due to the leakage flux inherent in the assembly lower the value of this precise method to a large extent.

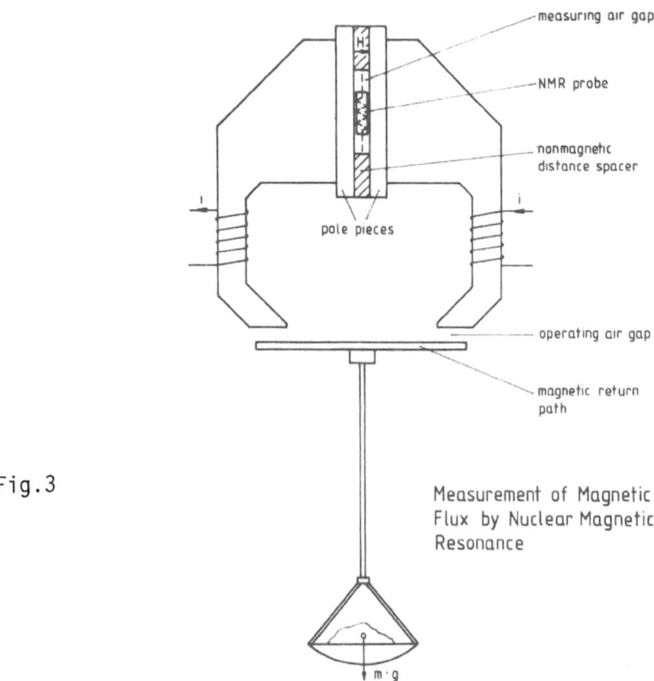

Fig.3 Measurement of Magnetic Flux by Nuclear Magnetic Resonance

3. PRINCIPLE OF THE DIRECTLY WEIGHING SUSPENSION BALANCE

There exists another possibility of force and mass determination in the suspended state,requiring only a minimal expenditure (Fig.4):

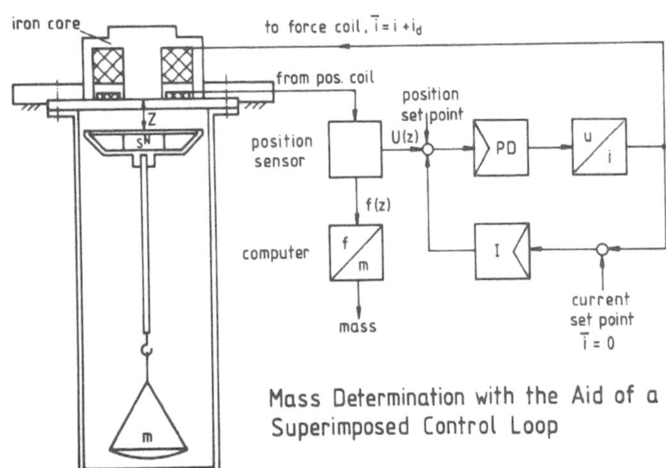

Fig.4 Mass Determination with the Aid of a Superimposed Control Loop

The position of the suspended magnet is measured by a high frequency eddy current sensor,consisting of a flat coil on a glass partition and an aluminium disc mounted on top of the suspended part.The partition supports the electromagnet,while the sample is suspended inside the closed reaction chamber to allow weighing under controlled atmospheres.As to be seen,the mechanical assembly is very simple.The device operates as follows:

The position sensor generates a frequency f(z),dependent on the distance z.A voltage u(z),dependent on f,is fed via a PD-controller to the electromagnet.The dynamic disturbances are superposed as fluctuations on a certain average value of the coil current: $i = \bar{i} + i_d$.$\bar{i}$ is a value for the additional exciting current in the control winding. This current increment is reduced to zero by a superimposed integral controller.For a given weight of the sample on the weighing pan,the distance z between the magnetic poles is thereby adjusted to a certain value which is determined by the force - distance characteristic of the magnetic system.If the weight increases,the distance decreases and vice versa.The system acts similar to a spring with negative compliance.Because the distance of the poles is correlated to the frequency of the oscillator,frequency and weight are correlated too.The frequency generated by the position sensor is consequently converted to the mass value.

Fig.5 shows the electrical block diagram of the position sensor:

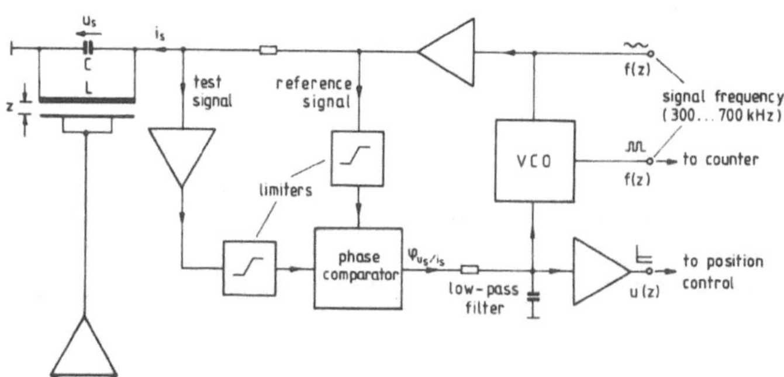

Fig.5 **Block Diagram of the Frequency Variant Position Sensor for a Magnetic Suspension Balance**

The phase difference between the voltage u_s and the current i_s on the resonant circuit is kept at zero by a phase locked loop.L and C determine the frequency of phase resonance,which is the measured frequency. Application of precision wideband operational amplifiers provides high stability of the distance - frequency characteristic.The control voltage of the oscillator serves also for the position control as signal.

Fig. 6 illustrates a special feature of the method:

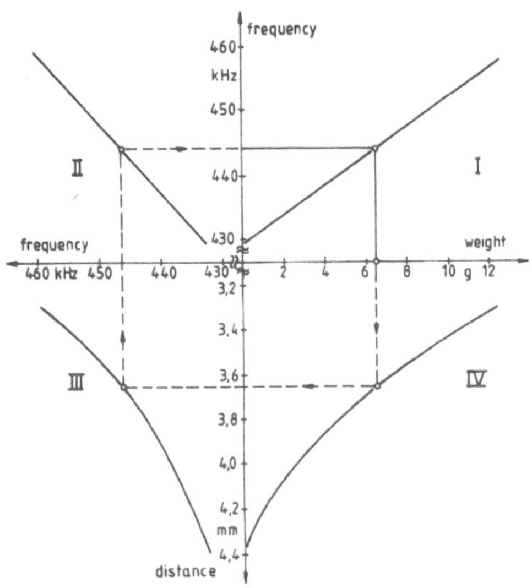

Fig.6 **Relations between Weight, Distance and Frequency in a Suspension Balance**

The IV. quadrant shows the force - distance characteristic of the magnetic poles,the III. quadrant presents the frequency - distance characteristic of the eddy current sensor.In the II. quadrant,the frequency values are projected to the mass - frequency characteristic,i.e. the weighing characteristic shown in the I. quadrant.It is obvious,that the nonlinear characteristics III and IV are bent in an opposite manner,which ultimately leeds to a linear weighing characteristic of frequency versus weight (Fig.7).

Fig.8 shows some properties of magnetic materials.Rare earths-cobalt magnets possess a remanent magnetism of 900 mT and a maximum energy product of about 120 kJ/m^3.Lately,neodymium - iron - boron magnets are available,which show a remanent magnetism of 1.3 T and a maximum energy product of 280 kJ/m^3.With rare earths - cobalt magnets,used in the described measuring assembly,load capacities of 100 g are realizable,while the net weight of the magnet is 3.5 g and the air gap width is 2 mm.The load range of 100 g corresponds to a change in distance of 4 mm,while the measuring counter runs through 10^5 steps,corresponding to a utilizable resolution of 10^{-4}.

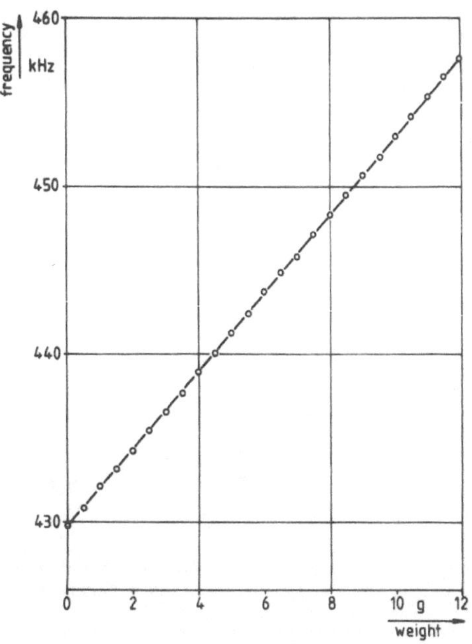

Fig.7 Weight-Frequency Characteristic of a Magnetic Suspension Balance

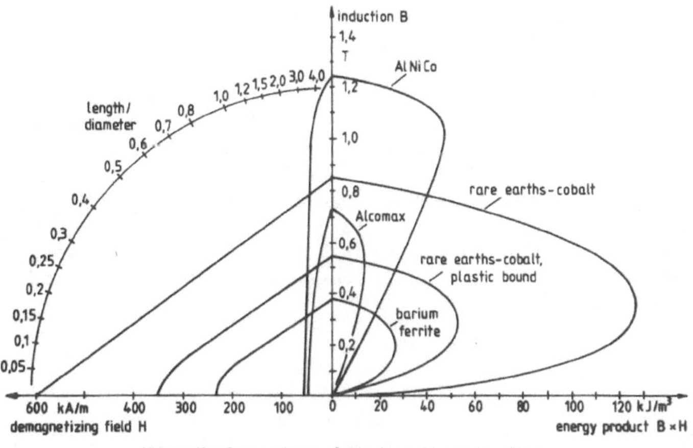

Fig.8 Magnetic Properties of Various Magnetic Alloys

MICROPROCESSOR-BASED SYSTEM FOR BELT-SCALES DOSING OF MIXED FEED COMPONENTS

A.I.Farfel
F.S.Galperin
V.A.Okun

Dosing in weighing and mixing has found an extensive use in a number of industrial processes including the processes of preparing grain mixtures in the mixed feed industry /I/. Continuous flow production processes, which ensure a significant growth of the technical-and-economic indices, have brought about the following complement of processing equipment for mixed feed components in a continuous dosing and mixing section (DMS):

-Dosing belt scales (DBS) based on strain-gauge weighing (Fig.I);

-Continuous (flow) mixer providing for the uniformity of the grain mixture at the outlet of the DMS;

-Weight verifying device (WVD) based on strain-gauge weighing principle;

-Dosing automatic control system (DACS).

Set forth in this paper are the aim of the functioning and the tasks of the DACS, along with the formulation of the DACS system of functions providing for the fulfilment of these tasks. Shown also are the DACS structures of hardware (HW) and software (SW) supports, as well as some peculiar features of the DACS software synthesis on the basis of the aggregate weighing and dosing software system (AWDSS).

<u>Key Abbreviations</u>:

DMS - dosing and mixing section;
DBS - dosing belt scales;
WVD - weight verifying device;
DACS - dosing automatic control system;
AWDSS - aggregate weighing and dosing software system;
DDC - direct digital control;
CCC - control computer complex.

The aim of the continuous dosing and mixing section (DMS) is the production of a standard quality n-component grain mixture in compliance with the assignment Z :

$$Z = (Q, R, R), \qquad (I)$$

where Q= given overall capacity of the DMS;
$R=\|r_i\|$, $i=I,n$=vector of given parts of mixture components as determined by the recipe;
$R=\|r_i\|$, $i=I,n$ =vector of permissible deviations of actual parts of components from given values, as determined by the standard

specifications.

In the mixed feed industry a standard grain mixture is one for which the following expression holds true:

$$(R-\Delta R)\sum_{i=1}^{n} qi^{B}(t) \leq \|qi^{B}(t)\| \leq (R+\Delta R)\sum_{i=1}^{n} qi^{B}(t) \qquad (2)$$

where: $\|q_i^{B}(t)\|$, i=I,n=vector of instantaneous flow rates of grain mixture components at the outlet of the DMS.

The main processing aim of the DACS is to provide for the maximum output of a standard mixture M_c,the qauntity of which is determined by the equation:

$$M_c = \int_0^{T_c} Q_{CM}(t)\,dt \qquad (3)$$

where: T_c= total duration of time intervals,which is valid for the inequality (2),for the DMS operating time(T);

Q_{CM}(t)=grain mixture flow rate at the outlet of the mixer.

Since $Q_{CM}(t) \approx Q_\Sigma$,then:

$$M_c \approx Q_\Sigma K_{TH}\, p_c\, T_э \qquad (4)$$

where: K_{TH} = DMS utilization factor /4/;

p_c =$T_c/K_{TH}\,T_э$ =probability of standard mixture output.

Q_Σ, $T_э$ and K_{TH} are the indices which stand for the output, durability and reliability of the DMS production equipment. Therefore, the IMMEDIATE TASK OF THE DOSING AUTOMATIC CONTROL SYSTEM IS THE MAXIMIZING OF p_c.

Let us take the following structured model /2/ of the continuous dosing and mixing section (Fig.2) consisting of n elements type $D_i=(\mathcal{E}_i^{o}, T_{pi}, Q_i)$, which is a dosing belt scales (DBS), and one element type S=($\mathcal{T}_{CM}$),which is a continuous mixer.

Here, $\mathcal{E}^{o}_i$=reduced error of DBS characterizing its static accuracy;

T_{pi}=mathematical expectation of the time interval during which the integral error of maintaining the given capacity of DBS is equal to zero, which characterizes the dynamic properties of DBS;

Q_i= maximum capacity limit of DBS;,

$\mathcal{T}_{CM}$= effective mixing time characterizing the smoothing properties of the mixer.

Let us also take the following mathematical models of the processes at the outlet of the structured model elements:

(I) Acting at the outlet of the elements D_i is an ergodic centered random process:

$$qi(t) = \overset{o}{q}i(t) + \Delta qi(t), \qquad (5)$$

where: $q_i^{o}(t)$= given flow rate of mixture components at the outlet of the element(coordinating action);

$\Delta q_i(t)$= instantaneous deviation of flow rate of mixture components $\Delta q_i^{o}(t)$, being a random process , the cross section of which at any fixed moment of time $\Delta q_i^{oo}(t)$ is a random value distributed according to the normal law (standard deviation of $\Delta q_i^{oo}(t)$ is determined by the value of Q_i

and ε_i, with its expectation being equal to zero).

(2) Acting at the outlet of S is a normal process centered with respect to Q_Σ, which, considering a hypothesis on the uniform mixing of the grain mixture components/I/ and the sm smoothing properties of the mixer/3/, may be expressed as follows:

$$Q_{CM}(t)=\frac{1}{\tau_{CM}}\sum_{i=1}^{n}\int^{t+\tau_{CM}} q_i(t)dt \qquad (6)$$

The accepted DMS model and the models of the acting in it processes (5),(6), as well as the determination of a standard grain mixture, bring about a decomposition of the DACS immediate task into TWO SUBTASKS FOR THE MAXIMIZATION OF p_c, viz.:

(I) DETERMINATION OF COORDINATING ACTION

$$q_i^o(t)=f[Q_\Sigma, R, q_i(t), q_1(t)\ldots q_1^{(m)}(t)\ldots q_i(t)\ldots q_i^{(m)}(t)\ldots q_n(t), q_n(t)\ldots q_n^{(m)}(t)] \qquad (7)$$

on in the simplest case:

$$q_i^o(t)=r_i\sum_{i=1}^{n} q_i(t) \qquad (8)$$

(2) STABILIZATION of $q_i^o(t)$ with an accuracy ensuring the execution of the inequality:

$$\left|\int_t^{t+\tau_{CM}} \Delta q_i(t)\right| \leq \Delta r_i \int_t^{t+\tau_{CM}} Q_{CM}(t)dt, \quad i=1,n \qquad (9)$$

Such a decomposition of the DACS task brings about its two-level structure. Placed at the upper level is a central coordinator realizing the first subtask, and at the lower level- "n" local stabilizing systems for the DBS capacity, which realize the second subtask(Fig.3).

Apart from the fulfilling its main technological aim, the dosing automatic control system should provide for the accounting of the used material and for the output, read-out of the processing data (including print-out data), as well as for the minimizing of losses caused by probable failures of the DMS equipment. In connection with this, the following system of the DACS functions has been established:

- Measuring the parameters of the flows of loose materials (loading and speed) and their digital data processing, which provides for greater accuracy of the DACS measuring channels;
- Capacity setting of the dosing belt scales and total quantity of the dosed grain mixture and its components;
- Setting of capacity relationship of the dosing belt scales included in the DACS;
- Computing the capacity of the dosing belt scales, flow rate of mixture and its components;
- Formation and output of controlling actions to the electric drives of the dosing belt scales and the WVD;
- Accuracy checking of the DBS by means of the WVD;
- Input and execution of operator commands;
- Display of operational data on the video display terminal;
- Print-out of systematized data.

The above functions are realized with the dosing automatic control system running in three operating modes. Direct digital control (DDC) of the DBS complex is effected in the dosing mode. Automatic adaptation of the DACS parameters to the actual conditions of the DBS complex is brought about in the setting-up mode. Direct digital control of the WVD and the DBS,as well as the accuracy control of the DBS, are accomplished in the verification mode.

The basic hardware should provide for the reception of DC signals of up to 5 mA,variable frequency signals of up to 2 kHz, and positional signals of 24V logic "I" level,as well as the forming of DC signals of up to IOV and positional signals of 24V logic "I" level. To realize fully the established system of functions, the DACS hardware support should include developed operating and control equipment, as well as means for the development and setting up the software support and diagnostics of the DACS.

These requirements are most completely met with by the control computer complex (CCC) CMI803.03, which is one of the family of the specified models of CCC CM I800 based on the microprocessor KP580IIK80. The CMI803.03 computer complex includes modules for the input of analogue,discrete and digital-pulse signals, and for the output of the analogue and discrete signals. The complex is equipped with a symbol display unit BTA2000-32 and a character-synthesizing printer DZM-I80. The overall software support of the CMI800 computer complex includes a two-way assembler, translator from PL/M and BASIC languages, editor,and control program for a magnetic -disk library.

The selection of the CMI803.03 control computer complex as basic hardware,as well as rather rigid requirements as to the cost of the DACS, have brought about its synthesis as a centralized system of processing control, whereby the two- level structure of the DACS is realized through its special software. The DACS special software ensures the software support for the functioning of the hardware and does solve both the inherent DACS problems (computing the instantaneous values of linear loading of dosing scales conveyor belts,belt speeds, DBS capacities, flow rate of material and amounts of released mixture, determination of coordinating and controlling actions, and also some additional problems (correction of non-linearity of transfer characteristics of the linear load measuring channels and computation of capacity, forming of averaged and smoothed estimations of the running values of linear loadings, conveyor belt speeds and DBS capacities,compensation for the temperature drift of the zero of the weighing cells and compensation for the weight of the conveyor belts). The realization of the additional problems should provide for greater accuracy of dosing by weight and for more accurate account of the used material and released product.

The synthesis of the algorithmic support of the DACS is provided on the basis of the aggregate weighing and dosing software system (AWDSS).

The AWDSS is comprised of a set of algorithms A:

$$A = \{M, V, Q, T, U, K\},$$

where: M= subset of algorithms for servicing the measuring channel for linear loading;
V= subset of algorithms for servicing the measuring channel for conveyor belt speed;
T= subset of algorithms for time measuring;

Q= subset of algorithms for servicing the computing channel for capacity and flow rate;

U= subset of algorithms for servicing the controlling action forming and output channel;

K= subset of coordinating algorithms.

Every one of the six subsets contains realization algorithms for the imperative and additional tasks. For instance, subset M:

$$M = \{m_I, m_2 \ldots m_7\} \tag{IO}$$

where: m_I = linear loading input code algorithm;

m_2 = algorithm for conversion of linear loading code into a format with a floating point;

m_3 = algorithm for correction of non-linearity of conversion characteristic of measuring channel for linear loading;

m_4 = algorithm for averaging instantaneous values of linear loading;

m_6 = algorithm for smoothing down the running estimations of linear loading;

m_7 = algorithm for compensation of weighing cell zero temperature drift.

The imperative algorithms are thereby only m_I and m_2. Every one of the algorithms is characterized by three parameters, vis.:

$$m_i = \{\alpha m_i, \beta m_i, \gamma m_i\} \tag{II}$$

where: αm_i = relative error reduction factor, which is determined by the relationship between the relative error of the linear load measuring channel, functioning under the support of the algorithm, and its error without the support;

βm_i = time interval requored for the execution of the algorithm;

γm_i = storage space required for the arrangement of the programs realizing the algorithm m_i.

Such an organization of the AWDSS enables synthesis of the DACS software with the given relative channel errors $\mathcal{E}_z$ to be ensured under the restrictions of the execution time for the programs realizing the algorithms of their servicing, and of the storage space γ which is necessary for the arrangement of the said programs.

If the hardware of a DACS channel (for instance, the channel for measuring linear loading) has a relative error $\mathcal{E}_o$, then to provide for the given value of $\mathcal{E}_z$, the software support for the channel should possess a relative error reduction factor:

$$\alpha = \frac{\mathcal{E}_o}{\mathcal{E}_z} \tag{I2}$$

The task of designing such software support consists in determining the combination of algorithms $m_i \in M$, the products αm_i of which should be greater than or equal to α. The possible combinations of algorithms thereby are defined by the following relationship:

$$S(M) = \{m_I m_2 S(m_3, m_4, m_5, m_6, m_7)\} \tag{I3}$$

where: S(M) = set of combinations of algorithms ofs set M; $S(m_3, m_4, m_5, m_6, m_7)$- set of combinations of algorithms $m_3, m_4 \ldots m_7$.

Then, a subset $S_\alpha(M) \in S(M)$ may be formed, so that for each element $S_j \in S_\alpha(M)$ the following inequality is valid:

$$\alpha_j \geqslant \alpha \tag{I4}$$

The combination of algorithms corresponding to the elements S will ensure the given relative channel error. From among such combinations those should be chosen which will have $\beta_j \leq \beta$, $\gamma_j \leq \gamma$. If there are a number of such combinations, the one should be chosen which has a minimum value of either β_j or γ_j.

In a similar manner algorithmic support for all the DACS channels can be formed. Since the development of software support is the most labour-consuming stage of designing the dosing automatic control system, the use of the AWDSS and the above-mentioned procedure for the synthesis of algorithmic support enables the time for a concrete realization of the DACS to be substantially cut down.

Thus, the use of the DACS on the basis of the microprocessor CCC CMI803.03 allows for the DMS aims to be successfully achieved. The task of the DACS is thereby the coordination of the DBS and stabilization of their capacities in the DDC mode. Sufficiently large computing capabilities of the CCC CMI803.03 enable the performance of the DBS to be corrected by the DACS to provide for the maximum probability of output of standard grain mixture. The developed operational and dispatch equipment of the CCC CMI803.03 allows the processing data to be displayed in the most convenient form for the operators.

Fig.I Information Model of Dosing Belt Scales

I - feed hopper, 2 - conveyor belt, 3 -starter, 4 -auxiliary relay, 5 -speed transducer, 6 - weighing cell, 7 - strain-gauge transmitter, 8 - belt break transducer, 9 - belt run-off transducer, IO -electric drive, II - control unit.

Fig.2 DMS Structured Model

Fig.3 Hierarchial Structure of DACS
Process of dosing and mixing

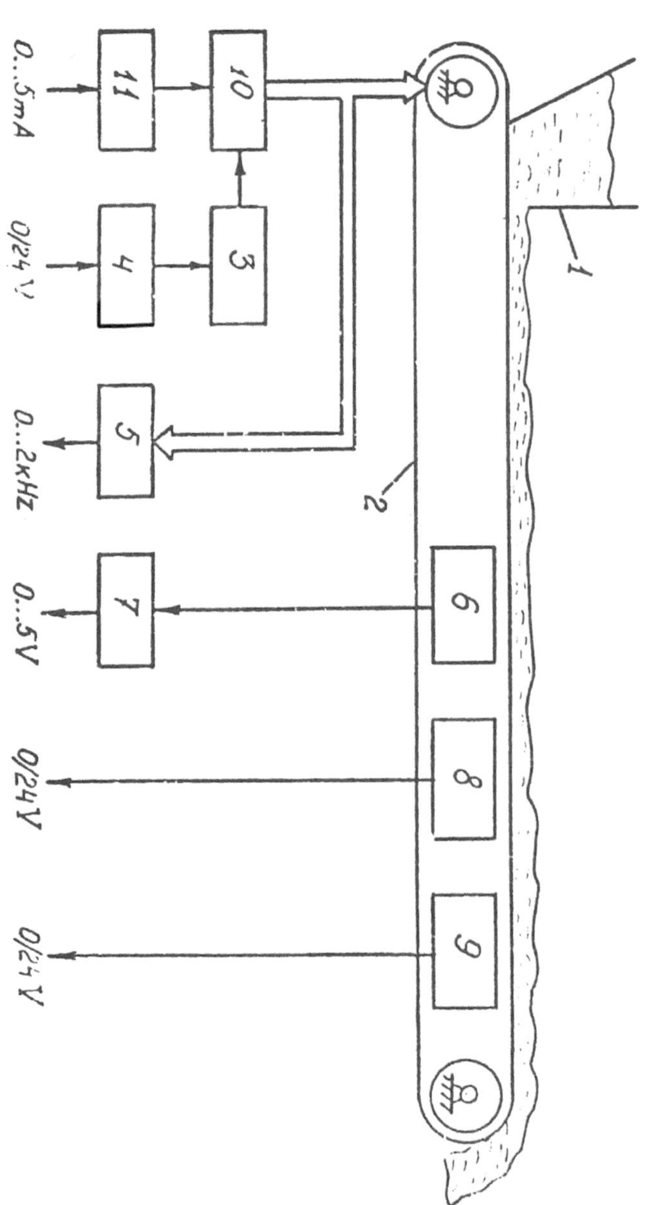

1
2
3
4
5
6
7
8
9
10
11
0...5mA
0/24V
0...2kHz
0...5V
0/24V
0/24V

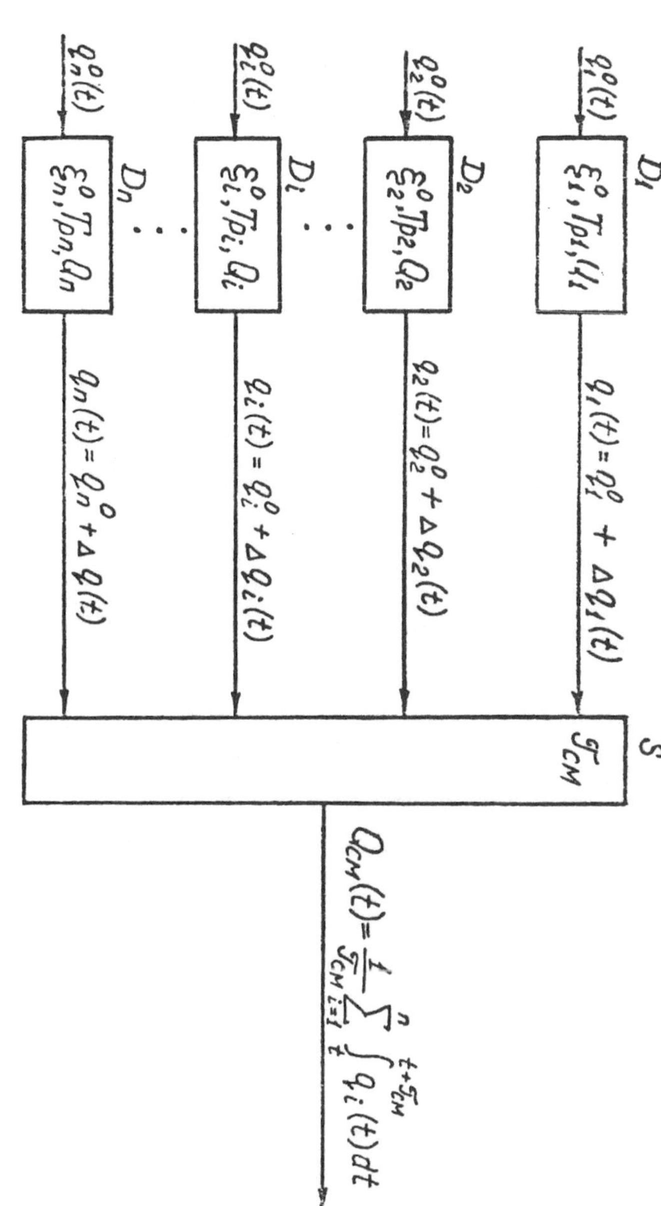
$q_1^0(t)$
D_1
ξ_1^0, T_{p1}, Q_1
$q_1(t) = Q_1^0 + \Delta Q_1(t)$
$q_2^0(t)$
D_2
ξ_2^0, T_{p2}, Q_2
$q_2(t) = Q_2^0 + \Delta Q_2(t)$
$q_i^0(t)$
D_i
ξ_i^0, T_{pi}, Q_i
$q_i(t) = Q_i^0 + \Delta Q_i(t)$
$q_n^0(t)$
D_n
ξ_n^0, T_{pn}, Q_n
$q_n(t) = Q_n^0 + \Delta Q(t)$
S
T_{CM}
$Q_{CM}(t) = \frac{1}{T_{CM}} \sum_{i=1}^{n} \int_t^{t+T_{CM}} q_i(t)\,dt$

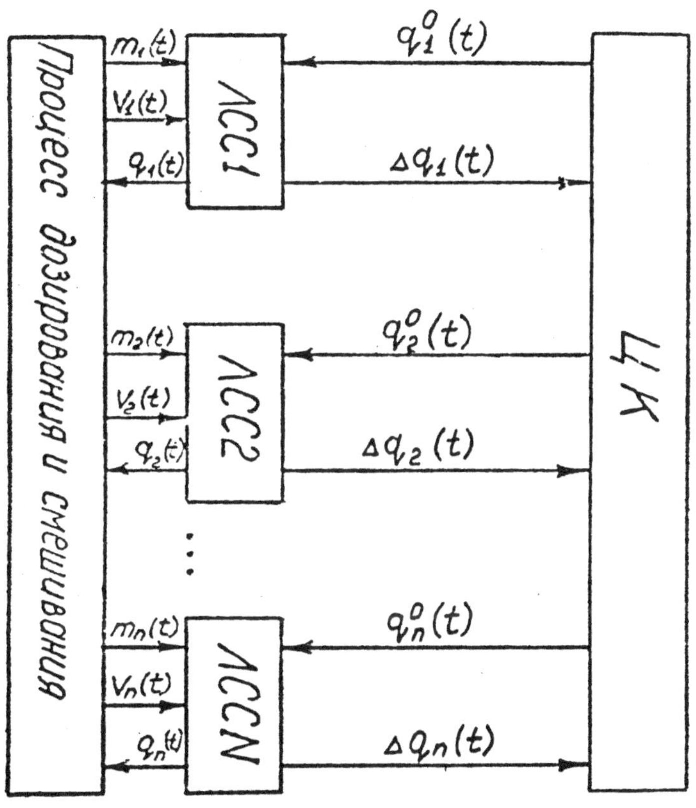
Процесс дозирования и смешивания
m1(t)
V1(t)
q1(t)
ЛСС1
q1⁰(t)
Δq1(t)
m2(t)
V2(t)
q2(t)
ЛСС2
q2⁰(t)
Δq2(t)
mn(t)
Vn(t)
qn(t)
ЛССN
qn⁰(t)
Δqn(t)
ЦК

A CONTRIBUTION TO THE ANALYSIS OF CONVEYOR BELT WEIGHING ERRORS.

KNUT FRISTEDT

S.E.G. INSTRUMENT AB, SWEDEN

1. INTRODUCTION

Conveyor belts are used to move bulk material between industrial processes and to load bulk materials into e.g. trucks and cargo ships. Continuous weighing of conveyed material can be carried out if one or more of the conveyor's idlers are replaced by weigh-idlers. A belt scale system often consists of a strain gauge-based load cell, a sensor which measures belt travel and a signal conditioning unit with sufficient computer capacity.

A belt scale's total systematic measurement error is influenced by a number of factors. However, we will only consider the most significant influences, i.e those contributed by the belt's pre-tension and bending stiffness when the scale idler is misaligned in relation to surrounding idlers. The measurement error appears as a change in the zero and a change in the sensitivity of the signal from the load cell. In practice nowadays, the zero is automatically controlled by update operation when the belt runs empty, but sensitivity must be checked by calibration of the system with a known quantity of material.

For many years theoretical models have been available which attempt to explain the dependence of current errors on the idler misalignment. As a result of the approximations involved, the models produce tare error but no sensitivity error. In the present study,these models were compared to one another and to a model based on superposition of elementary cases of beams in bending. The basic formulas for the latter can be found in many engineering handbooks. Calculated results were also compared to the results of tests from a conveyor whose scale idler was intentionally misaligned.

Only systems with one weigh-idler were considered.

2. SYMBOLS

P_0 = force on scale idler

$P_1, P_2 \cdots$ = forces on idlers on either side of the scale idler

F_0 = change in scale idler force due to scale idler misalignment

$F_1, F_2 \cdots$ = changes in forces on idlers on either side of the scale idler due to scale idler misalignment

F_{OA} = change in idler force due to belt load. Total error minus tare error

Q = belt load

L = idler pitch

n = number of idler pitches

D = scale idler misalignment

M, M_1 = bending moments

T = belt pre-tension

Wieringa, H. (ed), Mechanical Problems in Measuring Force and Mass.

ψ = $F/(\frac{E \cdot J \cdot D}{L^3})$ = error parameter

ϵ = $F_{0A}/g \cdot Q \cdot L$ = relative error
g = acceleration due to gravity
B = belt width
t = belt thickness
t_v = reinforcement thickness
J = second moment of area of belt cross-section
E = Young's modulus
E_v = effective Young's modulus
K = belt stiffness factor due to Colijn
δ = linear deflection
θ = angular deflection

3. THEORETICAL ERRORS

The scale idler load can be expressed as

$$P_0 = g \cdot Q \cdot L + F_0 \quad (1)$$

As previously mentioned,two parameters, i.e. belt pre-tension and belt bending stiffness, cause measurement error when the scale idler is misaligned. In theoretical consideration of the problem some approximations are introduced, such as the assumption that idler diameter is negligible and that the influences exerted by the different load parameters can be superimposed. Superimposition means that only first order effects are considered.

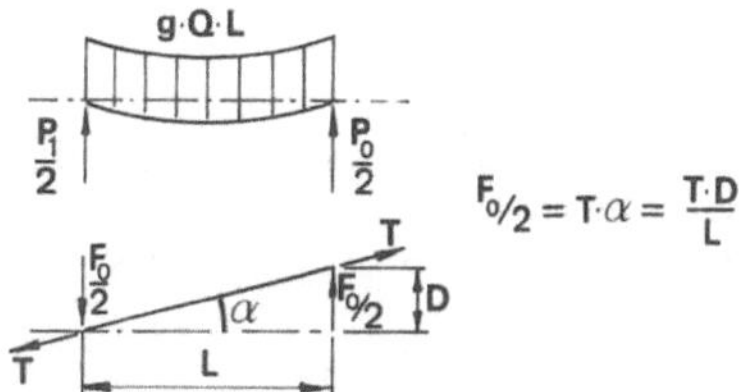

Fig 1: Force error due to misalignment of a pre-tensioned belt with no bending stiffness.

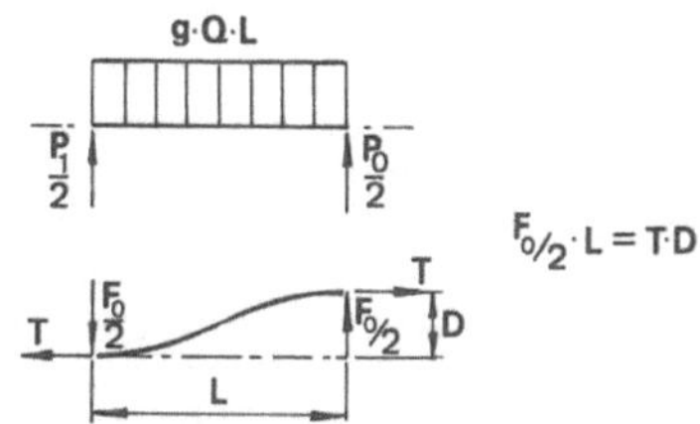

Fig 2: Force error due to misalignment of a pre-tensioned belt with bending stiffness.

Fig. 1 illustrates how error in scale idler force arise when the belt's bending stiffness is zero. Belt draping around the idler changes when the idler is misaligned. When the load is distributed as shown in Fig. 1, the error can be expressed with the equation

$$F_0 = 2\,\frac{T \cdot D}{L} \quad (2)$$

The problem is considered in /1/.

If the belt is represented by a beam as shown in Fig. 2, the error due to the pre-tension can still be described with (2).

The error due to the belt bending stiffness is more complicated and will be considered in the following two chapters.

4. EARLIER ERROR MODELS

Colijn's model /2/ postulates that a belt can be represented by a beam, fixed at both ends and with one end misaligned for a distance D. He utilizes the differential equation of the deflection curve, and both the pre-tension T and the distributed load gQ are included in determining the bending moment of the beam.

Thus:

$$F_0 = \frac{2T \cdot D}{L\left[1 - \frac{1}{\frac{L}{2}\sqrt{\frac{T}{E \cdot J}}} \cdot \tanh \frac{L}{2}\sqrt{\frac{T}{E \cdot J}}\right]} \qquad (3)$$

Colijn expresses (3) as

$$F_0 = 2\,\frac{K \cdot T \cdot D}{L} \qquad (4)$$

and concludes that the beam's bending stiffness amplifies the error because of belt pre-tension.

The forces on the most adjacent idlers become

$$P_1 = g \cdot Q \cdot L - F_1 \qquad (5)$$

with

$$F_1 = \frac{K \cdot T \cdot D}{L} \qquad (6)$$

The forces on the remaining idlers are not affected.

Hyer /3/ made a strain energy analysis of a belt supported on two idlers and applied Castigliano's Theorem. The loading corresponded to a beam, fixed at both ends, which affected by misalignment of one idler, rotaded the adjacent idler. The bending moment in the beam was determined as in the previous example.

Thus:

$$F_0 = 24\,\frac{E \cdot J \cdot D}{L^3} + 2\,\frac{T \cdot D}{L} \qquad (7)$$

The first term on the right side represents error due to the belt's bending stiffness, and the second term represents the error due to belt pre-tension.

A comparison of (2) and (7) shows that the effect of pre-tension on error is the same in both cases.

If the "tanh" in (3) is replaced by an equivalent series expression limited to three terms, and if a conjugate extension is then used (3) is transformed into

$$F_0 = 24\,\frac{E \cdot J \cdot D}{L^3} + 2.4\,\frac{T \cdot D}{L} \qquad (8)$$

A comparison of (7) and (8) shows that the influence of belt bending stiffness is equal and that the influence of pre-tension is virtually identical in the two cases.

Hidden /4/ assumed that a belt can be represented by an infinitely long beam, supported on an infinite number of idlers, with the scale idler misaligned for a distance D. Pre-tension is included in determination of the bending moment, and he uses the differential equation of the deflection curve. /4/ contains equations describing force errors on the scale idler and surrounding idlers. Hidden also compared his results with those of Colijn and Hyer and showed in a diagram that F_0 becomes roughly 2/3 of corresponding calculated errors when (3) or (7) are empolyed. However, present author was unsuccessful in his attempt to compare /4/ with the results in the following chapters.

5. SUPERPOSITION OF ELEMENTARY CASES

The following models are based on superposition of elementary cases for beams in bending. As noted in the previous chapters, only the error forces F are determined. Necessary deflection equations will be found in many engineering handbooks, e.g. /5/.

Two different approaches were used. A fixed beam supported on a number of idlers and misaligned at one end (see Fig. 5) represented the first model. A beam simply supported on a number of idlers and misaligned at one end (see Fig. 6) represented the second model.

A beam of length L, fixed at both ends and misaligned at one of the ends, was considered,when the first approach was compared to /2/ and /3/. As shown in Fig. 3, the model is described by a system of 3 linear equations, two obtained from the deflection conditions and one from the vertical force balance. The terms in the equations are due to the elementary cases cited and thus:

$$F_0 = 24\frac{E \cdot J \cdot D}{L^3} + 2\frac{T \cdot D}{L} \tag{9}$$

$$F_1 = 12\frac{E \cdot J \cdot D}{L^3} + \frac{T \cdot D}{L} \tag{10}$$

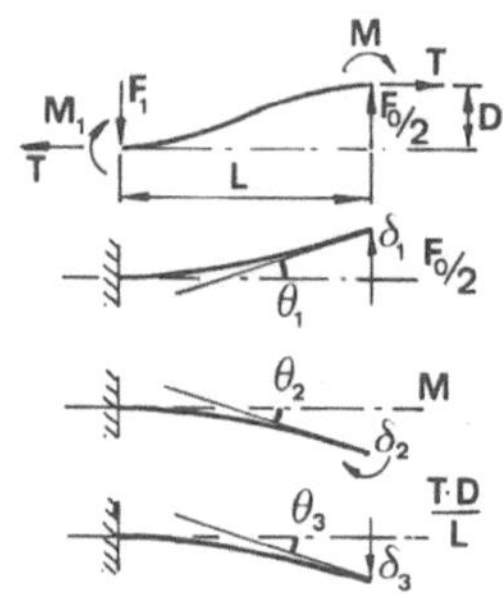

System of equations

$$\begin{cases} D = \delta_1 - \delta_2 - \delta_3 \\ 0 = \theta_1 - \theta_2 - \theta_3 \\ 0 = F_0/2 - \frac{T \cdot D}{L} - F_1 \end{cases}$$

Fig 3: System of elementary cases describing a fixed, misaligned beam of semi-length L.

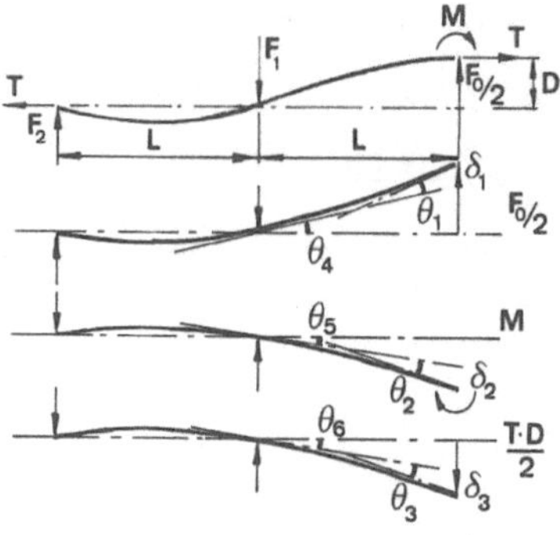

System of equations

$$\begin{cases} D = \delta_1 - \delta_2 - \delta_3 + L(\theta_4 - \theta_5 - \theta_6) \\ 0 = \theta_1 - \theta_2 - \theta_3 + \theta_4 - \theta_5 - \theta_6 \\ 0 = F_0/2 - \frac{T \cdot D}{L} - F_1 + F_2 \\ 0 = F_0 L - M - 2T \cdot D - F_1 \cdot L \end{cases}$$

Fig 4: System of elementary cases describing a simply supported, misaligned beam of semi-length 2L.

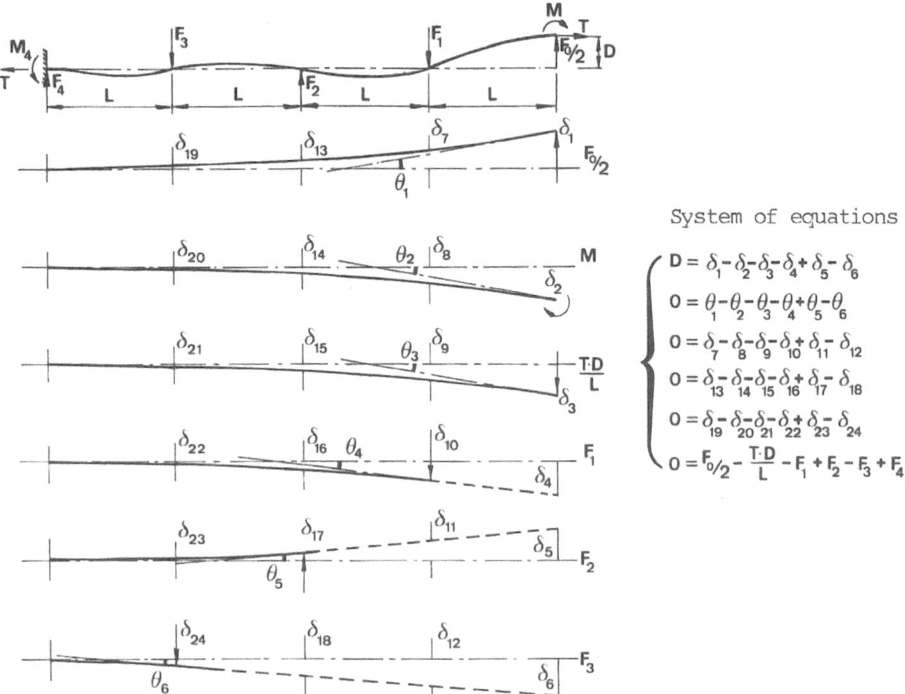

Fig 5: System of elementary cases describing a fixed, misaligned beam of semi-length 4L.

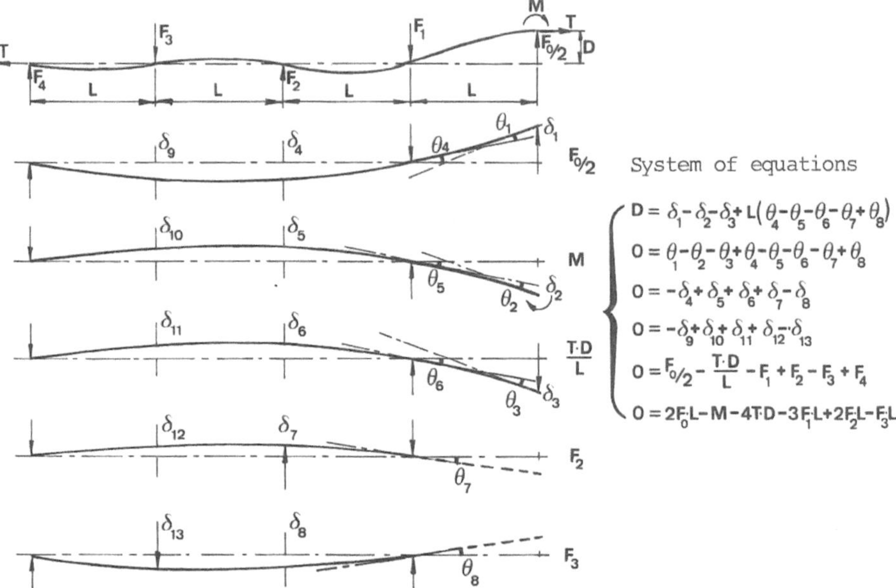

Fig 6: System of elementary cases describing a simply supported beam of semi-length 4L.

(9) agrees exactly with (7) and approximately with (8). (7), (8) and (9) overestimate the value of F because the beam is prevented from rotating the idler adjacent to the scale idler.

A lower limit is obtained for F_0 if the beam is instead simply supported as shown in Fig. 4. In this model, which represents the second approach, the system of equations is derived from two deflection conditions and from the force and moment balance.

The results obtained can be expressed as follows:

$$F_0 = 13.71\frac{E \cdot J \cdot D}{L^3} + 2\frac{T \cdot D}{L} \qquad (11)$$

$$F_1 = 9.48\frac{E \cdot J \cdot D}{L^3} \qquad (12)$$

$$F_2 = 2.47\frac{E \cdot J \cdot D}{L^3} + \frac{T \cdot D}{L} \qquad (13)$$

If an increasing number of idlers on either side of the scale idler are considered, then the number of equations in the system increases. Figs. 5 and 6 depict the situation when the semi-length of the beam is 4L. However, solving the system of equations remains simple.

The influence of T in the expression of F_0 is independent of the number of idlers involved and always has the value indicated in (9) and (11).

Fig. 7 shows the values of F (expressed as ψ), depending on bending stiffness, when the number of involved idlers increases. The semi-length of the beam (belt) considered is nL.

Cantilever beam, with transverse load

n	ψ_0	ψ_1	ψ_2	ψ_3	ψ_4
1	24	-12			
2	15	-12	4.5		
3	14.4	-10.8	4.8	-1.2	
4	14.36	-10.71	4.5	-1.29	0.32

Simply supported beam, with transverse load

n	ψ_0	ψ_1	ψ_2	ψ_3	ψ_4
1					
2	13.71	- 9.43	2.57		
3	14.71	-10.62	4.15	-0.69	
4	14.35	-10.7	4.45	-1.11	0.19

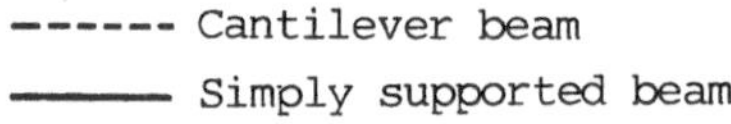

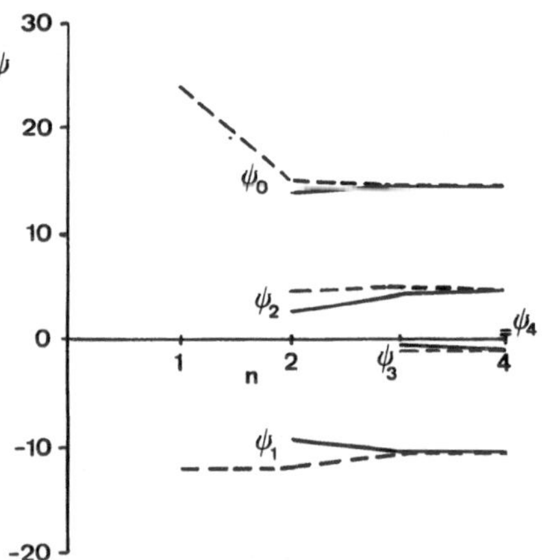

Fig 7: Summary of theoretical bending stiffness in idler forces, when the scale idler is misaligned and beam semi-length varies.

The following conclusions can be drawn:

a) If the semi-beam length is 4L or more, then it does not matter whether a fixed beam or a beam simply supported at the most remote idler is considered.

b) The error F_0 approaches the value of

$$F_0 = 14.35\frac{E \cdot J \cdot D}{L^3} + 2\frac{T \cdot D}{L} \qquad (14)$$

when a beam length of 4L or more on either side of the scale idler is considered. Coljin's and Hyer's expressions predict higher values. (14) indicates the importance of having a belt with low bending stiffness, low pretension and a large pitch.

c) Misalignment of an idler in a row of idlers influences the reaction forces exerted on other idlers in the row. Idlers located at a distance of 4L or more from the misaligned idler are only marginally affected.

d) When the idlers of a conveyor are aligned, inclusion of 3 or 4 idlers on either side of the scale idler is consequently necessary, depending on the measurement accuracy desired.

6. THEORY VS. EXPERIMENT

The fertilizer manufacturing company Supra in Landskrona, Sweden, a division of Norsk Hydro, kindly allowed us to conduct a number of tests with one of their conveyor belt scales equipped with a permanent test weigh-hopper. Its weighing unit is an S.E.G.,model BE with a computerized System T console instrumentation. The aim of the test was to compare calculated and measured errors arising after deliberate misalignment of the scale idler. All idlers were carefully aligned before the tests, and the automatic tare compensation was disconnected. The conveyor was fed from the hopper with its precision scale. In each test, 2000 kg of granulated fertilizer in the hopper were deposited onto the conveyor belt. The tare of the running belt was recorded before and after each test . The belt's characteristics are listed in Fig. 8 which also contains the evaluated test results and calculated values according to (7) and (14).

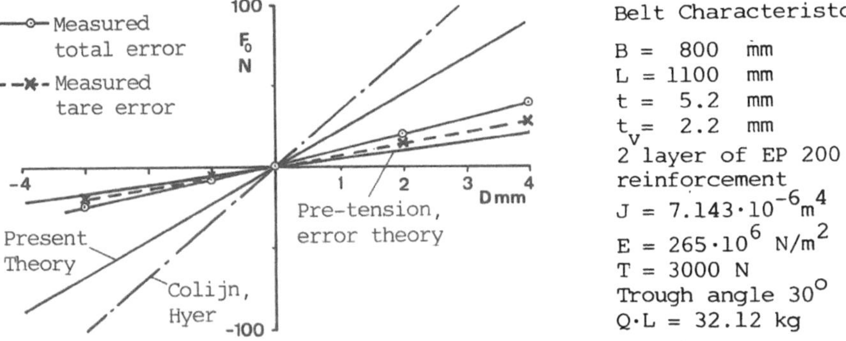

Fig 8: Measured and calculated scale force errors vs. scale idler misalignment.

Young's modulus was estimated from information supplied by the belt-manufacturer /6/.

$$E = E_v \frac{t_v}{t} \tag{15}$$

The stiffness of the rubber is negligible compared to the stiffness of the reinforcement. Each experimental point in Fig. 8 is the mean-value of 3 tests. Scale idler misalignment caused by the maximum load was 0.2 mm.

The following conclusions can be drawn from Fig. 8.

a) The theoretical model in Chapter 5 yields errors which exceeded the measured errors by approximately 135%. This was probably due to local buckling phenomena which made classical calculation of the second moment of area of the belt cross-section unreliable. The current belt is also thin. Fig. 9 shows the running belt during one of the tests. The arrow indicate a visible local dent.

Fig 9: The conveyor belt scale tested at Supra, Landskrona.

b) The theoretically calculated error due to belt pre-tension was somewhat smaller than the tare error measured. If the calculated error is assumed to be correct and is compared to the measured tare error, the order of magnitude of the calculated error due to the belt's bending stiffness could be wrong.

c) The theoretical model produces tare error but no inherent error caused by the load. Figs. 8 and 10 show that experimental errors caused by the load exist.However, these errors were considerably smaller than the tare errors.

d) The results in Fig. 11, which were calculated from Fig. 10 and also pertain in principle, to partial loads, indicate that the present balance with misalignment less than D = 1.6 mm could fulfil the requirements for class 1 scales in OIML No. 50.

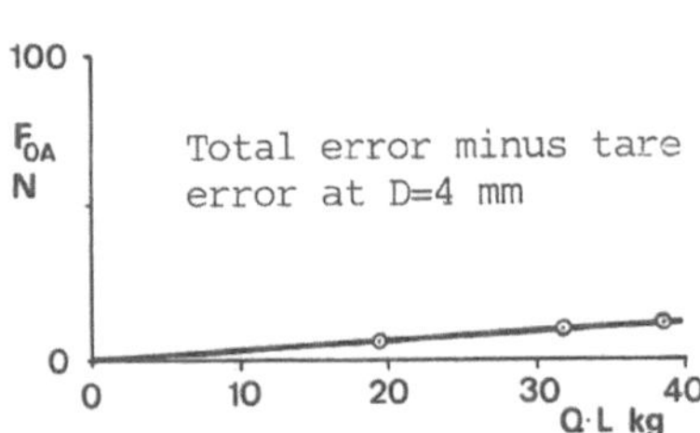

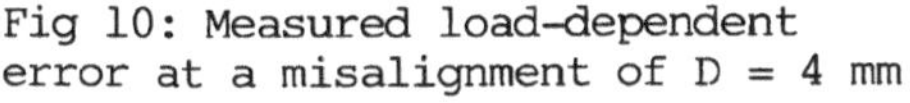
Fig 10: Measured load-dependent error at a misalignment of D = 4 mm.

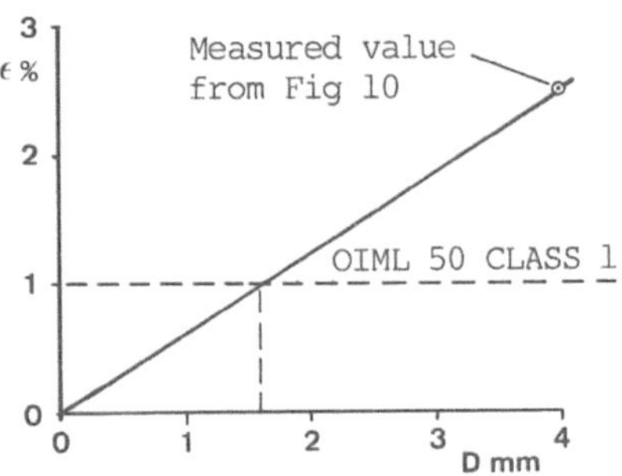

Fig 11: Maximum load dependent error as a percent of current load with differing misalignment, extrapolated from the value measured at D = 4 mm.

7. CONCLUSIONS

a) The theoretical models presented are mathematically simple but do not permit reliable numerical calculation of errors due to belt bending stiffness in connection with scale idler misalignment, at least not as regards thin belts.

b) Tests with thicker conveyor belts must be performed before a final assessment can be made of the theoretical models.

c) Aligning the scale idler and 3 to 4 idlers on either side of the scale idler is sufficient to prevent the present type of weighing errors.

d) The tests performed showed that tare error dominates and that automatic uppdating of the zero when the belt runs empty is of great importance.

e) A conveyor belt scale must be calibrated with a known quantity of material for determination of load dependent error.

f) The present tests showed (Fig.11) that single idler scales may comply with class 1 requirements in OIML No. 50 if provided with automatic zero (tare) compensation.

REFERENCES

1. van den Berge H et al: "Belt Weighing" Science and Industry No. 9, 1977
2. Colijn H: Appendix B of "Weighing and Proportioning of Bulk Solids" Trans Tech Publications, 2nd Ed, 1983
3. Hyer F S:Appendix A of "Weighing and Proportioning of Bulk Solids" Trans Tech Publications, 2nd Ed, 1983
4. Hidden A E:"Errors in Conveyor Belt Weigher Systems" Measurement and Control, Vol 6, 1973
5. Roark J R:"Formulas for Stress and Strain" Mc Graw Hill
6. Bengt Hansson, Trelleborg AB, Sweden,Private communication.

FEEDING CONTROL OF AUTOMATIC BAGGING SCALE

Takeyoshi Nagao

Yamato Scale Co., Ltd.
5-22, Chaemba-cho, Akashi-city, 673, Japan

ABSTRACT

This papar describes the feeding control of automatic bagging scale which weighs and discharges the bulk material.

We have incorporated a new concept in its construction, and have adapted the original algorithm to the feeding control. Consequently weighing time could be shortened remarkably, while maintaining high accuracy.

KEYWORDS : automatic bagging scale, dynamic programming, weighing time, high accuracy

1. INTRODUCTION

In such fields, where bulk materials are generally disposed of, as chemical, pharmaceutical or food stuff industries, automatic bagging scale are being used, which are designed to fill sacks or containers with a proportioned quantity of final products at the last production stage. The capability of an automatic bagging scale originally is dependent on the weighing speed beside the weighing accuracy. For a bagging scale it is essential to accelerate the weighing speed in view of the economy of production.

As the feeding rate of a bagging scale increases, the weighing time will certainly be shortened. However, merely increasing the feeding rate would degrade the weighing accuracy since a wider dispersion of value will occur in the proportioning operation.

To overcome such difficulties the conventional bagging scale incorporates the 2 or 3 staged feeding system, i.e. material will be supplied at a higher rate at the first stage to accelerate the weighing speed, and subsequently additional material will be supplied at an appropriately lowered rate to satisfy the requirement in weighing accuracy. However, due to inadequacy in the design concept, the requirement for higher feeding rate with higher accuracy is satisfied only limitedly in such case.

Through an experiment intended to improve such deficiency of conventional system with a new idea incorporated in the control of feeding operation, some successful results were obtained, which were found satisfactory in terms of shortening the weighing time as detailed later in this paper.

The second chapter in this paper describes the feeding process by conventional automatic bagging scale, and in the third chapter the construction of the newly developed automatic bagging scale.

A new feeding process is introduced in the fourth chapter with an explanation of "optimal control theory" which has helped us to determine an algorithm of the new feeding process. The experimental results of the new feeding process that can afford a higher weighing speed of automatic bagging scale, while maintaining high accuracy, will follow.

2. FEEDING PROCESS BY CONVENTIONAL AUTOMATIC BAGGING SCALE

The object of an automatic bagging scale is to obtain a target value

Wieringa, H. (ed), Mechanical Problems in Measuring Force and Mass.

quickly and accurately. In general, increasing of feeding rate inevitably brings about a proportional degrading of weighing accuracy, because of:

(1) a larger fluctuation in feeding rate of material to be weighed.

(2) a larger dispersion in values due to a larger feeding rate per sampling time, when mass is converted from analog to digital.

To achieve a higher accuracy the feeding rate must be reduced, which, however, is not allowed in view of the proposed speedy weighing. To cope with this problem the conventional automatic bagging scale incorporates a feeding process represented by Fig.1. According to Fig.1 showing 3 staged feeding system the feeding operates first at a higher rate of Q1. The feeding rate is then to be reduced down to Q2 and to Q3 each time when the pre-set masses M1 and M2 has been reached respectively. The weighing accuracy associated with feeding rate Q3 at the last stage must satisfy the requirement provided. The above process is accompanied by the following drawback.

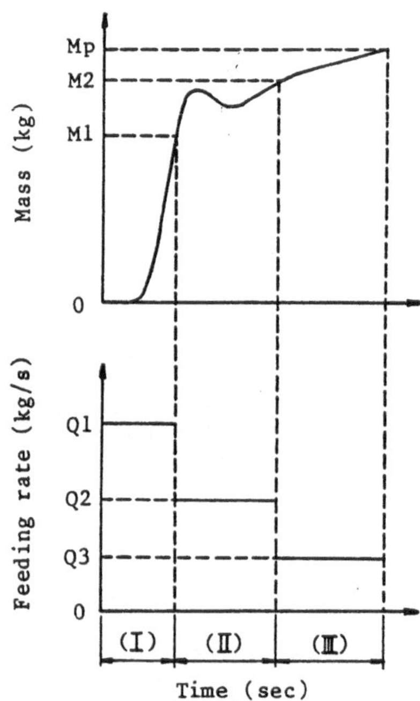

Fig.1 Feeding control of conventional automatic bagging scale

Whenever the feeding rate is changed over, an indented response appears due to a sudden variation of feeding rate. Particularly in the section (II) the feeding curve experiences a slight descending due to an adverse effect attributable to a shock load, in which case the weighing time will inevitably have to be prolonged. Also in the section (III) an indented response appearing at change-over will give an adverse effect on the weighing signals, resulting in an unwanted dispersion of weighing accuracy.

As a result the intervals for feeding operation have to be prolonged until the indented response ceases. This is why the conventional type is not able to reduce the weighing time.

In view of the foregoing a key factor for being able to shorten the weighing time depends on how to suppress the indented response taking place at each change-over of feeding rate.

3. CONSTRUCTION OF NEW SYSTEM

An outward-view of the new automatic bagging scale is shown in Fig.2 and a schematic diagram of new system in Fig.4. The machine is designed as an automatic gravity bagging scale. The mass of bulk material supplied into the weighing hopper is measured by a strain gage type load cell which directly supports the weighing hopper. The measured mass is then turned into a digital signal through strain gage amplifier and A/D converter and then sent to a micro computer. In response to the said signal an operational signal is calculated by an algorithm being programmed in the micro computer, and is to be sent to the drive section through the D/A converter. The drive section will provide an arbitrary gate positioning subject to the operational signal produced, since it consists of servo circuit for position control.

An outward-view of the control device is shown in Fig.3. This type of

control device equipped with 16 bit A/D converter (sampling time = 2msec) has as high an accuracy as the analog. In addition, it comprises every function required for an automatic bagging scale such as automatic zero compensation, automatic over-feed compensation, etc. In fact, it is a compact designed unit.

The block diagram of control object is shown in Fig.5.

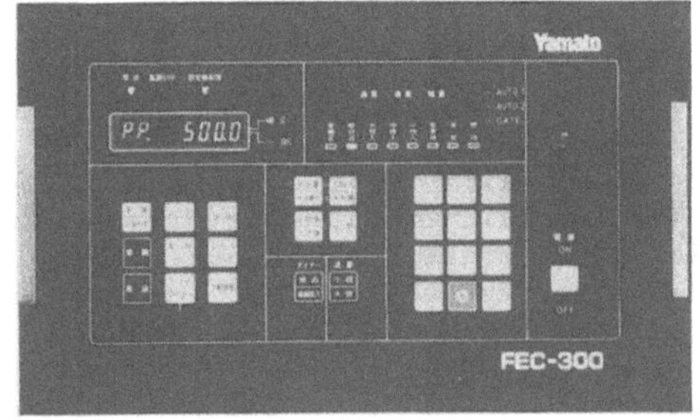

Fig.2 New automatic bagging scale

Fig.3 Control device

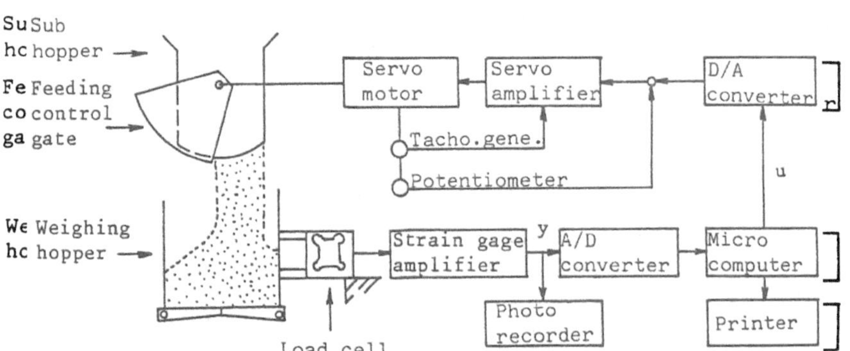

Fig.4 Schematic diagram of new system

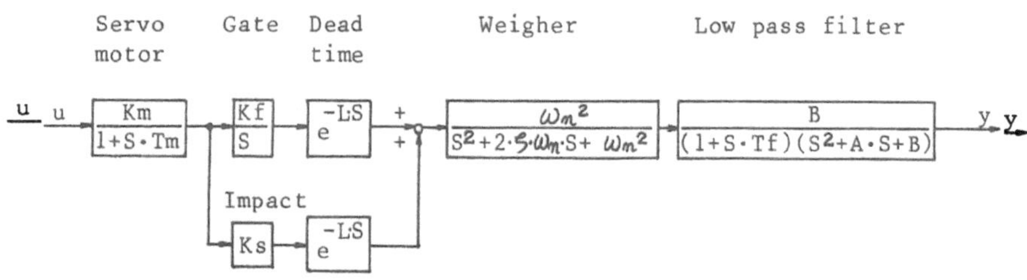

Fig.5 Block diagram of control object

where :

- Km,Tm : gain and time-constant of servo motor
- Kf : gain of feeding control gate
- Ks : impact coefficient
- L : dead time
- ω_n,ζ : natural frequency and damping coefficient of weigher
- Tf,A,B : constant of low pass filter

4. DETERMINATION OF FEEDING CURVE

An ideal feeding curve implies that no indented responses occur at any change-over instance while in feeding operation and that the target value is attained earliest possible. The problem to determine an ideal feeding curve can be regarded as the optimal control problem entitled "how to determine a series of operational input which will make the arbitrary criterion function the smallest of all". One of the keys available to solve the optimal control problem is the "dynamic programming". In an attempt to find out "which kind of operational input is required to provide the optimal feeding curve", a simulatoin was performed by the dynamic programming using a low degree model approximating the new system.

4.1. Dynamic programming

Generally the optimal control problem is considered to be a multi decision process. The dynamic programming is therefore defined as a procedure available to solve the optimal control problem of multi decision process through converting it into a series of single decision process. This conversion procedure incorporates a meaningful theory, the principle of optimality. In practice, the differential equation representing the controlled system is to be converted into the discriminating equation which is used to solve, by numerical calculation, the functional equation obtained through the principle of optimality.

The processing of the dynamic programming algorithm using a general purpose computer for a generalized high degree system inevitably presuppose an enormous amount of memories and calculation time which will make this processing scarcely feasible.

A reasonable compromise in this case will be an approximation to a low degree system at the cost of accuracy. The system of the automatic bagging scale was a compromise using approximation in Fig.6. As the time-constant of servo motor in this case is of extremely small value, the transfer function of the motor is appoximated with Km. The approximation employed in Fig.6 reflects the fact that the pole of the low pass filter is found lying extremely close to the imaginary axis when compared with that of the weigher, and that the system response will almost be dominated by the low pass filter.

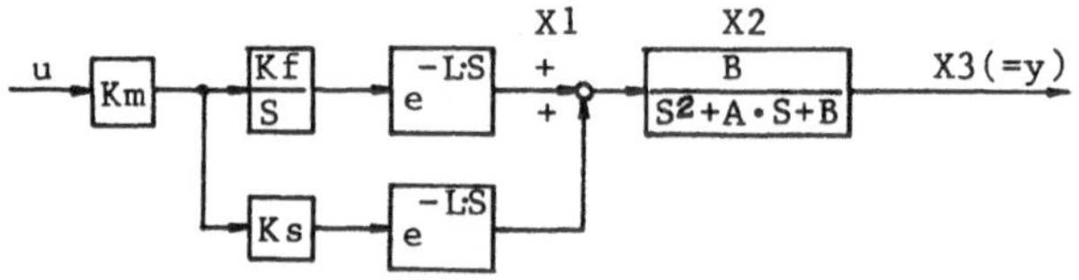

Fig.6 Approximate block diagram of control object

3 kinds of state variables, as seen in Fig.6, X1: mass through the gate (kg) ; X2: rate of mass variation (kg/sec) ; and X3: mass (kg), are used to establish the following state equation expressed in a matrix form:

$$\begin{cases} \dot{X}(t)=G\cdot X(t)+H\cdot U(t-L) \\ y(t)=C\cdot X(t) \end{cases} \tag{1}$$

where :

$$G=\begin{bmatrix} 0 & 0 & 0 \\ B & -A & -B \\ 0 & 1 & 0 \end{bmatrix} \qquad H=\begin{bmatrix} Km\cdot Kf \\ Km\cdot Ks\cdot B \\ 0 \end{bmatrix} \qquad C=\begin{bmatrix} 0 & 0 & 1 \end{bmatrix}$$

Here a comparison between experimental value and computer-simulated value to examine the authenticity of this approximated model is shown in Fig.7, and pulse input was given for this. As a result, the resemblance in Fig.7 between the two data can justify the authenticity of computer-simulation in relation to the existing system.

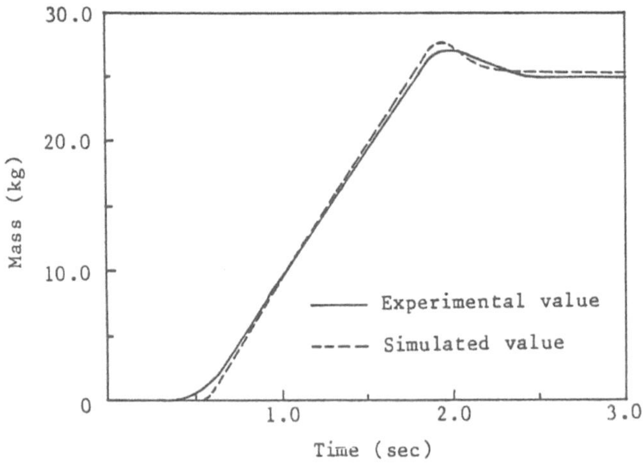

Fig.7 Pulse response of new system

The equations (1) can be transformed into following form.

$$\begin{cases} X(k+1)=E\cdot X(k)+F\cdot U(k-Ld) \\ y(k)=C\cdot X(k) \end{cases} \tag{2}$$

where :

E : state matrix discreted
F : input matrix discreted
Ld : dead time discreted

The criterion function to solve this optimal control problem is defined as:

$$J=\sum_{k=0}^{N-1} [[Mp-y(k+Ld+1)]^2+j\cdot U(k)^2]\cdot h \tag{3}$$

where :

Mp : target value
j : constant applicable to limit the input
h : sampling time

Under the principle of optimality a functional equation can be derived as:

$$V(k)=\min_{U}\left[\left[\left[Mp-y(k+Ld+1)\right]^2+j\cdot U(k)^2\right]\cdot h+V(k+L+1)\right] \tag{4}$$

The functional equation shall be solved backward in terms of time. In other words, a series of optimal operational input is to be solved backward from the future to the past successively.

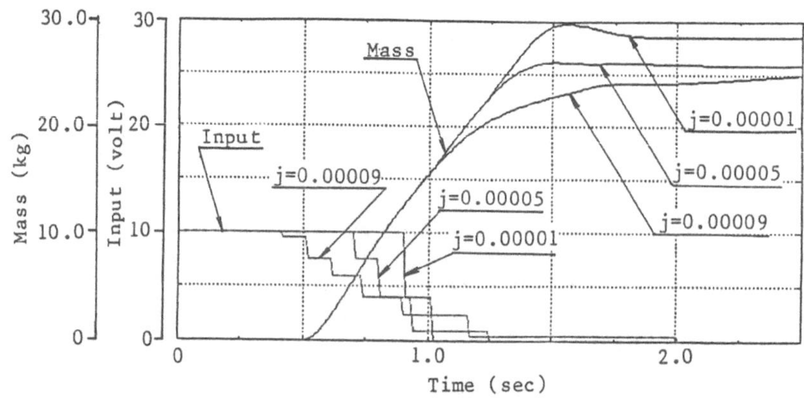

Fig.8 Optimal reponse pattern of system (by dynamic programming)

The result from a computer simulation under the condition covering a target value of 25kg, sampling time for digitalization h of 0.01sec and the limitation of operational input U=0v to 10v is depicted in Fig.8, where the response curves representing j=0.00001 to 0.00009 are plotted in contrast with the input pattern. On the assumption that the ideal response curve should not cause any indented response and the target value shoud be attained earliest possible, the response curve representing j=0.00005 would be the most suitable.

Altogether, it is surmised that the optimal response curve can be determined by such operational input pattern as will make it possible to provide the maximum operational input for the specified initial interval and to reduce the operational input from the relevant point on so that possible indented response could be prevented.

4.2. Determination of feeding control algorithm

In paragraph 4.1., a pattern of operational input (equivalent to the feeding rate in a practical automatic bagging scale system) which provides the optimal feeding curve of an automatic bagging scale was determined.

In view of a pattern as such, an optimal feeding curve is given by such an operational input series, which makes it possible that the feeding in the initial stage, as seen in Fig.9, is made at a higher rate Ql and then from an arbitrary point A the feeding rate is successively reduced, so that no indented response would take place. Based on such a pattern an original feeding contol algorithm which can be calculated by micro computer has been sought. A basic concept of feeding control algorithm thus obtained is schematically shown in Fig.10. In this case the feeding is made in the section (I) at a higher flow rate of Ql (bulk feeding); from the point A to

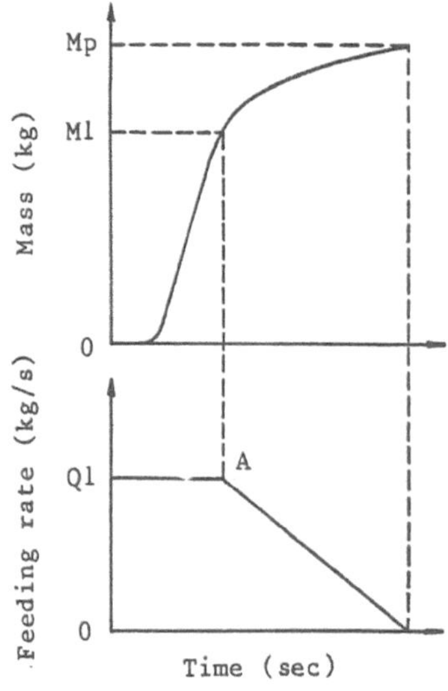

Fig.9 Optimal feeding curve and input pattern

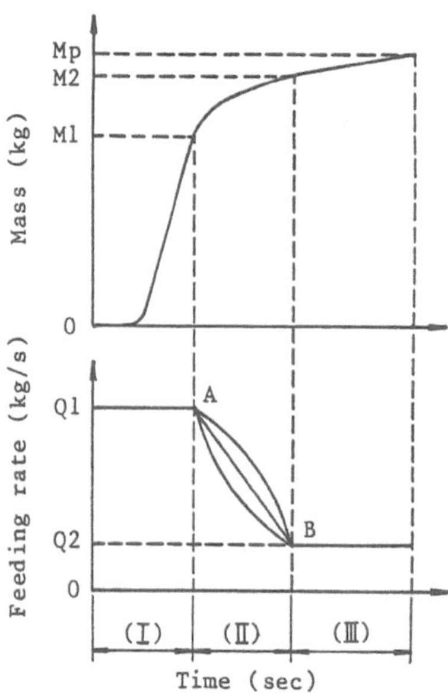

Fig.10 Feeding process of new system

the point B the feeding is made under such an aribitrary function which suppresses a possible indented response; and in the section (III) the feeding is made at a lower rate of Q2 (dribble feeding) which will satisfy the requirment in accuracy. The reason why the section (III) has been supplemented is the fact that where the feeding control continued up to the moment of final cut-off as seen in Fig.9, a degrading of the stability in weighing time due to a fluctuation of feeding rate would occur. Determination of the points A and B, where the feeding rate will have to be changed over in Fig.10 is made possible through the comparion between the mass of feeding material and the pre-set masses M1 and M2.

With Mx as variable the feeding rate Qx at an arbitrary point can be given by:

$$Qx=\begin{cases} Q1 & : \ 0 \leq Mx \leq M1 \\ (Q1-Q2)\cdot f(Mx)+Q2 & : \ M1 < Mx \leq M2 \\ Q2 & : \ M2 < Mx \leq Mp \end{cases} \qquad (5)$$

boundary conditions : f(Mx)=1 where Mx=M1
f(Mx)=0 where Mx=M2

where :

- Qx : feeding rate at an arbitrary point
- Q1 : higher feeding rate (bulk feeding)
- Q2 : lower feeding rate (dribble feeding)
- Mp : target value
- M1 : mass at the end of bulk feeding
- M2 : mass at the strat of dribble feeding
- Mx : mass at an arbitrary point
- f(Mx) : arbitrary function in the section (II)

Considering of boundary conditions, the arbitrary function f(Mx) is derived as:

$$f(Mx)=\left(\frac{M2-Mx}{M2-M1}\right)^{p} \tag{6}$$

where p is a constant.

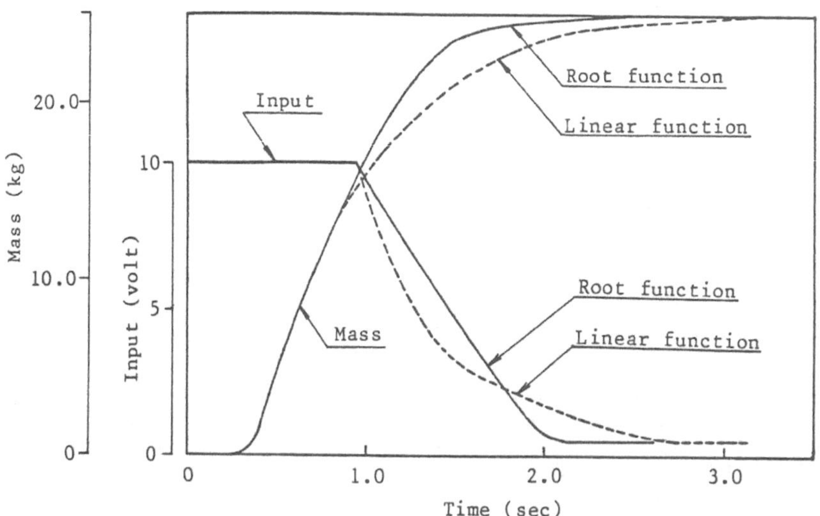

Fig.11 Simulation results of new feeding process

For the determination of the said p was made a simulation of the feeding curve using a computer. For this purpose a model used in Fig.6 was again put to use here. In Fig.11 the simulation results are represented with the case where p=1, i.e. f(Mx) is the linear function and the case where p=0.5, i.e. f(Mx) is the square root function. Through a comparison, it is evident that the square root function shows a faster response and the target value is attained earlier.

In view of the foregoing we have decided to make a choice of the square root function as the feeding control algorithm for the automatic bagging scale.

5. EXPERIMENTAL RESULTS

A test on the new system was carried out, with the algorithm determined in the proceeding paragraph incorporated into a micro computer. For this purpose polyethylene pellets were put to use which showed a better fluidity (grain dia.; 0.3mm, density; 0.5g/cm^3). The results obtained are depicted in Fig.12. Here, (a) represents the square root function and (b) the linear function. It appeares that in case of (a) the target value is attained earlier, which is similar to the results obtained in the simulation. Fig.13 shows a feeding curve of a conventional system of automatic bagging scale. It appears that an indented response occurs due to a shock load, resulting in a loss of time.

A comparison of each system performance is shown in Table 1.

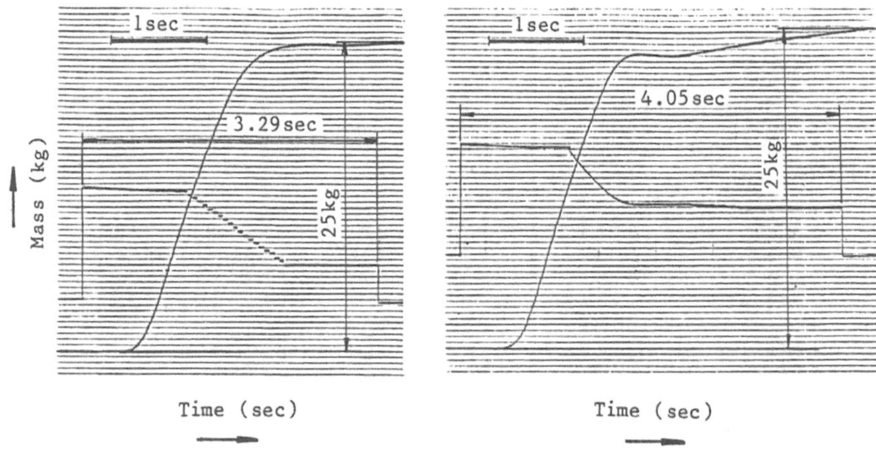

(a) Square root function (b) Linear function

Fig.12 Experimental results of new system

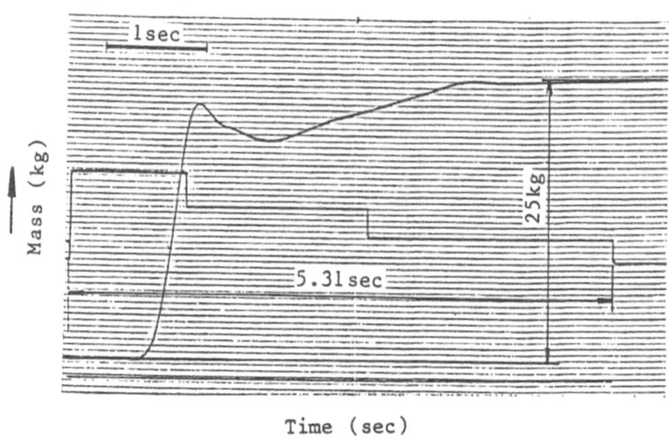

Fig.13 Experimental results of conventional system

Table 1. Comparison of system performance

	Conventional system	New system	
		linear	root
Accuracy (g)	±10	±10	±10
Standard deviation at sigma 2 (g)	±5.4	±5.3	±5
Feeding time (sec)	5.31	4.05	3.29
Weighing time (sec)	7.11	5.85	5.09
Weighing speed (baggs/h)	500	600	700

The feeding time in table 1. means an interval from the starting point of feeding to the final cut-off point. The weighing time is equal to the total amount of the feeding time, evaluation time (1sec) and discharging time (0.8 sec).

It is evident from the table 1. that the square root function mode is superior to the others. And the feeding time could be shorter than 3sec, with some compromise to be made in terms of weighing accuracy.

6. CONCLUSION

A successful feeding control, based on algorithm enabling the target value to be attained earlist without any indented response at a change over of a feeding rate, has been carried out on a automatic bagging scale. As a result the weighing time has been shortened by as much as 30% in comparison with the conventional bagging scale.

In spite of the above success a procedure of "try and error" inevitably has to be involved in some part of determination procedure for correcting pre-set parameters. To overcome this drawback a development of an automatic on-line adjustment or a self error compensation function is being planned within the framework of the projects in future.

We would like to express our sincere thanks to Prof. Shinzo Kitamura of the Department of Instrumentation Engineering of Kobe University and his colleagues for the valuable advice and assistance.

REFERENCES

1. K. Wada, N. Hayano and G. Uetaki: Weight Control of Hopper. Trans. of SICE, Vol. 20, 1984.
2. K. Ogata: Dynamic Programming. Baifukan, 1973.
3. H. Kimura, H. Maeda, U. Inoue, Y. Misaka and R. Takahashi: Programming for Control System Design. Nikkan Kogyo Shinbun Co., Ltd., 1985.

WEIGHING IN ROADVEHICLES

W. Jeuken

Dutch Metrology Service, weighing machines division
PO Box 654, 2600 AR Delft, The Netherlands

1. INTRODUCTION

Weighing in vehicles is a well-known procedure in the case of determination of loaddistributions or rough quantities. For these purposes a very high accuracy is not necessary. For trade purposes quite different requirements apply, however. Lately in The Netherlands a grewing the need arised for weighing at the recipients place of bulk materials as forage, bakers' flower and industrial gasses. Farmers want to know the weight of their fattened pigs before they are taken for slaughter. As not everybody has a weighbridge or another rather large weighing machine in his neighbourhood one looked for other solutions. Weighing in the delivery truck itself could give an answer to the problem. It appears that straingauge measurement and electronics together offer rather simple possibilities.

2. THE TECHNICAL APPLICATIONS

Nowadays, a great number of weighing methods are used in The Netherlands. Only few of these methods are permitted for trade purposes by the legal metrology authorities. Other methods are still used in the area of rough quantity determination, whilst some methods have not yet grown out of the development stage.

2.1.The free hanging cattle cage. The lorrie used for collecting cattle at the farmer's is equiped with a movable rail in the top of the cargo compartment. An electric hoist can be moved in horizontal direction along the rail. A load cell is suspended from the hoist, supporting four chains, one to every edge of an enclosured platform; this platform - the cattle cage - is the load receptor.
The measuring system is used as follows. At the collecting place the cattle cage is lifted out of its resting position on the tail board, transported enough out of the vehicle and sunk to the bottom. After being loaded with cattle, the load receptor is lifted untill it is free from the bottom. The weighing procedure can be carried out. If equiped with sufficient accurate and approved load cell and indicator, this system is allowed to be used for trade purposes in The Netherlands, because of a correct vertical load introduction.

2.2. The hydraulic tailboard used as a weighing machine. This tail board principally consists of a very stable frame and a rather slack board, which is the load receptor, mounted to the frame by means of four load cells. In horizontal position of the tail board it is necessary to choose a balanced position for the load receptor. This will be possible by fitting on each side (to the stable frame and to the slack board as well) of the load cell a free swinging construction.

In figure 1 an existing hanging unit with a load cell of the bending beam type and double ball bearings is showed. Ideally the friction between ball and housing will be negligible. Moreover, if the free space S between load cell unit and frame and the tolerated swinging in the ball bearing is wide enough, good weighing is possible even without any other precaution of leveling the vehicle and the load receptor. A vertical load introduction is guaranteed. For this reason it is allowed to use the system for trade purposes in this country.
At departure the tailboard of course is in a vertical position again; the slack board is secured against involuntary movements. This is necessary because of the possibility of damaging the load cell.

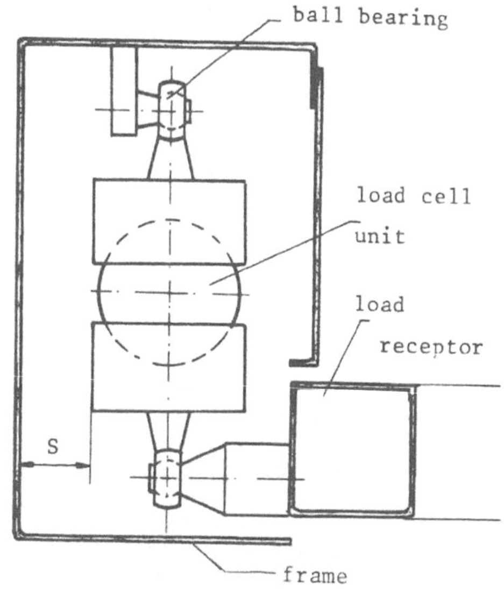

fig. 1

Although a fixed connection between load cell and frame on the one side and/or load receptor to the other side is imaginable this will give a lot of trouble for a correct vertical load introduction in all cases. Because of the introduction of horizontal forces into the load cell it will not be possible to perform measurements with a great accuracy. Special arrangements would be necessary to reduce the uncertainty.

2.3. <u>The cargo compartment (or bulk tank) used as a loadreceptor</u>. Several constructions can be distinghuished.

2.3.1. The cargo compartment or tank has a rigid connection to the chassis of a lorrie via a number of strain gauge load cells, so the weighing system is an integrated part of the vehicle.
The load cells are mounted between subframe and cargo compartment and must be very robust. The special dimensions of the load cells will in general cause a small resolution of the measuring system. In principle only in a rather leveled position weighings can be carried out. To meet a horizontal position special precautions, as fitting four independent lifting systems at the four edges of the vehicle, should be taken. Because of the very small resolution of the load cells it is not possible in this moment to use the measuring system for more than a rough quantity determination.

2.3.2. A special appliance of the above mentioned weighing system in a lorrie is that one, which is built in a semi-trailer. In that case the load cells are fitted between the tank and the running gear on the one side and between the semi-trailer-coupling-tray and subframe of the truck on the other side. In some existing constructions the measuring signal of the running gear load cells is replaced by a pneumatic signal, produced by the air feeded spring system of the semi-trailer. In our service we do not have available any experience with mass measurement by using a pneumatic signal in this moment.

Both of the above mentioned applications of mass measurement in vehicles have some common disadvantages. An important one is the necessary flexibility of every vehicle. This flexibility causes torsions and horizontal forces into the load cells, which will give unpredictable influences on the measuring results.
Problably the most advanced method of mass measurement in vehicles from a technical point of view is the one, where it is possible both to eliminate the influence of not leveled subsoil and to unload the cells during driving I already mentioned the free hanging cattle cage and the special hydraulic tailboard. To perform measurements with bulktanks on vehicles other answers have to be found. The next examples are known in this country, nowadays.

2.3.3. The bulk tank has a rigid connection to a chassis cab.
Hydraulic jacks are mounted on four spots. This jacks will lift the whole lorrie. They are independently controled so that an uneven subsoil does not have any influence upon the leveling of the lorrie. The load cells are fitted in the jacks in such a way that they are not loaded until the pistons touch the subsoil. With this system the influence of crossforces is small indeed, but a corrects mass measurement is not guaranteed, because of the indifferent frictional resistance in the gaskets between pistons and housing of the jack. As a result of a not quite horizontal subsoil considerable cross-forces in the lower gasket f.i. can cause a force in the same (vertical) direction as the force to measure.

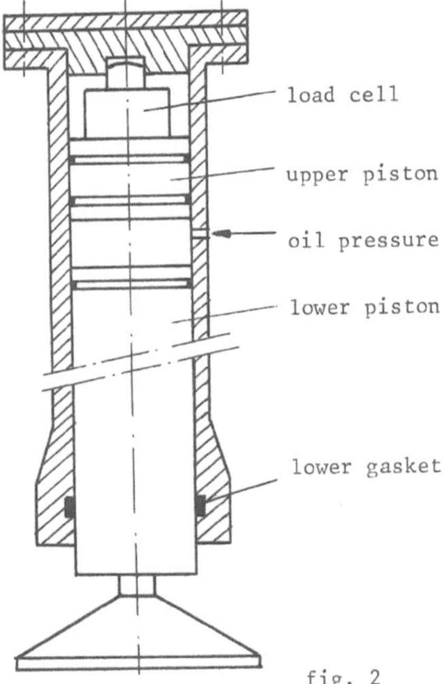

fig. 2

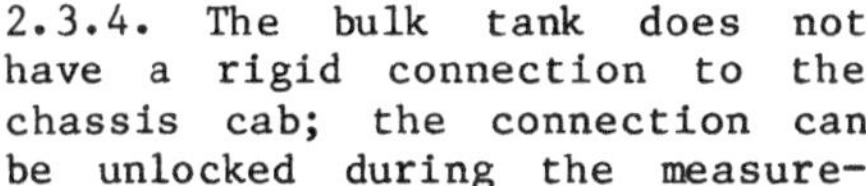

2.3.4. The bulk tank does not have a rigid connection to the chassis cab; the connection can be unlocked during the measurements. In this case, only the tank is lifted by the jacks and put horizontal on the load cells fitted in the jacks. This design looks like the in clause 2.3.3. described one, but in this case the unlocked situation minimalizes the affects of cross-forces mentioned there. A few other things earn special attention now. First of all special precautions have to be taken to secure the tank on the chassis cab during transportation. Secondly, connections between pumps or compressors, which, mostly are mounted to the chassis, should not cause to much extra uncertainty in measurements, when they are driven by the power take off.

2.3.5. Lately a very special design has been made for the Dutch dairy-industry. On a chassis cab just behind the driver's cab one hydraulic jack is rigidly fixed on a subframe. At the end of the chassis a second jack is mounted on the same subframe not rigidly now, but turnable in one direction. Between the two jacks the tank is suspended and fixed during the weighing position. In this position the tank lies completely free on two strain gauge load cells which are mounted in the top of the jacks. Because of the turnable connection between jacks and tank on the one side and independent controling of the hydraulic system on the other side, it is possible to take measurements in a leveled situation. At departure the load cells will be mainly or completely unloaded again. The tank is locked against involuntary movements. This tank is to be used for collecting milk at the farmers in portions of even 200 up to 800 liters.

2.4. An external device as loadreceptor
For the sake of completeness we mention some other constructions, which are more or less comparable. The differences mainly are found in the construction of the load receptor and the load introduction. At present, in The Netherlands two additional systems are available, but neither of these is allowed for trade purposes, i.e:

2.4.1. A load cell in the hydraulic jack system of a waste container truck. In this case the container functions as a load receptor.

2.4.2. A load cell in the hydraulic jack system of a fork lift truck. A pallet acts as the load receptor.

3. INFLUENCE FACTORS AND DISTURBANCES.

3.1 Vibrations and shocks. The effects of vibrations on the reliability of weighing machines can be very great. At this moment in The Netherlands no regulations have been formulated regarding this kind of influence factors, with respect to pattern approval or initial verification. Also in the most recent documents of the OIML working group SP7-SR2 such regulations fail.

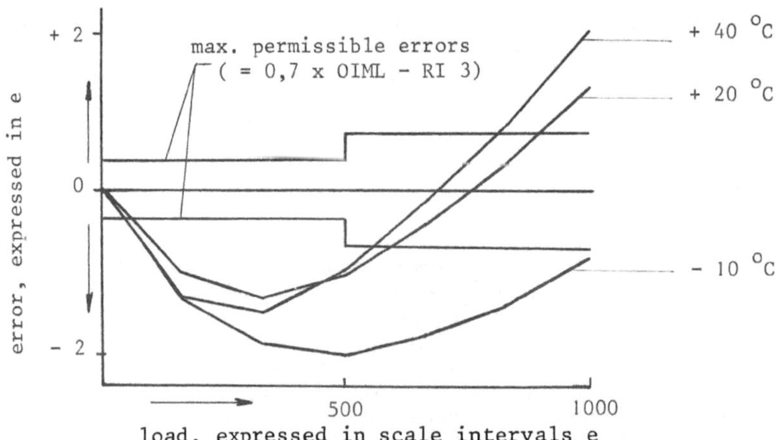

fig. 3

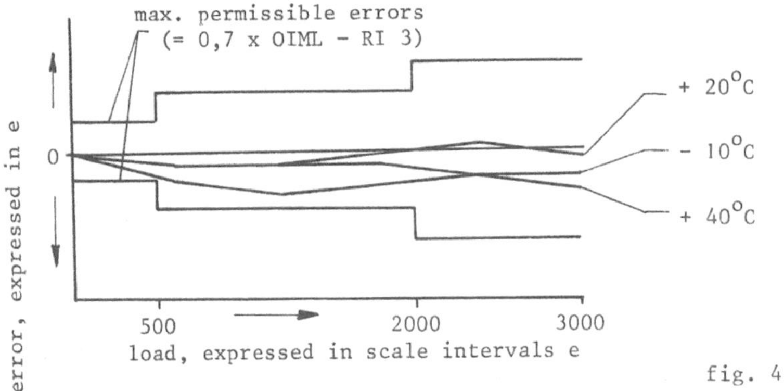

fig. 4

Probably there is no need for such kind of regulations, since a disturbance of an electronic weighing system caused by vibration generally will not result in any weight indication at all. Specially in road transportation, strong and continuous vibrations will occur. In that case it is reasonable to expect that a suitable weighing machine does not break down as a result of these vibrations. These systems should be able to cope with very heavy shocks caused by normal road surfaces. Electronic parts such as print boards and connecting pins should be able to resist such extreme vibrations and physical shocks. It is advisable for various reasons (a.o. temperature and humidity aspects) to create a location for the electronics inside the driver's cab to prevent vibration as much as possible.

The load cell itself presents a special problem. The load cells referred to under 2.3.1, 2.3.2 and 2.4.2 must be of a very robust construction. During transportation they must absorb all the swinging forces, produced by the load receptor and by the cargo. In the weighing po-position, the elastic deflection of the measuring element (the body carrying the strain gauge) is too small to produce a proper resolution for measuring the load. In diagram 3 some test results of such a robust load cell are shown. A resolution of 1000 parts within a reasonable accuracy is not possible. Perhaps it is not the robust construction only, which is responsible for the bad results. The vulcanized rubber pads between the moving parts, to protect the load cell against the swinging forces, might contribute as well. For comparison in diagram 4 the results of a "normal", not overdimensioned load cell are shown. This kind of load cells, which

Max. Cap. 5,6 tons
Resolution 100 parts
Safe overload 300-600 %

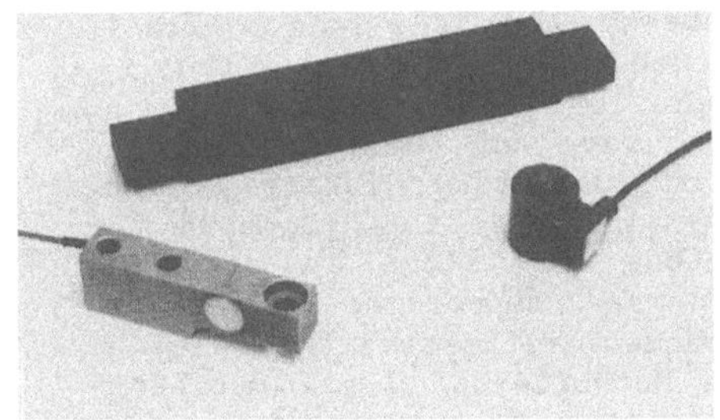

Max. Cap. 5 tons	Max. Cap. 25 tons
2500 parts	3000 parts
safe overload:	safe overload:
150-300 %	150-400 %

fig. 5

have a considerable lighter construction, can only be applied if the load cells are not under load during transportation.
In those cases where during the vehicle's movement the load cell is not in principle under load, still adequate measures must be taken, to ensure that the load cells own mass and that of the subsidiary construction required for the proper load introduction do not produce inpermissible high forces. A plastic deformation and damaging of the load cell could easily result in unacceptable uncertainties of measurement. Diagram 5 shows 3 load cells, two of them virtually with the same maximum capacity, to illustrate how great the physical differences may be for different applications. The largest one is intended to be loaded during a drive.

3.2 The load introduction. A faulty load introduction may result in considerable inaccuracy. This can be traced back to incorrect methods of incorporation in the vehicle, resulting in lateral forces and/or to torques. Also the positioning and load distribution of the vehicle may cause inaccuracies in some applications.
We already saw in 2.1 and 2.2 situations where this is not so important. In this application there is little chance for lateral forces or torques.

In the applications 2.3.4 and 2.3.5 a vertical load introduction easily can be obtained. However, special attention has to be paid to the friction aspects in the rotation points.

In all other mentioned applications more severe uncertainties are to be expected when taking measurements on any location of the vehicle. In practice this source of errors has proved the most important one, so we will discuss this here in more detail.

What, can be the effect of a 10% slope on the weighing result? The play of forces shown in diagram 6 may give some insight. The force N, measured by the load cell will be reduced to L.cos α ; thus, a slope of 10% results in an inaccuracy of nearly 2%. The extra inaccuracy, because of the lateral force Q, depends on various provisions in construction and cannot be simply determined arithmetically.
The Netherlands' legal metrology authorities stipulate that weighing machines incorporated in vehicles may be used only if the vehicle is horizontal, unless a position at a slope up to 10% does not affect the weighing result. This implies, that provisions must be made to realise a horizontal position and to prevent the reading of weighing results if the vehicle is not level. Leveling usually is realised

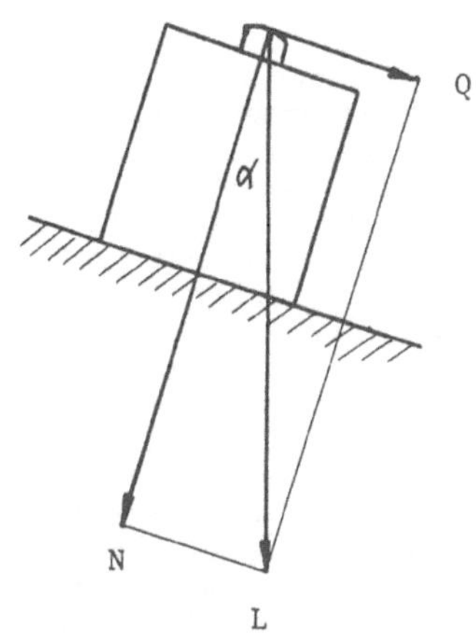

fig. 6

by a hydraulic system below the weighing machine or below the vehicle itself.

Prevention of weight indications, when not level may be realised by means of a klinometer of sufficient accuracy built into the weighing machine. The construction of the klinometer can be quite simple. A plumb line construction with reed contacts seems an acceptable solution.

3.3 Temperature and humidity aspects. Compared with weighing machines located indoors, the ones incorporated in vehicles are in a rather bad environment. A stabilized temperature of the complete weighing machine seldomly occurs. The sun only heats one side of the vehicle, the influence of which hardly can be controlled.
In particular high demands are made to the functioning of the essential components as load cells and indicators. Also the load cells are exposed to strong temperature changes. Splashing water may moisten the load cells and/or the indicator. In winter the salty water used to fight the slipperiness (in The Netherlands only?) is an additional problem, since zerodrift and sensitivity both are resistance dependent factors.
For that reason, the only load cells that can be used in practice are hermetically enclosed ones, unless their incorporation is such that the possibility of penetration of moisture can be ruled out. As already mentioned, it is advisable to house the electronics in the driver's cab. Although a virtually hermetically closed casing at the outside of the vehicle may seem acceptable, we strongly recommend against. As weighing machines in road vehicles are only shortly used in relation to the periods of movement on the road, the temperature effects, resulting from inadequate warming-up time, cannot be neglected.

3.4. Electric power supply.
Normally, a weighing system will be fed from the public grid supply of 220 V. Sometimes the supply is carried out via an adaptor, whilst usually the 220 V supply is transformed in the indicator. However a weighing system built into a vehicle is in a completely different situation. Only a 12 V or 24 V DC supply is available. All the acceptable systems known at present are fitted with a 220 V indicator, which gets its power from the DC system of the vehicle by way of a convertor with trapezoidal output. Problems with the trapezoidal output instead of the usual sinusoidal output of the public voltage power system are not known at present.

3.5 Supervision of the weighing operation.
In The Netherlands it is stipulated that the weigh master must be able to see what he is doing. Obstacles, which might influence the measurements should be noticed. In the case of weighing with a vehicle as a weighing machine he cannot be at the two sides of the vehicle at the same time. So, before taking measurements he has to ascertain that all the conditions for good weighing practice are fulfilled. It may be expected that the driver of the vehicle is a good weighmaster in this respect. Therefore, no special stipulations apply.

3.6. Effects of wind, rain, draughts and the like.
Weighings in the open air should have to be made with the same accuracy as indoor weighings.
Mostly a stabilizing or oscillation integrating device has to be used in this kind of weighing machines. The problems, well known of weighing with a

weigh bridge, are also present when weighing with a vehicle. In the latter case the effects of the natural elements even can give some more difficulties. The vehicle may catch more wind and rain. Measurements themselves sometimes have to be read under conditions a lot worse than with a weigh bridge. Because of the use in all kind of weather, the metrology authorities recommend to use no more than 3000 divisions in the open air. In view of the special construction of a weighing system integrated in a vehicle there should not be any opportunity for using a multi-interval instrument.

4. UP-WEIGHING OR OUT-WEIGHING.

Some weighing machines are intended only for out-weighing, in particular for weighings with a bulk vehicle delivering granular or liquid products to various customers. This may be done by subtracting two gross weighing results or, after zerosetting, considering the negative quantity indicated as the quantity delivered. At present, no adequate tolerances exist for method. After all, tolerances for weighing machines relate to the mechanism itself and not to the product weighed with it. We think that good work may be done with a variant of the tolerances which are stipulated in the clause 7.1 of the international OIML-recommendation No. 3 for non-automatic weighing instruments:

For weighings in class III the maximum permissable error at initial verification in the nett quantities delivered, expressed in verification scale intervals e, should not exceed:

0,5 e — for a delivered quantity corresponding with a value of max. 500 e

1 e — for a delivered quantity corresponding with a value of more than 500 e not exceeding 2000 e

1,5 e — for a delivered quantity corresponding with a value of exceeding 2000 e.

These tolerances make the requirements for this kind of weighing machines more severe. Even to the extent that a weighing machine approved in

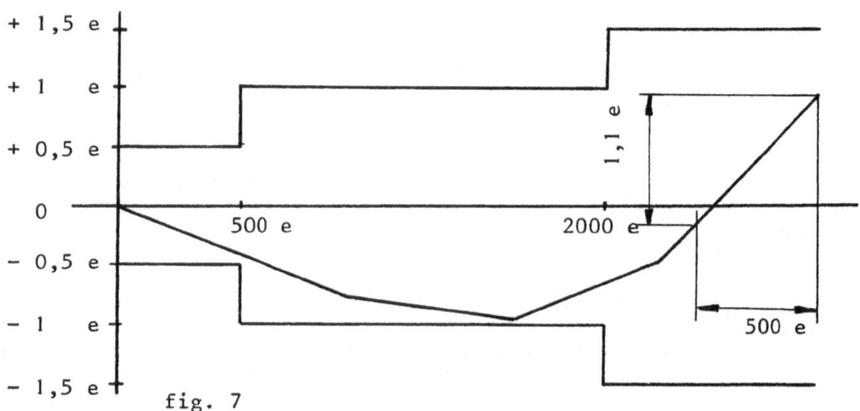

fig. 7

accordance with the present OIML recommendations might be rejected, when subsequently tested against the requirements formulated above. This is elucidated with fig.7. The weighing machine, with the fictitious error curve shown will be approved according to the present requirements. However, when used in weighing out from the maximum load, it would show an error of 1,1 e over the subsequent 500 e and consequently, would have to be disqualified according to the above tolerances. Since this is a realistic problem and some relation should exist between the quantity delivered and the properties of the weighing machine, this problem is considered within our weighing machine division at the moment. In practice the above stipulation means that whatever the starting point in the error curve, the variation must never exceed 0,5 e per 500 e (in cases of initial verification).

5. TEST FACILITIES

First of all, the general basic requirement that "the examination should bear relation to practical use" must be honoured. Clearly, a weighing platform incorporated in a loading platform intended for weighing live cattle may be loaded with mass standards in the conventional manner. Equally, all tests such as for dependability, movability, sensitivity, the effects of excentric loading, the effects of being placed at a slope, the course of faulty indication, etc. must be carried out as with any other weighing machine. With bunker shaped load carriers (tank bulk vehicles) used for weighing a high fluidity product, unlike that weighed with the plateau in the loading platform, loading with mass standards may be more difficult. The examination, to correspond with the practical use, ought to be with the accent on one off-loading operation per water level adjustment. For some bulk tanks it should be admissible to hang two yokes in the form of saddles about the tank. On each side of the tank could be fitted a bar with chains to contain sufficient standard weights. This method, which is in preparation now in The Netherlands is only applicable to tanks with dividing walls. Other tanks could succumb to the load. Another method is to place the standard weights on large stable platforms below and/or next to the tank. Whatever method is chosen, we consider it essential that the weights are placed in position by other means than manpower for maximum capacities over ten tons. In The Netherlands the mechanical transfer of test loads is used for weights over 2 tons. Mechanical handling of standard weights does not necessarily refer to weights of 10 or 20 kilograms. In this country standard loads of 200, 500 and 1000 kg are quite normal. When building a weighing machine, due regard must be given to problems that might arise in testing it, with particular relevance to the accessibility of the weighing machine for mechanical aids and the possibility for placing bulky mass standards.

Applications with a load receptor rigidly connected to the chassis or to the running gear of a vehicle do not seem to be out of the developping stage. The verification technology is quite another matter. We may wonder whether solid standards of mass will continue to be necessary. If the weighing properties of weighing machines could be verified with the aid of liquid or granular masses that have been measured elsewhere, and which are then deposited in the bulk tank, external provisions for applying test loads would be superfluious. Some research in this direction is currently carried out in The Netherlands, albeit at a very small scale. However, sensible and technologically indisputable results are not yet available.

6. EPILOGUE

In the foregoing, a brief outline was presented of the present technological situation in the field of weighing in and with road vehicles. Much will yet have to be done in this wide field of applications before we arrive at metrologically acceptable weighing results. The stress between the heavy constructions required in international road traffic on the one hand and movable and sensitive weighing techniques on the other hand, probably present many problems which will prove hard to overcome in the near future.
The applications 2.1 and 2.2 and more or less the application 2.3.4 and 2.3.5 present quite nice and acceptable measuring systems, which can meet all the present requirements.

IN-MOTION WEIGHING SCALE AND SOME PROBLEMS OF ITS METROLOGICAL EVALUATION

Georgi F. MALIKOV, Yuli M. SERGIENKO, Leonid K. TIMOFEEV,
Vladimir A. CHUKHNO.

The obvious advantages of the coupled-in-motion railroad track scale motivated rapid development of this branch of scale making. The scales of the world-known companies, such as AVERY, RAILWEIGHT, STREETERAMET, etc., are now of such a high accuracy which allows them to be used in commercial operations.
Taking into account intensive railroad goods traffic and serious losses caused by demurrage and difficulties in static weighing we have every reason to say that in-motion weighing is now the only possible method of weighing such bulk loads as coal, cement, etc.
The aim of the development was to manufacture a scale complying with the following requirements:

- the weighing mechanism must be close to statically determinate systems;
- the scale must be designed to provide a possibility of checking the laod cells by independently loading each cell with standard weights within the assembled scale;
- the value of the output signal of every load cell should not depend on the load position on the load receiving unit;
- every operation of the scale adjustment must be independent;
- the scale must be assembled and certified in statics before delivery.

The natural design, which meets these requirements, is based on the well-known elastic parallelogram principle which is widely used in static low-capacity scales. However, it proved impossible to simply transfer this principle to a heavy-duty scale. This can be explained by the fact that inevitable physical imperfections of the elastic parallelogram which only negligibly affects low-capacity scales (which allows to consider this mechanism to be ideal) reach intolerably great values in heavy-duty scales thus resulting in errors of one to two per cent.
Thus the problem of the physical elastic parallelogram appeared. This problem was studied theoretically and a simple adjustment device was proposed, which makes it possible to compensate the mentioned imperfections and to reduce the cost of the scale manufacture and adjustment.
A very important advantage of the 1959 TC-200B scale consists in its design, which allows to make the main operations of its manufacture and adjustment independent of each other. Here are some examples. All of the following three operations - elimination of influence of the load position on the scale, zero and span adjustments - are independent. This makes it possible to attain the specified acccuracy within a very short period of time by fulfilling simple operations without subsequent adjustments by repeatedly placing heavy weights. No one of the known types of the heavy-duty scales provides for such a possibility.

Wieringa, H. (ed), Mechanical Problems in Measuring Force and Mass.

Each of the two measuring channels is manufactured and adjusted individually; the adjustment is carried out by direct and independent loading of each load cell with the scale units being assembled.
As it is known, no scale now features such a property. It allows to reduce the cost of the load cell manufacture since in this case there is no need (without deterioration of the load cell quality) in any "rigid" adjustment of the cell sensitivity coefficient to an accuracy of 0.02 to 0.03 per cent.
The necessary preventive maintenance and repair procedures are greatly simplified since the operation of the load cell replacement which was recently very complicated, has become very simple.
The dynamic adjustment of the scale is independent of the static one. The compensation of some dynamic errors by means of electronic and mechanical elements does not change the static characteristics of the scale.
Such exceptional features of the 1959 TC-200B scales as their modular design, possibility of checking by independent direct loading of each load cell with standard weights and independent adjustment operations make it possible to reduce labour consumption in the course of their manufacture and to organize their modern intensive production.
The up-dated in-motion weighing scale calls for expensive foundation and construction of approach rails since its accuracy is directly dependent on the quality of the understructure. The cost of the foundation is far in excess of the cost of the scale proper. As to the 1959 TC-200B scale the developed methods of adjustment and weighing contribute to the reduction of the foundation cost and make the requirements on the rail track profile not so severe and all this along with the increased accuracy as compared with the scales manufactured by many well-known companies.
This scale together with the developed method of weighing, allows the trains with rigid automatic coupling, to be weighed to an accuracy which is not lower than that obtained in weighing trains with flexible automatic coupling. The requirements on the scale foundation herewith are not so stringent.

When evaluating the accuracy of the scale intended for weighing trains in motion some serious difficulties arise, which are seldom when dealing with static scales.
The accuracy of the in-motion weighing depends on the length of the train, type of the automatic coupling, design of the bogie, profile of the track, quality of the drive and so on. Under the action of these factors the errors of measurement are greater than the error of the scale proper. Therefore the results of the scale tests for compliance with the standardised value of the accuracy are substantially dependent of the testing program and of the method of evaluation of the test results. For example, the test results obtained when weighing a train having a total mass of 1000 tons differ 1.5 to 2 times from the results obtained with 2000- or 3000-ton trains. Taking into consideration that the main part of the error is of a random character and cannot be predicted, some methods allow 5 of every 100 test results to be beyond the maximum permissible values.
In spite of some variety of the testing methods all of them can be divided into two groups:

1 Verification tests when real conditions of the scale exploitation are simulated to maximum approximation. For example, if a scale is intended for weighing trains with a total mass of 2000 tons, it is just a

2000-ton train the scale is to be tested with. During operation such a scale is not allowed to be used for weighing trains with a greater mass, for example 3000 tons, because it is not tested with such a mass and hence its accuracy in this case remains unknown.
This method is accepted in the USSR with none of the test results admitted to be beyond the maximum permissible error.

2 Testing the scale with the use of a uniform train consisting for example of 10 cars. The scale tested in such a way may be used for weighing trains with any total mass. This method is cheaper and easier. As may be seen the aim of the test carried out by this method differs from that described above and may be interpreted as to obtain comparative metrological characteristics which give the possibility to compare the accuracy of different scales. In this case it is allowed that 5 per cent of the test results are beyond the maximum permissible error but not any result should exceed its doubled value.

Application of various test methods is very inconvenient. Lack of uniform standard test methods hampers comparison of the accuracy of different types of in-motion weighing scales. In such circumstances the user can come to a faulty opinion about the scale accuracy.
Now it is necessary to speed up the work within the limits of the International Organization of Legal Metrology on the elaboration of some recommendations aimed at the standardization of the parameters of the in-motion weighing scales and their verification.

6-COMPONENT LOAD CELL FOR AUTOMOBILE WHEEL

Takeshi Yoshida and Kazuo Nakayama

Yamato Scale Co., Ltd.
5-22, Chaemba-cho, Akashi-city, 673, Japan

ABSTRACT

It is indispensable to know the characteristics of the tire when discussing the movements of an actual car, but the forces to be generated within the footprint area of the tire are complicated and are hard to be detected.

To investigate the characteristics of the tire, various testing machines have been developed in the past, and used by tire makers and car makers. Not only the testing machines for the tire static rigidity measurement but also the tire running test machines for measuring the forces while running on a rotating drum have greatly contributed toward the improvement of tire technology.

However, they are only the pure data in a research laboratory, and it would be very interesting to compare them with the data obtained on actual cars. In other words, it is essential in evaluating the movement characteristics of actual cars to investigate the relation between the inertia, acceleration, deceleration, spring characteristics of each wheel and the forces to be generated within the footprint area of the tire.

We have so far manufactured and offered various 6-component load cells for use in tire testing machines and actual car wheels. However, because such load cells were of the fixed type, installation of them in the wheels required space, so the car had to be modified.

In this connection, we have developed a rotary type 6-component load cell for exclusive use of actual car wheels and a digital data memory for storing its signals.

The 6-component load cell is installed between the wheel flange and the wheel for measuring the 6-component of forces to be generated within the footprint area of the tire. The 6-component signals generated from the load cell are put into the digital data memory mounted in the car through the slip rings. The signals will be subject to interference correction operation and coordinate conversion operation after A/D conversion, and will be put into memory as the 6-component.

KEYWORDS : rotary type 6-component load cell, controllability, stability, tire testing machine, trailer testing car, slip ring, digital data memory

1. INTRODUCTION

It is approximately 100 years since the air-inflated tire has been developed. During this period, the quality and life of the tire have become remarkably improved in proportion to the familiarization, high-speed motorization and highly-improved performance of cars.

For the past years, there has been an increasing demand in reducing rolling resistance from the energy-saving viewpoint, and in securing high performance of the tire from the viewpoint of the reliable controllability and stability of the car.

Accordingly, a lot of testing machines have been developed to measure the

Wieringa, H. (ed), Mechanical Problems in Measuring Force and Mass.

tire characteristics. The material, construction, strength, rigidity, etc. of the tire have become improved. For example, a drum testing machine is usually used to measure the tire characteristics. The fatal disadvantage with the machine is that the drum itself has some curvature. So the tire grounding condition is different from that on the actual flat road surface, thus resulting in slight differences of characteristics measured.

Therefore, there have been developed such testing machines as the one with flat belt in substitute for drum, a trailer type testing car operated on an actual road, and a traction testing car or a bus equipped with the testing wheel, which have been used for experiment and research. They were developed only to measure the characteristics of tire, but it is essential to measure tire characteristics including other characteristics such as steering and suspension, to study the controllability and stability of the car.

This lead us to develop the 6-component load cell to be installed in the wheel of a car for the above purpose 10 years ago. The unit was designed exclusively to be installed in driven wheel, and it was not possible to measure the characteristics of driving/braking wheels with it.

Now, we have developed the 6-component load cell which can be easily installed in the driving wheel, and the digital data memory used to store the measurement data. Since the unit can be installed in the front wheels of FF-type cars which have become more popular these days, it is possible to measure driving/braking characteristics under steering condition.

It is also applicable to the 4-wheel-drive car and 4-wheel-steering car, and is expected to make a great contribution to the study of the controllability and stability of the car.

This paper, hereafter, describes the tire coordinate, tire testing machines, and further introduces a brief outline of the construction, basic performance of the 6-component load cell and the digital data memory.

2. TESTING OF TIRE CHARACTERISTICS

2.1. Forces working against the tire and its coordinate.

The characteristics of forces to be generated within the footprint area of the tire under side-slipping condition must be known for a research on the controllability and stability of the car. Side-slipping takes place when there is a difference in angle between the wheel plane and the wheel travelling direction when a running car takes a curve. The forces to be generated within the footprint area of the tire at side-slipping are complicated. Among them, forces which affect specially the controllability and stability of the car are "cornering force" and "self-aligning torque".

Fig.1 shows the tire coordinate.

Where,

α: Slip angle (SA)	: Angle formed by wheel plane and wheel travelling direction.
β: Camber angle (CA)	: Angle formed by wheel plane and Z-axis.
Fx: Rolling resistance(RR)	: Drive force, braking force and rolling resistance to the direction of wheel plane.
FX: Drag force (DF)	: Resistance force to wheel travelling direction.
FY: Side force (SF)	: Force generated against centrifugal force at the right angle to the direction of wheel travelling.
Fy: Cornering force(CF)	: Force generated against centrifugal force at the right angle to the direction of wheel plane.

FZ: Vertical load(VL) : Load applied to tire.
Mx: Over-turning moment (OTM) : Moment generated at a diviation of FZ from the origin toward y-axis.
My: Rolling resistance moment (RRM) : Moment generated at a diviation of FZ from the origin toward x-axis.
Mz: Self-aligning torque (SAT) : Moment generated at a diviation of Fy from the origin toward x-axis.

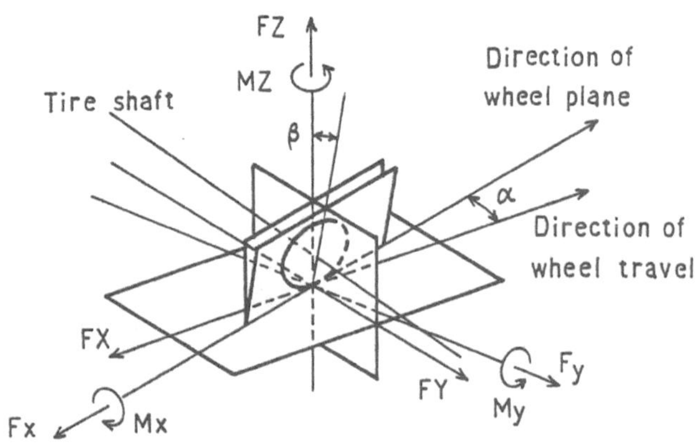

Forces and moments on the footprint area of the tire. Slip angle is formed by wheel plane and wheel travelling direction. Camber angle is formed by wheel plane and Z-axis.

Fig.1 Tire coordinate

2.2. Tire testing machine. First of all, the forces to be generated within the footprint area of the tire must be measured in order to obtain the tire characteristics. And there are various types of machines available for the measurement. The followings describe the features of the three typical testing machines used at present.

2.2.1. Indoor type drum testing machine. Fig.2 shows the outward appearance of the machine which is designed for specific measurement of the pure tire characteristics.

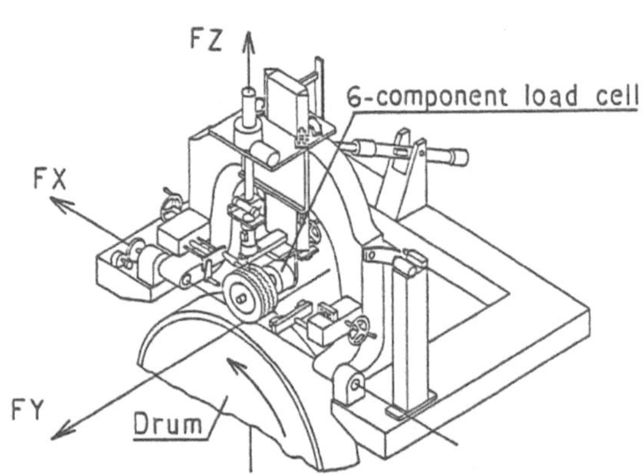

Slip angle is provided by the drum turning around Z-axis, camber angle by the main unit turning around X-axis. The 6-component load cell is installed as the tire shaft. The machine is capable of a precise control of the load, inflated pressure, speed, slip angle and camber angle.

Fig.2 Drum testing machine

The 6-component of forces are to be measured with the 6-component load cell. The machine is capable of a precise control of the load, inflated pressure, travelling speed, slip angle and camber angle, and provides the pure tire data for a research on the controllability and stability.

In addition, it is also possible to change the friction condition by replacing the drum surface material, and measure the uniformity of one rotation of tire with a precisely-machined true rim.

2.2.2. Trailer testing car. Fig.3 shows the outward appearance of the trailer testing car.

Fig.3 Trailer car

Each of the wheels is equipped with the 6-component load cell. The load can be adjusted by the number of the weight. Slip angle and brake are operated on the instrument panel in the tractor.

The testing car is an intermediary product of the indoor type drum testing machine and the 6-component load cell installed in an actual car. The machine is characteristic in that it is pulled by a tractor to travel on an actual road. Each of the wheels, right and left, is equipped with the 6-component load cell for the measurement of 6-component forces to be generated within the footprint area of the tire. The load to be applied to the tire can be adjusted by the number of the weight. The tractor is to be equipped with a variety of control switches, thus allowing an easy-control of the slip angle and of the brake on the right wheel and the left independently or at the same time.

It is also possible to adjust the spring characteristics by changing the internal pressure of the air spring or by replacing the shock absorber. Therefore, the machine can simulate features of various types of cars.

2.2.3. Fixed-type load cell installed in an actual car. Unlike the indoor type drum testing machine and trailer type testing car, it is a 6-component load cell to be fixed in an actual car, and is capable of measuring such tire characteristics including the characteristics of steering and suspension of the car.

Consequently, the data obtained comprise the pure tire characteristics to be given by the indoor type drum testing machine plus the characteristics of the car. Therefore, the data will help a lot in examining the controllability and stability. Fig.4 shows the fixed-type load cell installed in an actual car. The load cell is designed for use in driven wheel, and is a fixed-type to be installed in space of the brake which is to be taken off. Although the load cell can be used for the measurement of the cornering characteristics, it is impossible to measure those of driving/braking wheels. Therefore, it is unable to get the data of driving/braking wheels under cornering condition.

Fig.4 Fixed-type actual car load cell

Fixed-type actual car load cell is designed for use in only driven wheel, and is installed in space of the brake. This is the load cell installed on the right side of front wheel.
Therefore, the actual car has to be modified.
The data obtaind comprise the pure tire characteristics given by the testing machine plus the characteristics of the car, and help a lot in examining the controllability and stability.
Although using this load cell, we can not get the tire characteristics of driving/braking wheels.

3. ROTARY TYPE 6-COMPONENT LOAD CELL AND DIGITAL DATA MEMORY

3.1. The outward appearance and main specifications.

Fig.5 shows the outward appearance of newly developed rotaty type 6-component load cell and digital data memory. The hole(A) is prepared for installing the load cell in the car, and the hole(B) to attach the wheel to the load cell respectively.

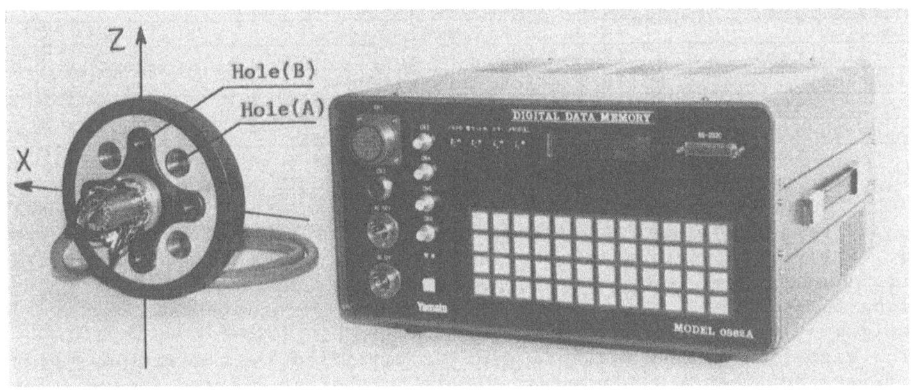

Fig.5 New load cell and digital data memory

The center point between the two holes(A) and (B) is to work as load cell, at 4 places in X-axis direction, and another 4 places in Z-axis direction, 8 places in all. The strain gages are fixed on each surface of these elastic places, 4 each at 8 places, amounting 32 pieces, which form 6 bridge circuits to measure each factor of 6-component.

Table 1 shows the specifications of the rotary type 6-component load cell. Since the practical accuracy required for each component of the forces in a measurement using an actual car is 3% ∼4%, the specifications can fully satisfy the requirement. Further, table 2 shows the specificaitons of the digital data memory.

Table 1 Specifications of 6-component load cell

Component	Fx	Fy	Fz	Mx	My	Mz
Capacity	±4.9kN	±4.9kN	±4.9kN	±590Nm	±590Nm	±590Nm
Accuracy	1 % of full scale					
Interference	1.5 % of full scale					
Outward dimension	Φ 200 x 50 mm					
Weight	55 N					
Subject car	Passenger car (weight is approximately 9.8 kN)					
Subject wheel	PCD Φ 114.3 mm, 4 holes wheel					

Table 2 Specifications of digital data memory

Analog input	6 ch.----±15 mV max. 4 ch.----±10 V max.
Amplifier	Non-lineality---0.02% of full scale Temp. characteristic--- 20 PPM/°C of sensitivity --- 0.2 μV/°C of zero
A/D converter	Approximate successive method Sampling time---0.5 ms/10ch. and less
Exciting output to load cell	10V DC(±5V), 200mA Stability---0.01% Temp. characteristic---0.005%/°C Ripple voltage---1mV rms
Digital input	1 ch.---Coordinate detection pulse 1 ch.---Sampling start pulse
Setting device	Full keyboard
Memory area	RAM 500k bytes, battery back-upped
Memory item	Data, Time, Set value and comment
Data output	Based on RS-232C
Working temp.	0°C ∼ 50°C
Display	16 figures, liquid crystal
Power source	12V DC ±15% or 100V AC ±10%
Outward dimension	400 x 200 x 370 mm

Fig.6 shows the block diagram of the digital data memory. It is capable of 10 channels analog input of such data as speed, steering angle, acceleration in lateral direction, acceleration in longitudinal direction etc., including the input of the force data given by the 6-component load cell, depending on each test purpose. It is also capable of 2 channels timing pulse input for detection of load cell coordinate and start of data sampling.

The signal sent from the load cell is amplified by the analog amplifier at first, and then A/D converted. It is further calculated in scale factor for physical value, and then stored through interference offset calculation and coordinate conversion calculation. The time required for these processes is less than 17ms/10 channels. The memory area has a capacity of approximately 500k bytes, and is capable of storeing the data of approximately 200,000 in all.

Upon completion of the test, the digital data memory is to be taken off the car and connected with the personal computer through the RS-232C to read the data from the memory area for data processing.

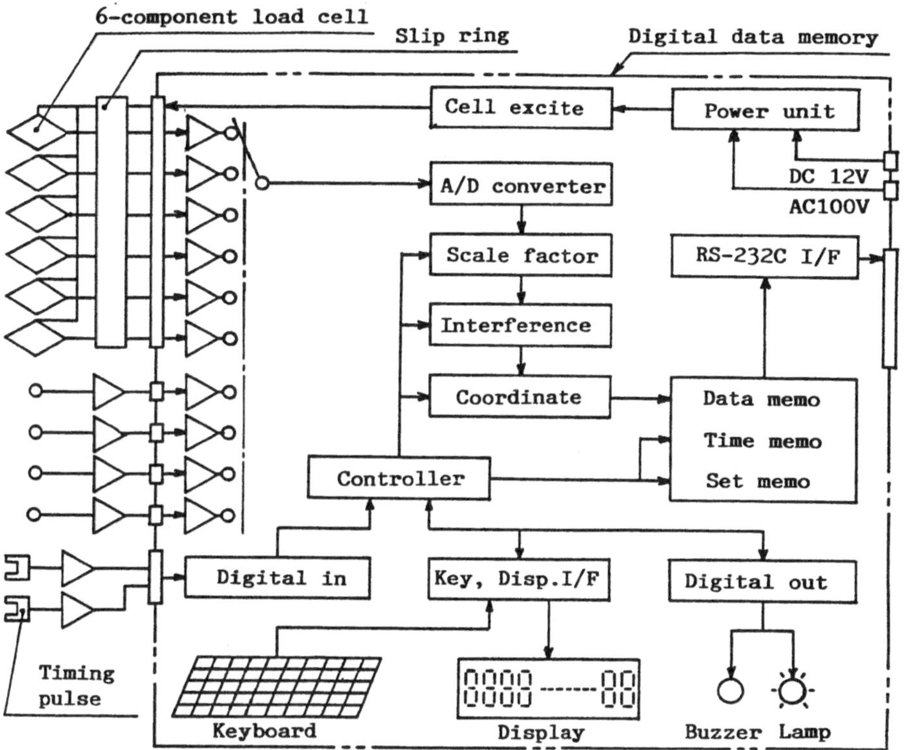

Fig.6 Block diagram of digital data memory

3.2. Operation.

The rotary type 6-component load cell turns together with the wheel. Therefore, it is greatly different from the ordinary load cell on the following two points.

3.2.1. Take load(VL), for instance, the VL is the Z-axis force at first. When the wheel makes 90° turn, however, the VL will be the original X-axis force. By another 90° turn, VL will be minus Z-axis force. If it further makes 90° turn, the VL will be minus X-axis force. The initial Z-axis force will not be the VL until the wheel makes another 90° turn.

In other words, the load cell is designed to sample the data only when the X-axis and Z-axis stand perpendicular against the road surface. Therefore, the data are taken 4 times for each component of the force while the wheel makes one turn.

As the load cell makes turn, the coordinate conversion calculation is carried out by the digital data memory. The rotating direction of the coordinate alternates when the load cell is installed either in the right wheel or in the left wheel respectively, and also when the car goes forward or reverses respectively. Accordingly, the rotating direction of the load cell needs to be put into the digital data memory.

3.2.2. As the load cell rotates, a slip ring is required to transmit signals. And extra care is required to decide on the selection of a slip ring used for signal transmission, as the level of singal output of the load cell is extremely small. Furthermore, the contact noise of brush and the thermoelectromotive force to be generated by heat at contact points should be small enough not to affect the guaranteed accuracy of the load cell.

Fig.7 shows how to install the load cell. First remove the wheel, install the load cell, and then the wheel to the load cell. In other words, the load cell is sandwitched between the wheel flange and the wheel.

If the original rim is used, the tire will project by the portion of load cell thickness. So the tire center must be placed in the original position by using rim with increased offset.

(a) Load cell to flange

(b) Wheel to load cell

Fig.7 Installation of 6-component load cell

Since the rotary type load cell can be installed not only in the driven wheel but also in the driving wheel of the car as described above, in addition to the characteristics under rectilineal travelling, it is also possible to measure the driving/braking characteristics under cornering condition.

4. RESULT

4.1. Load cell calibration by deadweight.

Table 3 shows calibration results on the rotary type 6-component load cell. The accuracy of each component force is shown in each cross of the same figures, moreover each cross of different figures show interferences which can not be compensated. The data satisfy the required accuracy of the load cell to be installed in an actual car.

Table 3 Calibration data of 6-component load cell

Each cross of the same figures show the accuracy of each component. And others show the interferences. The data satisfy the required accuracy.

Loaded component and capacity		Accuracy and interference (% of full scale)					
		Fx	Fy	Fz	Mx	My	Mz
Fx	± 4.9 kN	0.3	0.4	0.5	0.1	0.1	0.1
Fy	± 4.9 kN	0.3	0.8	0.2	0.1	0.1	0.2
Fz	± 4.9 kN	0.1	0.3	0.2	0.1	0.1	0.1
Mx	± 590 Nm	0.2	0.4	0.2	0.3	0.2	0.2
My	± 590 Nm	0.4	0.4	0.3	0.1	0.9	0.2
Mz	± 590 Nm	0.3	0.5	0.3	0.1	0.1	0.3

4.2. Experimental results on an actual car.

Fig.8 illustrates the experimental results on cornering force. The lines 1,2,3 show the relationship between slip angle and cornering force brought about by the indoor type drum testing machine, while values marked by □ are the experimental results by the load cell installed in an actual car. According to the data lateral acceleration increases in proportion to slip angle, resulting in the increase in cornering force. These are almost compatible with the data of the indoor type drum testing machine.

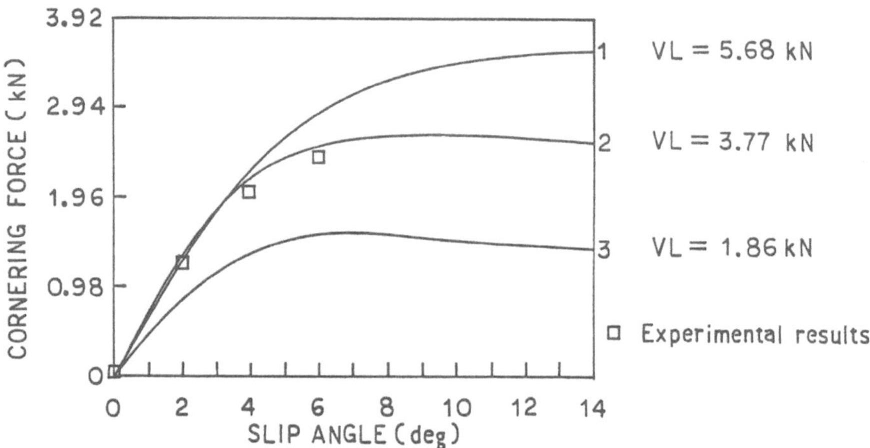

Fig.8 Relationship between cornering force and slip angle

Test car	:	TOYOTA COROLLA	Test wheel	:	Front, right
Test tire	:	165SR13	Pressure	:	0.17 MPa
Speed	:	Nearly 30 km/hr	Road	:	Asphalted

Experimental results are almost compatible with the data of the indoor type drum testing machine.

5. SUMMARY

5.1. We have obtained satisfactory data from the experiment carried out with a special note on cornering force.

5.2. The data shows no adverse effect of noise from slip ring.

5.3. Hereafter, we will also have to do a research with a special note on other component forces to obtain practical data under driving/braking condition. We think it would be quite interesting to carry out analysis of car movements with 4 load cells installed in each of the 4 wheels.

5.4. It is found from this experiment that the rotary type 6-component load cell and the digital data memory combines to provide very useful data to promote the performance of the car and the tire, and will contribute greatly to the development of new products of car makers and tire makers.

REFERENCE

1. T. Yoshida: 6-component force transducer and its application. Trans. of IMEKO, 1984.

MEASUREMENT UNCERTAINTY IN THE DETERMINATION OF WEIGHT AND CENTRE OF GRAVITY OF OFFSHORE MODULES USING STRAIN GAUGE LOAD CELLS

ing. P. Tegelaar and ir. L.J. Wevers
TNO-IWECO, P.O. Box 29, 2600 AA DELFT, the Netherlands

ABSTRACT

When measuring the mass of an offshore module accurate and reliable results must be obtained. Besides the mass also the position of the centre of gravity is of vital importance when the object has to be lifted or when transportation on a barge is necessary.
Reproducibility and measurement accuracy are of increasing importance to measurements for the offshore industry.

The various items that affect the measurement accuracy will be discussed. A method will be given for the determination of an overall uncertainty, including the standard deviation of the measured data.

A standardized weighing procedure will be described, underlining stable and reproducible loading conditions to obtain good statistical reliability.

The calculation methods will be discussed for a measurement system in use for field measurements.

INTRODUCTION

When weighing large structures for either onshore or offshore use, the intention is to determine the mass and the location of the centre of gravity.
No matter whether the weighing is for either financial purposes or that the results are needed for safety during transport or in active operation, reliable and accurate measurements are of increasing importance.

With a view to the importance of the results of the weighing, high demands are made upon the measurement accuracy and the reproducibility.
Load cells based on resistance strain gauges are nowadays of such high quality and reliability that the overall accuracy of the weighing is only partly determined by the properties and specifications of the load cells and the indicating instrument.

The authority carrying out the weighing should do so according to a well-reflected and defined method. They must be able to issue a document which describes the weighing procedure in detail and in addition a second document that defines the accuracy and the total measurement uncertainty, including its calculation procedure.

In this paper, full attention will be paid to the measurement uncertainty of the weighing whilst the possible inaccuracies will be distinguished as far as type and origin are concerned.
The errors of the load cells will then be discussed almost completely and quantified for a certain type of load cell. The rules of error summation laws will be applied to the various types of errors with the final result: an uncertainty for an estimated confidence level. This total uncertainty is supposed to be a reliable measure for the confidence limits of the measured quantity.

The method for weighing will be discussed, together with a detailed description of the weighing procedure. In this context, some variants, as occurring in practice, will be critically viewed regarding their effects on accuracy. It can already be stated here that prior to the weighing, a specification on the uncertainty of the load cell and the matching indicator should be available. However, only after the weighing itself, when the reproducibility and the standard deviation of repeated measurements are known, the final and total uncertainty can be determined.

METHODS OF WEIGHING

Various types of large structures are weighed, such as onshore constructions like movable cranes and offshore constructions usually in the shape of modules.

The structure is placed on a number of load cells and from the output of the load cells follows the mass of the structure. From the ratios of the load cell readings, the coordinates of the centre of gravity in the horizontal X-Y plane is calculated.

To support a construction, three supports are needed to make it statically determined. In practice, however, the minimum is four supports and often a much large number of them is used. In our series of weighings, 20 load cells was the maximum, supporting a mass of 2500 tonne (Figure 1, Underwater Manifold Centre (UMC) - Shell-Esso).

figure 1. Weighing a large offshore structure; a number of load cells and hydraulic jacks can be seen.

This means that we have in all cases a statically undetermined construction with all the difficulties involved. Depending on the flexibility of the structure, the stability of the foundation the load cells will bear a larger or smaller part of the total mass.

Levelling the construction has no influence on the mass distribution, but the procedure is followed to have a good reference.

Usually it appears that the positioning and levelling procedure must be repeated after the structure has been placed on the load cells for the first time. This caused by the setting of the soil and by possible plastic deformation in spreader plates and shims. When a repeatable zero reading and a repeatable output of each separate load cell is found then the set-up is stable. The weighing (load-unload) procedure is then carried out at least three times to achieve a reliable average value.

The errors and inaccuracies which may occur, are according to origin:

1. errors and uncertainties, related to the load cell;
2. errors and uncertainties, related to the indicator;
3. uncertainty related to the reproducibility of the measurement results.

DEFINITION OF ERRORS

Usually in error calculations, systematic and random errors are distinguished. We would prefer to follow a more instructive classification, based on the next two definitions (references [1, 2]).

1. "Uncertainty of measurement is that part of the expression of the result of a measurement which states the range of values within the true value is estimated to lie".
2. "The uncertainty of measurements is equal to the inaccuracy if all the systematic errors are corrected."

We now discern the errors that have to be known before defining the total uncertainty:

a. systematic errors that can be quantified and must be corrected (C-type);
b. random errors having a statistical (Gauszian) distribution and can be totalized according to statistical rules (R-type);
c. non-systematic errors that cannot be determined statistically and can be assumed to follow a square distribution (sigma-type);
d. standard deviation of the measurement results as a measure of inaccuracy resulting from the weighing procedure (S-type).

In the standard deviation is usually included a number of errors of the R-type. Between brackets symbols are given corresponding with the errors and uncertainties that will be discussed in more detail in the next chapter.

LOAD CELL AND INDICATOR ACCURACY

After 30 years of experience it is now possible to make an extremely reliable, stable and reproducible force measuring instrument. The matching electronic indicators are at least of the same quality.

Let us concentrate on a heavy type load cell, see Figure 2.

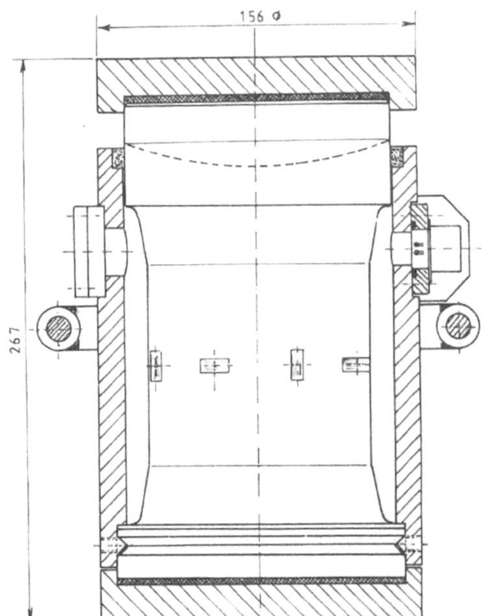

Fig. 2 Loadcell 2.5 MN

On the measuring cylinder there are, completely symmetric, 4 strain gauges in longitudinal and 4 in tangential direction. The four arm Wheatstone bridge thus arranged is balanced with correction resistors, corrected for temperature changes and adjusted for a specified nominal output [reference 4].
The concentrated load is distributed in a curved load plate in combination with a fully encapsulated rubber pad.

Errors that may occur are caused by:
1. the calibration, non-linearity, hysteresis and creep effects;
2. resistance of connecting cable;
3. temperature effect on zero reading;
4. temperature effect on calibration factor;
5. sensitivity to transverse loading;
6. the indicating instrument.

It is necessary to carefully evaluate the load cells available. A quantification of the errors mentioned, follows hereafter:

<u>Calibration, non-linearity, hysteresis and creep</u>

It is desirable and in many cases even compulsory that the load cell is calibrated by a certified authority which can issue a certificate that is traceable to national and international standards. For load cells that are used less than once a week, the British Calibration Service recommends a calibration interval of 4 years. According to the TNO weighing procedure in the Netherlands, the load cells are also checked before and after each weighing, i.e. calibrated on a less accurate, non certified loading machine.

The calibration is carried out in 6 series after 3 pre-load cycles up to 100% of nominal load. The load cell is rotated over 90° three times to compensate for possible effects of side loads. The calibration is carried out as follows:

Serie nr.		
1	0°	- position upwards
2	0°	- position upwards
3	0°	- position up and downwards
4	90°	- position upwards
5	180°	- position up and downwards
6	270°	- position upwards

The calibration authority indicates what accuracy is claimed, sometimes as a pretended accuracy without any further error definition (σ_1). More usual and required in many countries is a statement of the best measurement capability, expressed as an uncertainty and given for a stated confidence level (R-type, U_c).

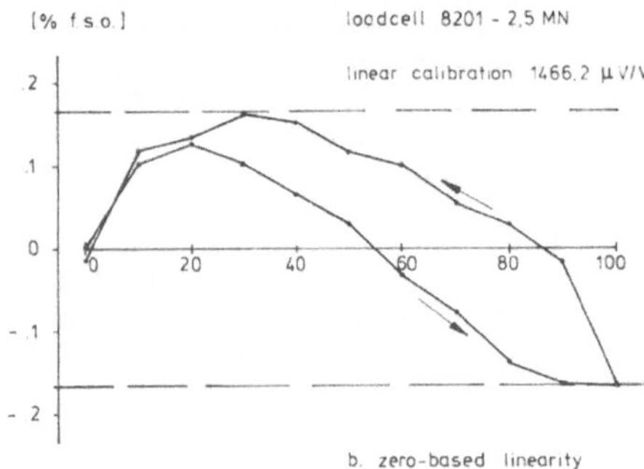

figure 3. Non-linearity of the 2,5 MN load cell.

In Fig. 3 the non-linearity is given for a 2.5 MN load cell; the best straight line (zero based linearity) is used as calibration value and the deviation from this line must be introduced as the possible error (σ_2, see reference [3]). Creep effects are small with a good quality load cell. Creep has a complex character caused by the type of steel used and by the strain gauges and their adhesive. For short term loading like weighing, creep effects are negligible [4].

Resistance of connecting cable

When cable compensation is used by means of a 6-wire connecting system, there is no influence of cable resistance on the calibration factor of the load cell. Any length of connecting cable is then possible theoretically.

With the usual 4-wire system, the resistance of two conductors is then in series with the resistance of the supply arms of the Wheatstone bridge. With normal cables (resistance per wire 50 to 100 m.ohm/m), there should be a correction for this resistance (C_2). As an example, we calculate for a 240 Ω load cell, for which the internal resistance in the supply arms is 260 Ω because of adjustment and correction resistors, a sensitivity reduction of 2.5% with a cable length of 100 m (R_i = 6.4 Ω).

Temperature effect on zero reading

A well-designed load cell is corrected for temperature effects both as regarding to zero and to output. Such a compensation is never perfect, but a zero shift of less than 0.1% of nominal output over a temperature range of 10°C to 30°C can be realized without too much effort. In weighing the time between the first zero reading, the output reading and the repeated zero reading is short, usually less than a few minutes.

Possible measurement errors caused by an unknown temperature change are random in character (R_2) and show themselves in a larger spread of the measured weighing results. When calculating the standard deviation, these errors are automatically included.

With a temperature change, attention should also be paid to the time constant of the load cell. This time constant is a measure for the time the load cell needs to stabilize after a temperature change.

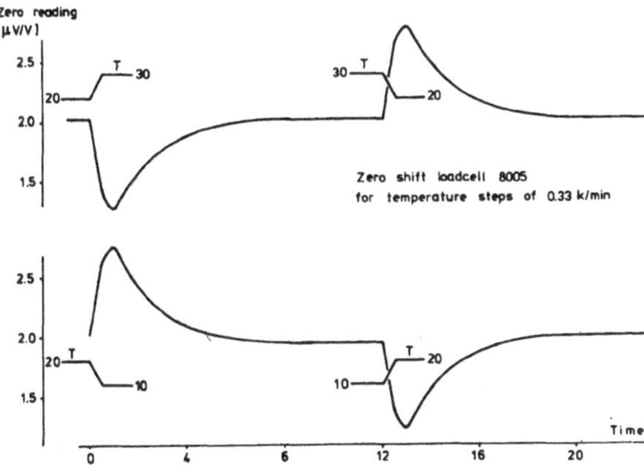

figure 4. Zero shift of a load cell caused by temperature steps of 10 K.

Fig. 4 shows the zero shift of a 2.5 MN load cell (mass 36 kg) at a temperature step of 10 K.
The disturbance caused by unequal heat transfer in the load cell has only disappeared after 8 hours.
Acclimatization of load cells, especially in calibrations, should be given ample sufficient time.

Temperature effect on calibration factor

Nearly all load cells have an electrical compensation for the change of Young's module with temperature (approx. -0.025%/K). As determining of the temperature effect on the calibration is a time consuming and complex matter, usually a calculated compensation resistance is used for the purpose. In practice it appears that the compensation is not always what it should be. For 16 identical load cells temperature coefficients in between +0.001%/K and +0.01%/K were measured.

When the average ambient temperature during the weighing is known, then the deviation which follows from the measured slope, can be corrected (C_3).

With respect to small unknown temperature changes, the same as for zero shift applies. The error thus caused has a random nature (R_3) and manifests itself in the spread of the measured results. In the standard deviation to be calculated, the error is also automatically included.

Effect of transverse loading

Most of the load cells are sensitive to transverse loading in certain directions. This sensitivity is caused by small inaccuracies when fixing the strain gauges onto the cylinder, in location as well as direction. The transverse sensitivity can be determined by placing the load cell in a measuring set-up between two wedges (Figure 5) so that a transverse load known in both direction and magnitude acts on the load cell.

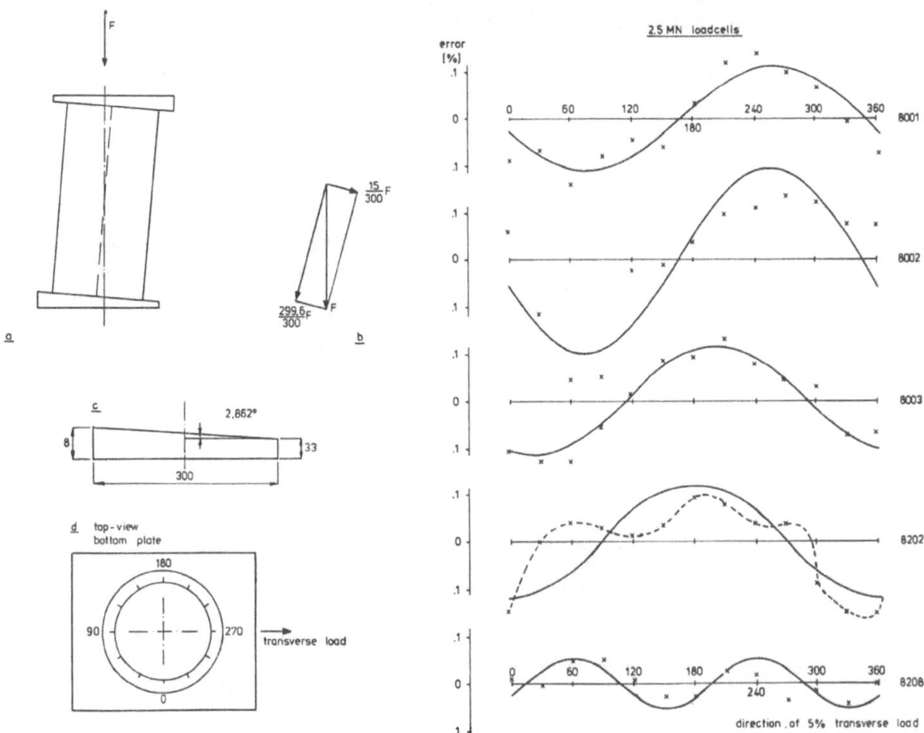

figure 5. Measurement set-up for transverse sensitivity.

figure 6. Error due to transverse sensitivity.

By rotating the load cell, the directional transverse sensitivity is determined. The effect for 5 load cells of the same type is shown in Figure 6.
With these load cells, with 4 strain gauges in lengthwise direction, a sinusoidal behaviour is explainable, whilst a first harmonic is also possible, see load cell 8208. Further study of the available measured errors due to transverse sensitivity shows the admissibility of a sinusoidal function with a superimposed first harmonic.

When during calibration and weighing the load cell is rotated around its central axis between successive measurements, the effect of the transverse load will average out.

If rotating is not possible, then a probable error must be taken into account. This error can be stated as the maximum error for a transverse load amounting to 5% of the nominal load (σ_3). For a high quality load cell, this error will be between 0.1 and 0.5%. The 5% side load seems to be rather arbitrarily chosen. Our experiences from a large number of weighings have shown that a possible transverse load during an expert weighing does not exceed the stated 5%. If the side load is supposed to be larger, the error must pro rata be increased.

Accuracy of indicators

For weighings carried out by TNO, a Hewlett Packard Data Acquisition System 3054A is used. The "accuracy" is stated to be be within + and - 1 micro-strain for a complete bridge configuration, temperature 23°C ± 5K over a period of 90 days.

For a load cell with a nominal output of 1.5 mV/V (or 3000 micro-strain), the inaccuracy is ± 0.3% (sigma type).

ERROR PROPAGATION LAWS

For a simple weighing, a structure is supported by 4 load cells, see Figure 7.
From the measurements of the 4 load cells, the total mass follows from the summation:

$$D = D_1 + D_2 + D_3 + D_4$$

The error propagation law states that for the expression:

$r + t + u + v + w$, the maximum absolute error is given by:

$dr = \pm (/dt/ + /du/ + /dv/ + /dw/)$ and the probable absolute error by:

$dr = \pm \sqrt{dt^2 + du^2 + dv^2 + dw^2}$

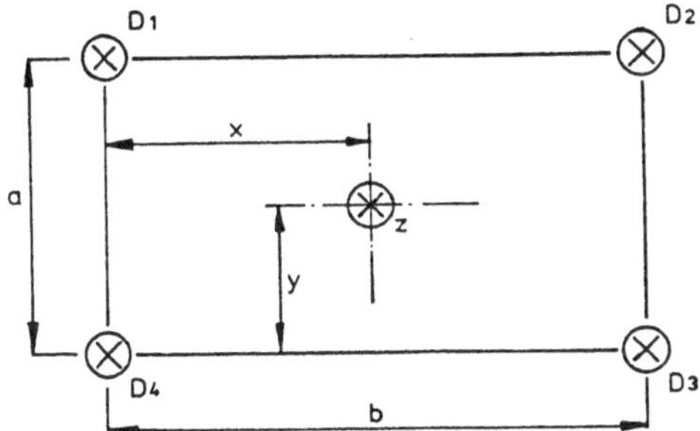

figure 7. Lay-out of load cell positions.

For identical load cells applied, the relative uncertainty for mass determination in the most unfavourable case (maximum error) is still equal to that of the single load cell and more likely twice as small (probable error). For a set of 9 load cells the factor is √ 9.

The co-ordinates of the centre of gravity in the horizontal X-Y plane calculated from:

$$\frac{x}{b} = \frac{D_2 + D_3}{D} \qquad \frac{y}{a} = \frac{D_1 + D_2}{D}$$

The errors (sigma type) of the measures a and b must be included in the final error summation for the c.g.

The standard deviation follows from:

$$s = \sqrt{\frac{1}{n-1} \sum_1^n (x_i - \bar{x})^2} \quad \text{with } \bar{x} = \frac{1}{n} \sum_1^n x_i \text{ is average value}$$

From statistics originates the concept of confidence level. This confidence level expresses a percentage of chance that the difference between the measured value and the actual value will be within certain limits:

$$\bar{x} - \frac{t.s}{\sqrt{n}} < \mu < \bar{x} + \frac{t.s}{\sqrt{n}}$$

with s = calculated standard deviation
n = number of measurements
t = a constant from the Student's table

The term $\frac{t.s}{\sqrt{n}}$ is a measure for the uncertainty. For n = 3 and a level of 95% the uncertainty becomes 2.5 s.

With the summation of all the errors and uncertainties, we follow the differentiation made before, according to C-type, Sigma-type, R-type and S-type [2].

- C-type errors are corrected and do not play a role in the error summation procedure
- Sigma-type errors are supposed to have a square distribution and have to be totalized according to:

$$\sigma = \sqrt{\frac{\sigma_1^2 + \sigma_2^2 \dots \sigma_n^2}{3}}$$

  The systematic uncertainty is calculated again using the Student's table for n = ~. At a confidence level of 95%, t ~ = 1.96 and the uncertainty U is:
  $U_\sigma = 1.96\ \sigma_s$
- R-type errors are totalized following:

$$R = \sqrt{R_1^2 + R_2^2 \dots + R_n^2}$$

  whilst the random uncertainty is calculated with:
  $U_R = t \sim .R$
- S-type error, the calculated standard deviation, gives the random uncertainty with:

$$U_S = \frac{t.s}{\sqrt{n}}$$

All calculated or given uncertainties have now to be summarized. Realising that only 50% of each error of one load cell has to be taken into account, the final error summation is given :

$$U_t = \sqrt{(\frac{U_\sigma}{2})^2 + (\frac{U_R}{2})^2 + (\frac{U_c}{2})^2 + U_s^2}$$

The summation of so many errors may easily give the impression that the total uncertainty becomes so large that the meaning of the weighing as such becomes questionable. A numerical example will show that this is certainly not the case.

Example total uncertainty

The example is about weighing a mass of 158 tonne on 4 load cells with an almost symmetric centre of gravity. After the necessary corrections for non-linearity, cable length, ambient temperature and systematic error of the indicator have been introduced, the following errors and uncertainties must be totalized:

calibration load cell, uncertainty	Uc = 0.1%
electronic indicator	σ_1 = 0.2%
transverse sensitivity (5% side load)	σ_2 = 0.7%
standard deviation mass	S_1 = 0.13%
standard deviation x/b	S_2 = 0.44%
standard deviation y/a	S_3 = 0.08%

After the summation follows for the uncertainty in the mass measurement U_m = 0.25% and for the two co-ordinates of the centre of gravity U_x = 0.71% and U_y = 0.18% respectively.

WEIGHING PROCEDURE

As the way in which the weighing is carried out is directly related to the accuracy and reliability obtained, an experienced weighing procedure is summarized here.

1. Inspection of the supports, both on the floor and at the bottom of the structure to be weighed. Attention should be paid to the flatness of spreader plates and shims.
2. The load cells are positioned with the construction jacked up. Cables are connected, cable length is measured.
3. The origin of the co-ordinate axes is chosen and from there the measuring locations are numbered. The co-ordinates of the locations are measured.
4. Zero readings of the load cells are taken.
5. The jacked up structure is lowered and carefully placed on the load cells. The structure in rest is levelled (up to 1 mm).
6. If necessary, the construction is jacked up again and shims are placed underneath the load cells so all load cells will carry a proportional part of the total weight. The procedure of putting down and jacking up has to be repeated if necessary.
7. Check if zero reading and output have become stable by the pre-loadings, otherwise the pre-loading cycle should be repeated.
8. First measurement: read zero, measure in loaded condition and after jacking up read zero again. When loaded check if the construction is free-standing and supported by the load cells only.
9. Measurement according to item 8 to be repeated a number of times, at least twice.
10. If possible, the load cells should be rotated around their central axis, e.g. over 120° for 3 measurements or in general over 360°/n for n measurements.
11. After each loading cycle, the output of the load cells must be converted to total mass to check the reproducibility of the measurement

and to avoid errors if the construction was not completely free-standing.

12. On request of the client sometimes, load cells have to be interchanged or it is requested to place reference weights. In these cases, the procedure must be repeated from item 5 onwards.

Finally, mention should be made about the use of an indicator with a micro-processor based data acquisition system. With such a system, it is possible to read all the load cells within a few seconds, which "freezes" the load distribution. Furthermore, it is possible by suitable programming to have the measurement results corrected, to determine the mean values, to calculate the centre of gravity co-ordinates and finally to calculate the standard deviations as well as the total uncertainty.

REFERENCES

1. Vocabulary of legal metrology - fundamental terms, International Organisation of Legal Metrology, Paris 1978.
2. The expression of uncertainty in electrical measurements, British Calibration Service, Document 3003, April 1977.
3. Terms used in metrology, British Standard BS 5233, April 1975.
4. P. Tegelaar and H. Wieringa, Some developments at TNO on load cells for weighing purposes, VDI Berichte nr. 137. 1970 (pp 29 - 40).

THE MEASUREMENT METHOD OF BIG TORQUE MOMENTS IN DRIVE SHAFTS

L.Kiełtyka
TECHNICAL UNIVERSITY OF CZESTOCHOWA, POLAND

1. INTRODUCTION

When a shaft diameter is big, and consequently a transmitted torsion torque is considerable, it is recommended that the shaft itself be used as an elastic element of a converter. There are a lot of methods of measuring a shaft torque (among others: an exensometric, inductive, phase-metric and capacitive one); in each of them a number of sensors can be used e.g. a strain gauge, string sensor, differential choke sensor, torsiometric sensor etc. However, specific operational requirements, in this case connected with the creation of big drive torques, sometimes decide not only the construction of a given equipment or measuring unit but also the choice of a physical principle of its operation, i.e. the choice of phenomena upon which the signal processing during the measurement is based.

2. THE ASSUMPTIONS OF THE NEW MEASURING METHOD

While working out the new measuring method, we were quided by the following assumptions (1):

a. the investigated shafts, being elastic elements of a converter, must be isotropic materials,
b. the proportionality of stresses in relation to deformations within the limits of the applicability of Hooke's law for the material investigated must be multiple for the sensor above the limit of proportionality,
c. considering both the homogeneity of the material investigated and the nature of the sensors used, there is no need to state places and character of stress concentrations and to determine their absolute value,
d. the research carried out has the character of dynamic measurements,
e. easy access to the shaft investigated (the facility of fixing the measurement sensors),
f. the sensors applied should be of light design considering the investigations of elements being in motion,
g. the accurate value of the moment measured should be casy to read off, to watch and to record,
h. the introducing of sensors onto the already working unit should go on without having to "slit" the shaft,
i. the system should be characteristic of considerable stability of work, particularly during long period of running.

The unit, which carries into effect the method described hereunder, is able to register the torque moment, power transmission by the shaft and the rotational speed of the shaft, also while working reversedly (2).

2.1. The Desciption of the Method

The suggested unit operates on the basis of the phase-metric method and the measured value of torque moment $\geq$ 100kNm. There are a number of known methods of phase displacement processing, among others such as:

- the method of frequency - time product processing,
- the method of time shift processing,
- the coincidence method.

The method applied by the author consists in filling factor processing (3). Figure 1 represents a block diagram of the unit, which besides measuring the torque moment also realizes the measurement of the power transmission over the shaft investigated and the measurement of rotational speed, in accordance with the assumptions presented above under item 2.

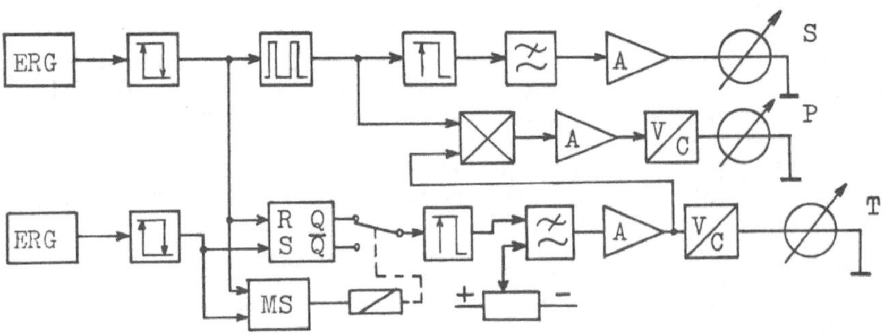

FIGURE 1. A block diagram of a system for measuring a torsion torque, the power transmitted by a shaft, and the rotational speed of a shaft, where ERG - electro-dynamic-reluctantive gauge, MS - the system which ensures the monomiality of a signal in a slotted line of a moment while operating reversibly (4); S, P, T - the instruments measuring: rotational speed, power transmitted by the shaft, and the torsion torque respectively.

The conception of solving the measurement of the torque moment as a prevailing value in the shaft investigated is based on the assumption of initial torsion being realized by proper arrangement of sensors ERG. Thus the measurement goes on by deviation method, and not by zero method. The measuring signal resulting from the initial torsion being assimilated by sensors, is corrected down to zero along the dotted line of measuring the moment after its coming out of the active filter.

2.2. The Accuracy of the Method

The exits of sensors ERG are connected with voltage discriminators with hysteresis. Taking into account the insensibility threshold of the voltage discriminators and the finite slope of curves E_1 and E_2 of the input signal (Fig 2.), ensues an error of time shift determined by the relationship:

$$\vartheta_{v\tau} = U_o \frac{(k_2 - k_1)}{k_1 \cdot k_2}$$

where: U_o - the insensilibity threshold of voltage discriminator in relation to mass,
k_1, k_2 - the slopes of input signals at the region of zero passage (taking into account the strengshening of input amplifiers).

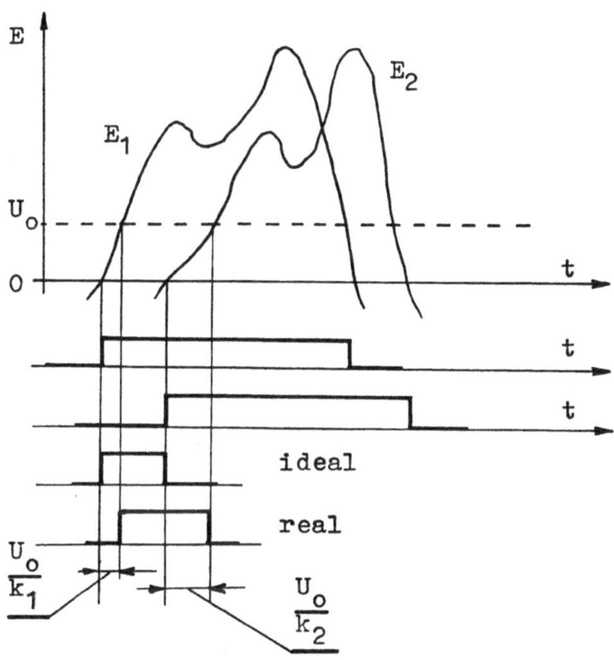

FIGURE 2. A diagram of error introduced by the formation system.

The maximum relative error, which can appear in the unit presented in fig.1, consists of all component errors. This error includes: the error introduced by sensors ERG being equal to 0.15%, the error resulting from the class of the instrument applied for the calibration of the new Unit being equal to 0.05% and the error ensuing from the graduation of slotted lines and meters S, P, T being equal to 0.4%. Finally, the maximum relative error of measurements is equal to 0.6%. Since the permissible value of the basic relative error is numerically equal to the grade of accuracy, the Unit can be regarded as meeting the requirements of class 1.

3. CONCLUSIONS

The suggested method of measuring big torque moments, according to the purpose it has been set, displays a lot of

advantages:
- it eliminates the contact transmission of the measuring signal,
- it eliminates the systems feeding the measuring sensors,
- insensitivity to humidity,
- low sensitivity to temperature changes (±0.2%/10K),
- multiple utilization of sensors while the converter being calibrated every time,
- the operation being independent of the position and direction of revolutions,
- relatively low cost of the construction of the converter,
- high sensitivity (the measurement error does not exceed 1.0%).

In comparison with alternative designs, used nowadays, of filling factor processing as being informative of the value measured, the present method eliminates the quantization error in the system and keeps full control over errors originating in the operation of the system that forms rectangular voltage signals. Besides, it eliminates the sources of errors resulting from the instability of standard frequency, which occur in methods applied nowadays.

LITERATURE

1. Kiełtyka L.: Doctor's thesis, 1984.
2. Kiełtyka L., Biernacki Z.: The measurement system of torque moment, power and rotational speed of the shaft. Patent application P-242515.
3. Kiełtyka L.: Modellierung des Fülfaktors als ein Messergebnis einer nichtelektrischen Grösse. 4 Wissenschaftliche Konferenz "Anlagenautomatisierung", Leipzig, 1983.
4. Kiełtyka L.: The processing system of the voltage sequence of alternative runs into a bivalent numerical signal. Patent PAT-24/4/7/82/86.

MOVABLE ELECTRONIC LOCOMOTIVE SCALE WITH MICROPROCESSOR

ZHU DING-MING, DUO JUN-QUAN, ZHAO HE-YUE, WU SHI-GUANG
NATIONAL INSTITUTE OF METROLOGY
BEIJING,CHINA

ABSTRACT

The Electronic Locomotive Scale of type ELS-12 has been developed in National Institute of Metrology,China. This type of scale can be used for determining the distribution of total weight of locomotives and vehicles on each wheel, and then determining the travelling safety by a vehicle on the railway.

This type of scale consists of shear cantilever beam load cells, self constrained-movable weighbridges and data acquisition and handling system with microprocessor.

It has high weighing accuracy: for total weight-0.02% F.S., and for each wheal weight-0.04% F.S. .

1. INTRODUCTION

For the travelling safety, many countries have made a rule of the weight distribution of the locomotive . In order to determine whether a locomotive of new product or after maintainance is accordance with the technical specification, we should use a weighing device to measure the wheel weights, axle weights and total weight of a locomotive.

The feature and performance of the ELS-12 scale are as follows:

a) It consists of 12 individual movable weighbridges suitable for weighing any locomotives and vehicles with 6 axles or less than 6 axles.

b) The application of the advanced shear cantilever beam load cell makes the construction of self constrained-movable weighbridge possible.

c) The use of the microprocessor has met the requirements of accuracy, function and weighing speed,etc. .

d) High weighing accuracy:for total weight-0.02% F.S., and for each wheel weight-0.04% F.S. .

2. GENERAL PLAN OF THE ELS-12 SCALE

2.1. Some significant problems considered in the general plan

a) In order to measure wheel whights, arrange 12 individual weighbridges for 6 axles measurement.

b) In order to make the scale movable and suitable to any different types of locomotives, the physical dimensions and self weight of the scale must be made as less as possible.

c) Because the scale is movable, the constraining of the platform of the scale can not use conventional foundation constraining method.

d) As 19 weighing data (12 wheel weights,6 axles weights and 1 total weight) must be provided simultaneously, and some additional requirements should be considered, the microprocessor is used here.

e) On the connection methods of the outputs of load cells, we do not use the series or parallel modes of the outputs in a conventional electronic scale. It is achived to replace analogue quantity addition by digital quantity addition from using the microprocessor. Therefore, it has not only improved the weighing accuracy but also eliminated the output standardziation compensation of all 48 load cells.

2.2 The operating system of the ELS-12 scale

The operating system of the ELS-12 scale is schematically shown in Figure 1.

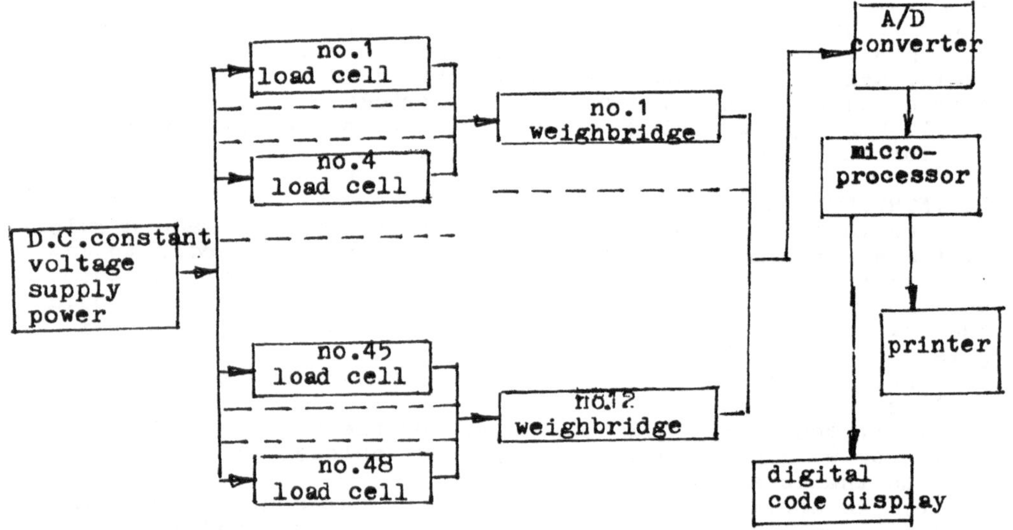

FIGURE 1. Type ELS-12 Electronic Locomotive Scale system scheme

3. TYPE JL-6T SHEAR CANTILEVER BEAM LOAD CELL

3.1. The feature of the load cell

An advanced type JL-6T shear cantilever beam load cell is developed for the ELS-12 scale. Figure 2 shows the schematic drawing of the JL-6T load cell.

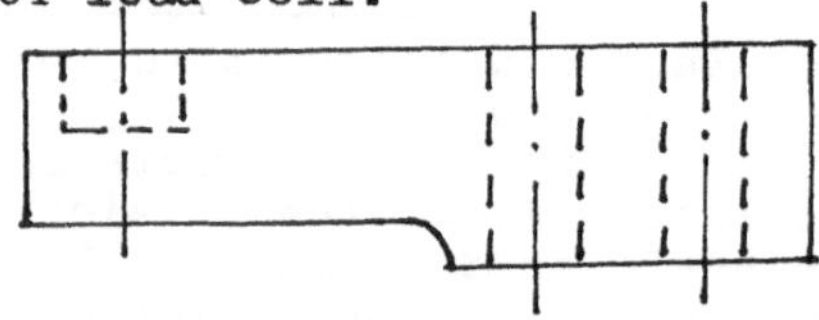

FIGURE 2. Type JL-6T shear cantilever beam load cell.

It features with high capacity of withstanding the side loads, good linearity, low profile, easy installation and simple construction.

3.2. Its major specifications are as follows:

Range: 6000Kg
Repeatability: 0.01% F.S.
Linearity: 0.01% F.S.
Hysteresis: 0.015% F.S.
Creep at 30 minutes: 0.02% F.S.

4. WEIGHBRIDGE CONSTRUCTION AND FOUNDATION

The foundation is a 30m long cast iron platform with two rows of T-shape grooves used for fixing weighbridges and rails.

The weighbridge construction, as shown in Figure 3, is composed of the following main parts, i.e., baseplate, load cells, scale platform, force delivery parts, transition rails, weighing rail and self constrained device.

The dimensions of the weighbridge are as follows:

Length and width of the baseplate: 602mmx260mm
Length and width of the scale platform: 358mmx540mm
Overall height of the weighbridge: 264mm

The self weight of the weighbridge is only about 200Kg.

This is a noval and unique weighbridge with self constrained-movable function. It gives up conventional constrained mode of weighbridge on the foundation and may be lifted by four hook screws on it to move to any position of the cast iron platform in order to meet the requirement for weighing various locomotives and vehicles with different axle-base. The dimensions and self weight of the weighbridge are much less than a conventional one.

5. SPECIFICATIONS OF THE ELS-12 SCALE:

Capacity: Individual weighbridge: 12000Kg
Overall: 144000Kg
Total number of weighbridge: 12
Total number of load cells: 48
weighing accuracy:
Wheel weight: 0.04% F.S.
Meets OIML to 2400 divisions
Total weight: 0.02% F.S.
Meets OIML to 7200 divisions
Resolution: 1Kg
Data record: Paper band

Locomotive weighing example: The ELS-12 scale has been installed in China and has been operating very well. For example, for weighing of a type BB-7 6-axles locomotive with the normal weight of 144000Kg, the repeatability is less than 0.02% F.S..

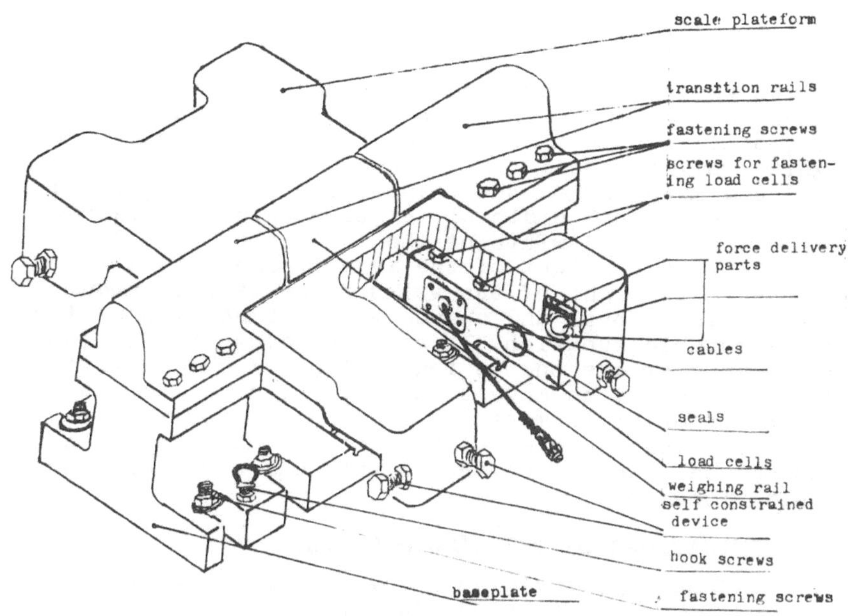

FIGURE 3: Construction of the individual weighbridge.

6. SUPPLEMENTAL FIGURES

Supplemental Figure 1. Type ELS-12 Electronic Locomotive Scale on the foundation (cast iron platform)

Supplemental Figure 2. The individual weighbridge.

Supplemental Figure 3. The data acquisition and management system installed in the instrument-control room.

IMPROVEMENT OF THE WEIGHING ACCURACY OF A 50 KG BEAM BALANCE

E. DEBLER, K. WINTER

Physikalisch-Technische Bundesanstalt, Braunschweig, Federal Republic of Germany

1. INTRODUCTION

Even today, beam balances are best suited for mass comparisons of the highest accuracy. Prototype balances used for 1 kg weights achieve relative standard deviations of $3 \cdot 10^{-10}$ (1).

Weights of more than 1 kg, however, must be determined by simpler beam balances (see Fig. 1) which mainly consist of the beam, the suspensions to carry the weights, a pointer and a line scale. These balances lack many characteristics which prototype balances possess to obtain high accuracy, such as constant load of knife-edges and planes, stable temperature conditions inside the case and sensitive readout assembly. Instead the simpler balances execute the change-over of the test specimen TS and the standard S (substitution method) by unloading the knife-edges and planes. Furthermore, the case must be opened several times while the masses are being compared.

When embodying a mass scale the lower accuracy of the high capacity balances must be taken into account. The more steps the weighing scheme includes to attain multiples of the mass unit, the more the uncertainties increase. To obtain as small a number of steps of different weights as possible when embodying the mass scale up to 5 t, the Physikalisch-Technische Bundesanstalt has manufactured one hundred 50 kg weights (2).

The following paper describes suitable changes made on a 50 kg beam balance; the results obtained and the effects on the mass scale up to 5 t are reported.

2. MODIFICATIONS TO IMPROVE THE WEIGHING ACCURACY

The modifications of a simple 50 kg beam balance are characterized by an automation of the weighing procedure.

2.1 Change-over from test specimen to standard

2.1.1. Bi-level arrangement. The most conspicuous modification of the balance which can be seen in the photograph (see Fig. 2) is the arrangement of test specimen TS and standard S one above the other. Another striking feature is the driving mechanism at the bottom side of the base (see Fig.3). This mechanism lifts and lowers bolts B_1 and B_2 which pass through bores in load receivers L_1 and L_2 of the suspension. As the motion of bolts B_1 and B_2 is oppositely directed, they either place test specimen TS on the upper load receiver L_2 of the suspension or the standard S on the lower load receiver L_1.

As a result of the bi-level arrangement the test specimen and the standard are situated at different heights at a distance of 0,4 m and that is why air buoyancy and gravity must be taken into account at two places. Both influences can be determined with sufficient accuracy. The bi-level

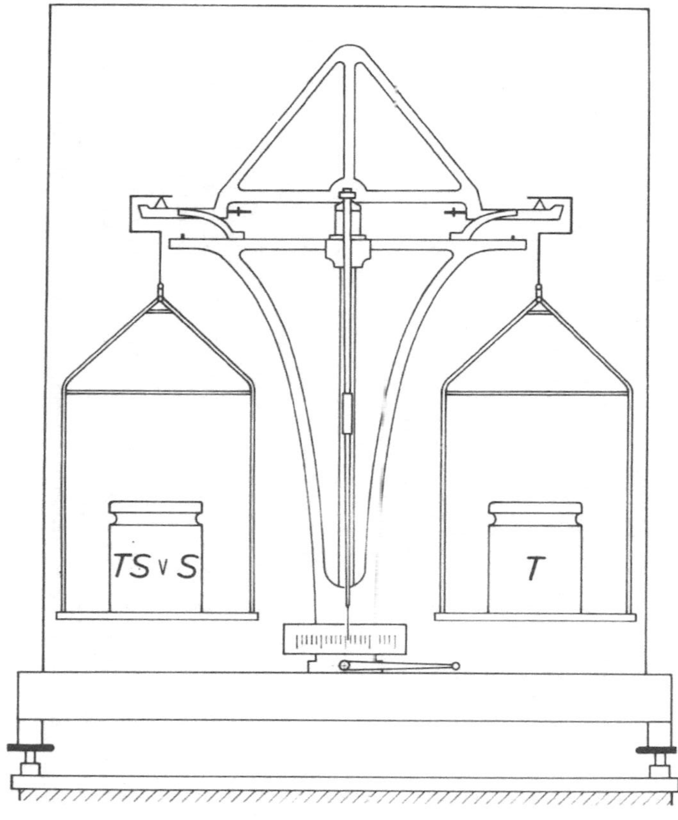

FIGURE 1. Simplified representation of a manually operated 50 kg beam scale. TS test specimen, S standard, T tare.

FIGURE 2. Modified balance, housing removed.

arrangement requires the least space. Another advantage is that the beam need not be arrested and that by careful operation of the change-over device, a smooth change-over can be achieved (3).

2.1.2. Driving mechanism. The manual change-over from test specimen to standard has been replaced by the computer-controlled positioning carried out by a driving mechanism. It consists of shaft SH rotated by electromotor EM via step-up gear SG (see Fig. 3). By means of steel strips wound around the shaft (not shown in Fig. 3), the shaft rotations are transmitted to bolts B_1 and B_2 so that these move in opposite direction. The tachometer generator T and the electronic control device CD allow the electromotor EM to be operated in a large speed range (1:3000). The desired speed is proportional to a voltage which the computer PC supplies via a digital-to-analog converter D/A. It is thus possible by a computer-controlled procedure to position the weights on the suspension with the lowest speed.

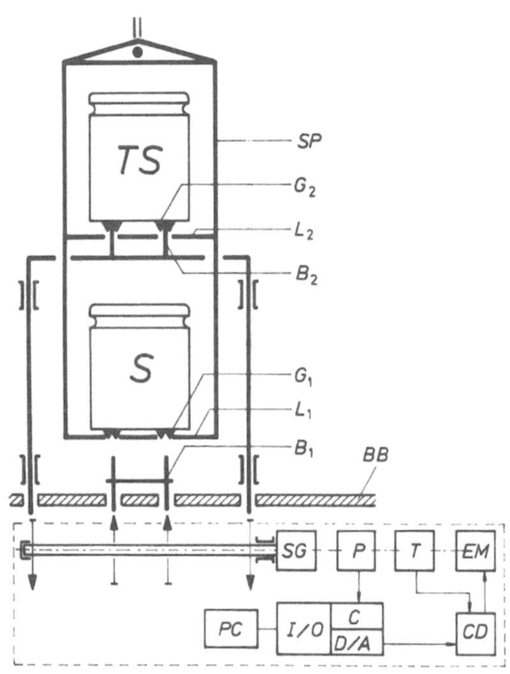

FIGURE 3. Device for the alternate positioning of test specimen TS and standard S on the load side. SP bi-level suspension, L_1 and L_2 load receivers, G_1 and G_2 guide pieces, B_1 and B_2 bolts, BB base of balance, SH shaft, SG step-up gear, EM electromotor, T tachometer generator, CD electronic control device, D/A digital-to-analog converter, PC personal computer, P pulse generator, B bi-directional counter. I/O input/output control.

Oscillations of the beam must be prevented, otherwise constant load of knives and planes cannot be attained.

2.1.3. Arrestment device. This device is meant to fix the beam in a specified position (see Fig. 4). The movable stop S arrests the oscillating beam. In order to avoid shocks it has been coated with flexible synthetic material SM. To ensure the safe resting position of the beam, a weight in the form of a wire D (2 g) is applied with the aid of an electromotor.

Even after wire D has been removed and stop S withdrawn, the beam remains in the adjusted position for a sufficiently long time and the unforced change-over from test specimen to standard is possible. Due to the low speed of bolts B_1 and B_2 the change-over can be carried out without shock.

2.1.4. Guide pieces. The guide pieces G_1 and G_2 are important constructional features. They carry the standard and the test specimen and are placed together with these on to the upper or lower load receiver. The conical shape of the guide pieces guarantees an unchanged position between the weights and the load receivers. It is not possible to dispense with the guide pieces, as without them the bores of the load receivers slide along the bolts hampering the free oscillation of the beam. The guide pieces must be taken into account in the evaluation on the basis of a preliminary determination of their mass.

2.2. Determination of the beam's inclination

In order to determine the inclination of the oscillating beam the balance is equipped with inductive transducers (see Fig. 5). Two ferrite cores are mounted on the beam whereas the respective coils are fixed to the frame. Sufficient play between core and coil has been provided so that the ferrite cores, together with the oscillating beam, can move on a circular arc about the line of contact between main knife-edge and main plane. The symmetric arrangement of the inter-connected coils ensures that the measured value is affected neither by horizontal nor vertical dislocations acting in the same direction. The signal is only modified by the oppositely oriented vertical displacement of the ferrite cores due to the inclination of the beam.

The inductive transducers, together with a carrier frequency measuring amplifier CFA, supply an analog voltage which corresponds to the inclination of the beam. By means of plotter P this voltage can be directly plot-

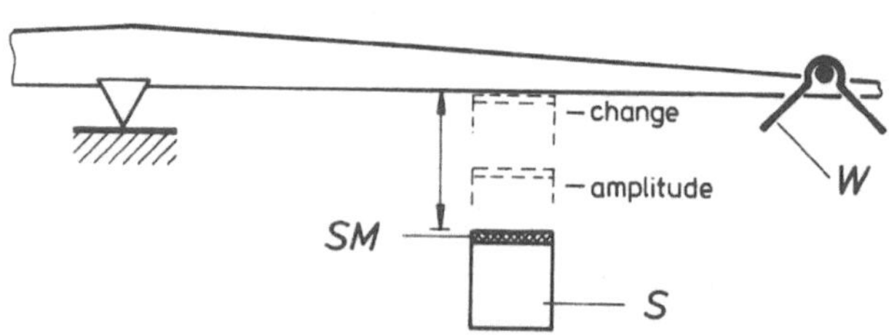

FIGURE 4. Arrestment device. S movable stop, SM flexible synthetic material, W wire.

ted as a function of time. A damped oscillation curve results as it is shown in Fig. 6 for a plotting period of 6 hours.

A digital voltmeter DVM subdivides the output voltage belonging to the range of inclination into 38000 steps (see Fig. 5). Compared with the visual reading of the pointer indication from the line scale, this means an almost 200-fold improvement of the sensitivity of the indication of the beam's inclination.

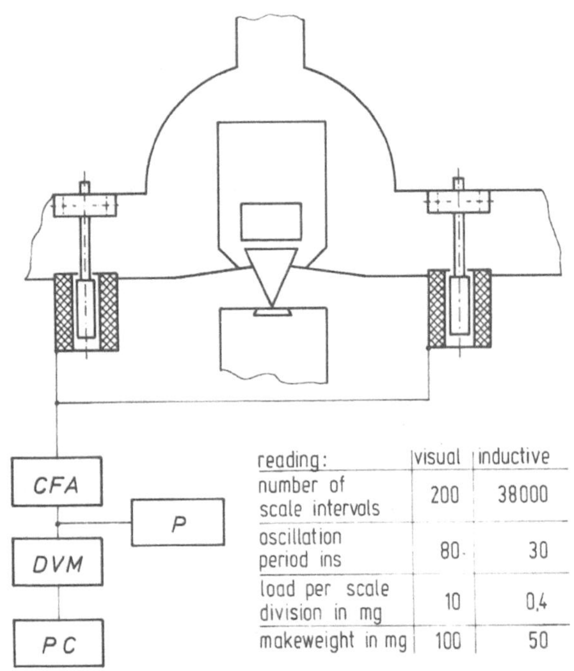

reading:	visual	inductive
number of scale intervals	200	38000
oscillation period ins	80	30
load per scale division in mg	10	0,4
makeweight in mg	100	50

FIGURE 5. Determination of beam inclination by means of inductive transducers. CFA carrier frequency measuring amplifier, P plotter for plotting the analog voltage, DVM digital voltmeter, PC personal computer.

Before the balance was modified the centre of gravity of the beam had to be placed in a high position to record the slightest mass differences. An oscillation period of 80 s had resulted. Because of the highly sensitive inductive transducers the centre of gravity can be lowered. The oscillation period is reduced to 30 s which makes shorter measuring times possible. The original scale interval was about 10 mg, whereas the more sensitive indication achieves 0,4 mg, which is a 25-fold improvement.

The personal computer PC takes over the measured values of the digital voltmeter DVM. Its calculating speed suffices to record and evaluate 10 measured values per second. The maximum and minimum values of the beam oscillation are determined, of which five each are used for calculating the centre of oscillation.

A linearization of the measured values is not necessary although inductive transducers do not work linearly over long displacement travels. There is also no linear relation as far as the arrangement used for transforming the circular movement into a voltage is concerned. A test carried out by means of an angle measuring device prior to installing the transducer in the balance showed that the coils and cores can be so adjusted that the deviations from linearity - converted into the unit of mass - amount to less than 1 mg over the total range of inclination. As in mass comparisons the centres of oscillation are apart by only fractional parts of the range of inclination and the beam oscillations are always adjusted to the same amplitudes by means of the arrestment devices, linearity errors are of no significance.

2.3. Makeweight

The makeweight for the determination of the sensitivity is applied in a computer-controlled procedure by means of an electromotor with gear. The makeweight is a wire 50 mg in mass, which a lever places on the sus-suspension.

2.4. Temperature

The change-over from test specimen to standard takes place without the balance housing being opened. There is no mixing of the air inside the housing with the ambient air. Instead, layers of air may develop. According

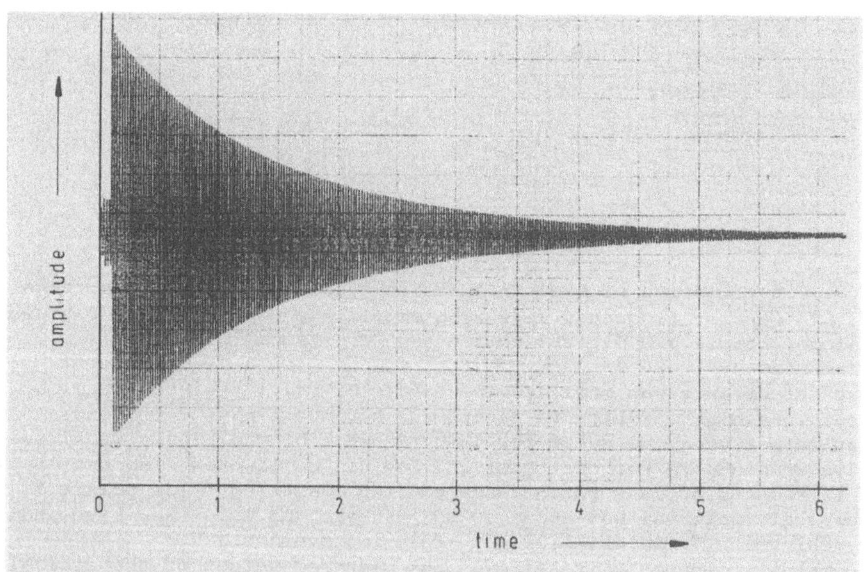

FIGURE 6. Damped beam oscillation.

to MACURDY (4) a horizontal temperature gradient inside the housing is detrimental, as it causes convection of the close air. As a result of convection, uncontrollable forces act on the load and tare sides and the dispersion of the weighings is increased. Vertical temperature gradients, on the other hand, have a stabilizing effect when the air temperature rises with the level. Different air densities in the vicinity of the two weights result. To these correspond the effects on the air buoyancy. Differences of the air temperature of up to 0,4 K were measured at test specimen and standard. In the case of 50 kg steel weights they result in differences of the buoyancy of 10 mg.

Measurements of the different air temperature at test specimen and standard are necessary. Two platinum resistance thermometers are used for this purpose which indicate the air temperatures in scale intervals of 0,01 $^{\circ}$C.

2.5. Pressure and humidity

The air pressure is measured at test specimen level; the air pressure at the test specimen is calculated from the difference in level of test specimen and standard. The air humidity is measured inside the balance housing. The instruments used for measuring temperature, pressure and humidity are connected with the computer.

3. RESULTS AND CONSEQUENCES

3.1. Automatic operation

All the weighings and their evaluation are made automatically. A weighing cycle for the determination of the mass of a test specimen comprises various determinations of the centres of oscillation. Standard S and test specimen TS are alternately positioned in the order S TS TS S. As a result those long-term changes of the balance which lead to a steady drift of the centres of oscillation are eliminated. ALMER (5) calls this type of a comparison by weighings the double substitution method. As each time a makeweight for the determination of the sensitivity is added to S or TS, a weighing cycle covers eight determinations of the centres of oscillation. It takes 20 minutes.

3.2. Standard deviation

The standard deviation as the characteristic number of the balance's quality can be derived on the basis of several weighing cycles. Fig. 7 shows standard deviations for 100 different 50 kg weights calculated from 10 weighing cycles. The deviations range from 0,15 mg to 0,8 mg; their average amounts to 0,32 mg.

The smallness of these values becomes evident in a comparison with the correction values which must be added due to the difference in level between test specimen and standard. The mass of the additional column of air causes differences in the air buoyancy which must be allowed for by 0,4 mg (in the case of 50 kg steel weights); with a difference level of 0,4 m, the effect of the different force of gravity (6) corresponds to 4 mg and is by a factor 10 higher than the mean standard deviation.

On an average, the relative standard deviation amounts to $6\cdot10^{-9}$. As can be gathered from Fig. 7, it rarely exceeds the value $1\cdot10^{-8}$ (s = 0,5 mg). Due to the measures taken, the repeatability at the weighings has become preciser by a factor 30 and thus reach the values until quite recently valid for prototype balances. A relative standard deviation of $1\cdot10^{-8}$ is also reported by UCHIKAWA et al. (7) for a 30 kg balance. They obtained this value in mass comparisons of 10 kg weights. Only for balances with maximum capacities of 1 kg have relative standard deviations of $1\cdot10^{-9}$ and less been published (1, 8 to 11).

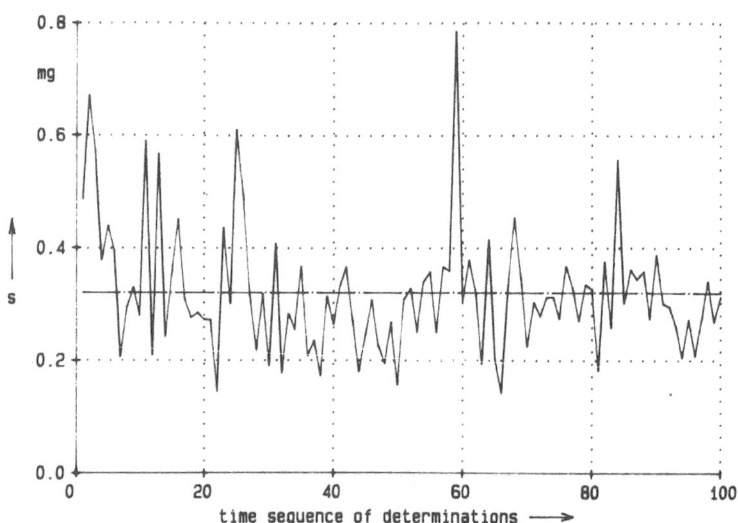

FIGURE 7. Standard deviation s of 100 weights à 50 kg determined from 10 weighing cycles each.

3.3. Mass scale up to 5 t

At the Physikalisch-Technische Bundesanstalt, the modified 50 kg beam balance serves to realize the mass scale up to 5 t with an accuracy higher than has hitherto been possible. The mass scale is embodied by one hundred 50 kg weights already mentioned. Each 50 kg weight is separately compared with a 50 kg reference standard which in turn has been checked against five 10 kg weights.

The two curves at the top of Fig. 8 show the influence of this improvement of the weighing accuracy upon the mass scale. By the comparison with the five 10 kg weights, the uncertainty of measurement of the reference standard is reduced by means of the improved balance from 15 mg to 13 mg; the relative uncertainty of measurement of the standard mass up to 5 t decreases from $3,1 \cdot 10^{-7}$ to $2,7 \cdot 10^{-7}$ (confidence level $1 - \alpha = 95$ %).

This improvement is a slight one. It is due to the mutiplication of the unit of mass from 1 kg to 10 kg, for which purpose balances with uncertainties of measurement of about $2,5 \cdot 10^{-7}$ are still used at present. The advantages of the 50 kg balance will only become effective when - as it has been planned - in every load step, balances with relative uncertainties of measurement of $2 \cdot 10^{-8}$ will be employed for multiplication. By this measure alone the relative uncertainty of the standard masses up to 5 t is reduced to $1,7 \cdot 10^{-7}$. With the improved 50 kg balance, an additional reduction of the relative uncertainty by a factor 2 to $8 \cdot 10^{-8}$ can be achieved (see Fig. 8, broken line).

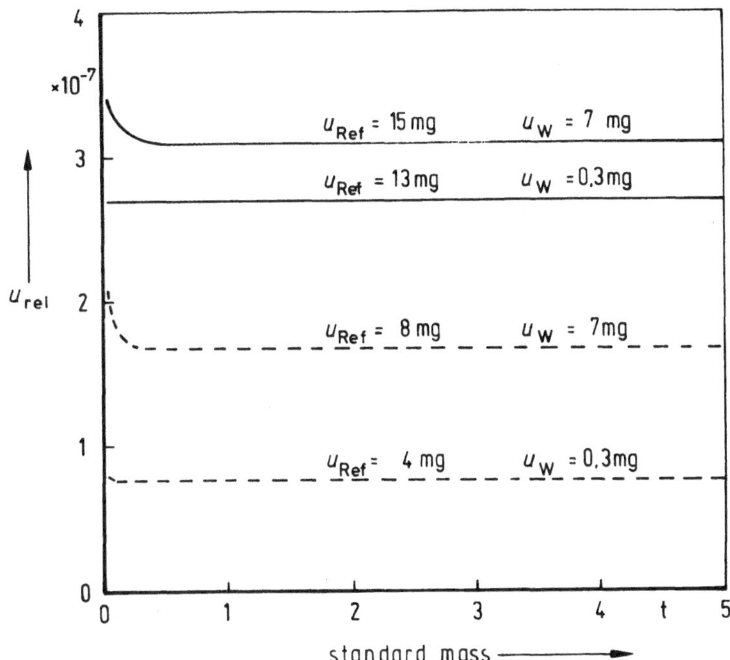

FIGURE 8. Relative uncertainties of measurement u_{rel} of the mass scale up to 5 t under different conditions for a confidence level $1 - \alpha = 95$ %.
u_{ref} uncertainty of measurement of the 50 kg reference standard,
u_W uncertainty of measurement of a weighing determined from s = 10 mg and 0,5 mg and 10 weighings.

The error estimates are based on a relative uncertainty of measurement of the national Pt-Ir kilogramme prototype of $5 \cdot 10^{-8}$. Long-term variations of the 1 kg prototype do not allow a lower value to be applied (12). Consequently, the uncertainty of measurement of $8 \cdot 10^{-8}$ of the standard masses up to 5 t which is only slightly higher, will probably have to be considered as the lowest value achievable.

4. SUMMARY

When it is possible - as has been described here - that balances with maximum capacities of 50 kg can reach uncertainties of $1 \cdot 10^{-8}$ ($1-\sigma$ limit), it appears possible that masses up to 5 t will be realizable with relative uncertainties of measurement of $8 \cdot 10^{-8}$ ($1 - \alpha = 95$ %).

REFERENCES

1. Kobayashi, Y. et al.: One Kilogram Balance (NRLM-2). Bull. NRLM, 33-2,7/18, 1984.
2. Winter, K.: Elektrochemisches Polieren und Abgleichen von 50-kg-Massenormalen. wägen + dosieren, November 82, p. 200-202.
3. Felgenträger, W.: Feine Waagen, Wägungen und Gewichte, Berlin 1932.
4. Macurdy, L.B.: Response of Highly Precise Balances to Thermal Gradients. J. Res. NBS 68C (Eng. and Instr.), July-September 1964.
5. Almer, H. E.: National Bureau of Standards One Kilogram Balance NBS No. 2. J. Res. Nat. Bur. Stand (U.S.) 76 C (Eng. and Instr.), Nos. 1 and 2, 1972.
6. Thulin, S.A.: The Local Value of g. Bulletin OIML No 94 - Mars 1984.
7. Uchikawa. et al.: Production techniques for standard gases by the mass-measuring method. Measurement Vol 1 No 2, April-June 1983.
8. Kibble, B. P.: A Modification to a Precision Balance Incorporating Substitution Weighing and Feedback Control which gives Improved Accuracy and Convenience. Metrologia 11, 1975.
9. Kobayashi, Y.: A Study on the Preciser Establishment of Mass Standard. Bulletin of NRLM, July 1981.
10. Spurný, R.: Metrological Balance for Comparison of Primary Standards with National Standard of Mass in CSMU. Report to CCM/81-22, BIPM 1981.
11. -: Rapport du directeur sur l'activité et la gestion du BIPM. Octobre 1982 - Septembre 1983. CIPM/83-3.
12. -: Rapport du directeur sur l'activité et la gestion du BIPM. Octobre 84 - Septembre 85. CIPM/85-1.

DECODING OF THE STATE OF SCALES EQUILIBRIUM BY MICRO-PROCESSOR SCALES TERMINAL

K.Grzywa, T.Olma, A.Zboinski

In basic version, designed and built microprocessor scales terminal consists of /Fig.1/: keybord, printer, display and single board microprocessor controller. The printer makes it possible to print 10 digits in line and some symbols or letters on the least significant position. The display includes 10 signs the height of which is 10 mm, the keybord consists of 12 numerical keys /10 digits, minus sign and space/ and 16 command buttons. In expantion vertion the device includes data transmitter and receiver which enable serial transmittion in standard V-24 /RS-232-C/ to/from superior computer. Superior computer collects data from many scales terminals

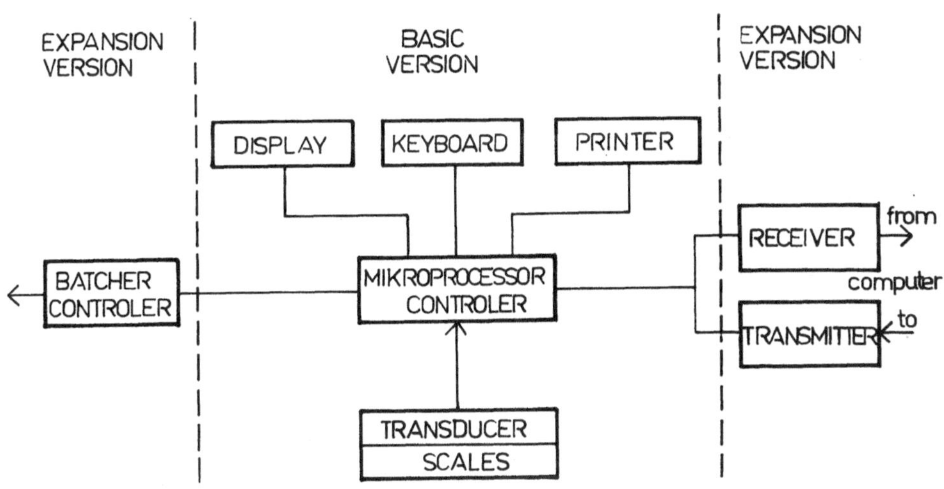

Fig.1. Microprocessor scales terminal-functional block diagram.

and may control scales terminals. The expantion version of microprocessor scales terminal for controlling the batchering of materials is elaborated.

The microprocessor scales terminal was designed with the following assumptions:

- low cost of production,
- small size /size of table calculator/,
- low energy consumption,
- easy both service and user's mannual,
- possibility of operation in combination with different types of scales transducers.

In hardware-software trade off the hardware was simplified as much as it was possible so as to meet the above assumptions.

One of the important functions realized by scales terminal is decoding of the state of scales equilibirum. This operation is sepcially important because it affects directly the accuracy of measurements and the time of settling the results of weighing. The necessary time for scales /special pendulum-cam scales/ to reach the final value is different and depends on many factors: individual parameters of scales, the value of mass being weighed, kind of mass being weighed /solid, liquid or loose material/, way of fastening on scales/on platform, hanging up hook etc/ and surrounding conditions such as shocks, vibrations of base /this happens in industry applications/. The simple method of time delay before reading the state of scales might be the reason of big errors.

A state in which two successive readings of scales state in an assumed time interval are the same might not lead to finding a value which is searched for /in the case of considerable vibrations of the base/.

In below considerations scales is treated as an second order oscillatory measurement transducer with concentrated elements characterized by:

- mass inertia or moment of inertia,
- viscous friction appearing independent from dry friction, gives force or proportional moment to velocity,
- returning moment or returning force proportional to angle or lineal displacement.

No matter if the exciting force being scales load is treated as:

- step function,
- product of velocity and step functions,
- forced pulse

the oscillatory scales system response in time domaine may be modeled by the function for every of these cases as follows:

$$y = A + Be^{-Ct} \sin (Dt + E) \qquad /1/$$

where: t - vary of time,

A,B,C,D,E - const coeficients.

From formal reasons different forced functions were assumed as the best model depends both on the type of scales and the manner of loading it.

There is usually a tendency that both angular freqency and damp vibrations of scales, in the range of oscillatory vibrations, would be the gretest in order to achieve the final value as quick as possible. Damp value in pendulum-cam scales can be easily fitted through the regulation of vibration damper. It is difficult to regulate angular frequency and besides it depends unproportionally on weighing mass. The consequence of increasing the dampvalue is decreasing of system sensitivity. It should be noted that measuring in equilibrium condition is accompanied by static friction resistance. The above assumptions lead to defining the final value of equilibrium state on the basis of dynamic signal measurements. Making use of lineal regression, considering its statistical features, seemd to be the best solution but the procedure of counting the final value for the accepted model is more complicated because of the shape of function it requires the application of new arguments of function.

It can not be performed in real time by simple microprocessor device which assumption have been already formulated. From this reason the counting method have to be simplified. If the analysis of the first expression is limited to its extremum the Eq.1. may be written as follows:

- for maximum

$$y = A + B e^{-k C} \quad , k = 1,3,5 \ldots p \qquad /2/$$

- for minimum

$$y = A - B e^{-n C} \quad , n = 2,4,6 \ldots w \qquad /3/$$

what in general can be written in the following way:

$$y = A - (-1)^{n} B e^{-nC} \quad , n = 1,2,3 \ldots r \qquad /4/$$

Sampling the output signal of scales /it is the electrical digital output signal from transducer which collaborates with scales and transforms lineal or angular displacements/ microprocessor controller seeks the extremum values: A_n, A_{n+1}, A_{n+2}, A_{n+3}... /Fig.2./.

The gathered datum A_i, for each natural i, supported by Eq.4. lead to a set of equations:

$$A_n = A - (-1)^{n} B e^{-nC} \qquad /5/$$

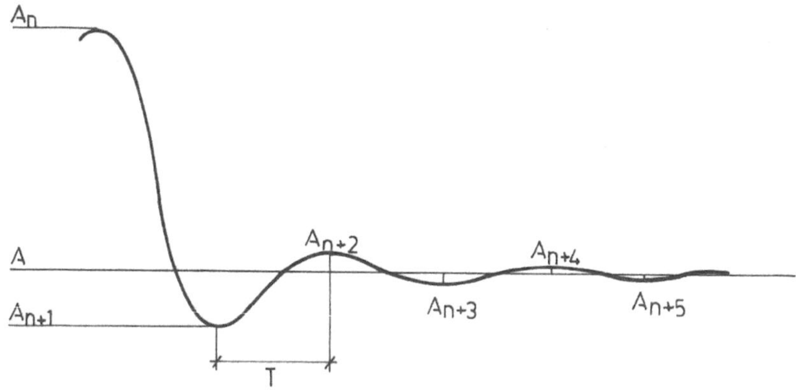

Fig.2. The chart of damp vibration of scales

$$A_{n+1} = A - (-1)^{n+1} B e^{-(n+1)C}$$
$$A_{n+2} = A - (-1)^{n+2} B e^{-(n+2)C} \qquad /5/$$

Solving of Eq. set /5/ determines the expression of the final value A

$$A = \frac{A^2_{n+1} - A_n A_{n+2}}{2 A_{n+1} - A_{n+2} - A_n} \qquad /6/$$

The set of equations for maximum is following:

$$A_n = A + B e^{-nC}$$
$$A_{n+2} = A + B e^{-(n+2)C} \qquad /7/$$
$$A_{n+4} = A + B e^{-(n+4)C}$$

its solution may be written this way:

$$A = \frac{A^2_{n+2} - A_n A_{n+4}}{2 A_{n+2} - A_{n+4} - A_n} \qquad /8/$$

The equation set for minimum may be set up in a similar form as the Eq. set /5/, it leads to a solution:

$$A = \frac{A^2_{n+3} - A_{n+1} A_{n+5}}{2 A_{n+3} - A_{n+5} - A_{n+1}} \qquad /9/$$

The equation set for two sequence maximum and minimum is following

$$A_n = A + B\,e^{-nC}$$
$$A_{n+2} = A + Be^{-(n+2)\,C} \qquad /10/$$
$$A_{n+3} = A - B\,e^{-(n+3)\,C}$$

and similar form for two sequence minimum and maximum:

$$A_{n+1} = A - B\,e^{-(n+1)\,C}$$
$$A_{n+3} = A - B\,e^{-(n+3)\,C} \qquad /11/$$
$$A_{n+4} = A + B\,e^{-(n+4)\,C}$$

The sets of equations /10/, /11/ may be solved by various methods /e.g. Cardan's method, the secant method, bisection method/ solution in this case has no as simple form as for Eqs /6/, /8/, /9/. 4/k-3/ values A /the final value/ may be derived from Eqs /6/, /8/,/9/, /10/ and for k /$k > 3$/ measuring extremum:

$$/A_1,A_2,A_3/,\ /A_2,A_3,A_4/\ \ldots\ /A_{k-2},A_{k-1},A_k/ \qquad /12/$$

$/A_1,A_3,A_5/,\ /A_3,A_5,A_7/\ \ldots\ /A_{s-4},A_{s-2},A_s/,$ s=k if k is odd number, s=k-1 if k is even number, /13/

$/A_2,A_4,A_6/,\ /A_4,A_6,A_8/\ \ldots\ /A_{s-4},A_{s-2},A_s/,$ s=k if k is even number, s=k-1 if k is odd number, /14/

$$/A_1,A_3,A_4/.\ /A_1,A_2,A_4/\ \ldots\ /A_{k-3},A_{k-1},A_k/,\ /A_{k-3},A_{k-2},A_k/. \qquad /15/$$

Statistic data processing is applied for values A searched in above way. The elimination of measurement reasults charged by excessive error could be performed e.g. by Chauvent's criterion. Verification whether A_i , i=1, 2...k is not charged by exessive error for k known extremum should by the computetion value A for the following combinations:

$/\underline{A_1}, A_2,A_3/$ – $/A_2,A_3,A_4/\ldots/A_{k-2},A_{k-1},A_k/,$

$/A_3,A_5,A_7/\ldots/A_{s-4},A_{s-2},A_s/,$ the restriction for s like in /13/,

$/A_2,A_4,A_6/\dots/A_{s-4},A_{s-2},A_s/$, the restriction for s like in /14/,

$\dots/A_{i-1},\underline{A_i},A_{i+1}/$ – $\dots/A_1,A_2,A_3/\dots/A_{i-3},A_{i-2},A_{i-1}/$

$/A_{i+1},A_{i+2},A_{i+3}/\dots/A_{k-2},A_{k-1},A_{k-1}/$

$/A_1,A_3,A_5/\dots/A_{p-4},A_{p-2},A_p/$

$/A_w,A_{w+2},A_{w+4}/\dots/A_{s-4},A_{s-2},A_s/$,

p,w are odd numbers, $p < i$, $w = p+2$ if i is even number, $w = p+4$ if i is odd number, the restriction for s like in /13/,

$/A_2,A_4,A_6/\dots/A_{p-4},A_{p-2},A_p/$

$/A_w,A_{w+2},A_{w+4}/\dots/A_{s-4},A_{s-2},A_s/$

p, w are even numbers, $p < i$, $w=p+2$ ifi is odd number, $w=p+4$ if i is even number, the restriction for s like in /14/,

$/A_{i-2},A_{i-1},A_{i-2}/$, $/A_{i-1},A_{i+1},A_{i+2}/$,

$\dots/A_{k-2},A_{k-1},\underline{A_k}/$ – $\dots/A_1,A_2,A_3/\dots/A_{k-5},A_{k-4},A_{k-3}/$

$/A_1,A_3,A_5/\dots/A_{s-4},A_{s-2},A_s/$, s is odd number and $s < k$,

$/A_2,A_4,A_6/\dots/A_{s-4},A_{s-2},A_s/$, s is even number and $s < k$,

Chauvent's criterion should be applied to every combination. The substantial feature of the method of searching extremum of damp oscillation is ability to recognition of disturbance. The case in which the frequency of disturbance is higher than the frequwncy of damp oscillations of scales is recognized when the local extremums exist. Then the counting of the final value is based on samplings of state without oscillation /after delay/ with disturbance. The disturbance where frequency is higher than frequency of damp scales vibration can be recognized during computation of value A for different combinations of samples /12/, /13/, 14/, 15/.

There is not enough room in this limited paper to discuss the full analysis of measurement accuracy and its dependence on disturbance.

The above consideration after a slight modification are substantial for the measurement of mass in batching, when static equilibrium state is difficult or impossible to achieve. In this case is necessary to approximate the current value from transient state. The dynamic error of measurement and the dealy being caused by the batcher ought to be taken into account.

APPLICATION OF A NEW BEAM TYPE LOADCELL TO MECHANICAL HAND

Yoshihiro Ochiai*, Koichi Kameoka**
Takashi Sugisako* and Toshiro Ono***

* Osaka Prefectural Industrial Research Institute, 2-1-53 Enokojima, Nishi-ku, Osaka 550, Japan
** Himeji Institute of Technology, Himeji, Hyogo 671-22
*** University of Osaka Prefecture, 804 Mozu-Umemachi 4-cho, Sakai-shi, Osaka 591

1.INTRODUCTION

Industrial robots have been widely used in various kinds of works such as welding works, painting works, and so on. Because of the features of the works, most of them are position-controlled type robots. Complicated works, e.g., assembly works or handling works of a brittle object, can be accomplished only by force-controlled type robots. In force-controlled types, a force sensor is one of the most important factors to determine the performance of the robot. Many force sensors have been developed and some of them are available now, which are mainly embedded in the wrist of a mechanical hand [1,2,3]. To make a robot do more complicated and delicate works, force sensors have to be embedded in the fingers of the mechanical hand. Such types of force sensors are eagerly expected to be developed [4,5].

The authors have made a new beam type loadcell, a force sensor, as a trial which can be applied to the gripper of a mechanical hand. The new feature of loadcell is to measure the horizontal and vertical forces and their acting points. An industrial robot equipped with these loadcells can grip a brittle object, check up the weight, and can detect the contact force and position of the gripper and the object. Thus the loadcell functions not only as a force sensor but also as a tactile sensor. The loadcell is small and slender in size, light in weight, to be embedded in the fingers of a mechanical hand.

In this paper basic properties in statics and dynamics of the loadcell are examined through several experiments. First, after the description of the measuring principle of a 2-component loadcell, a 4-component loadcell and a 5-component loadcell are proposed. Secondly, the experimental result is shown of the measurement in weight, gripping force and contact position, using a mechanical hand having two fingers in which 4-component loadcells are embedded. Finally the effect of noises on the measurements is discussed.

2.LOADCELL

2.1 Measuring principle

Fig.1 is the schematic drawing of the loadcell, which has an additional hole in the middel of a conventional beam type loadcell. To make the explanation of its measuring principle easily, let us consider the parallel beams as shown in Fig.2. The moment M at 1,2,3,4, and the force f at 5,6 can be written as follows;

$$M = \frac{LF}{2} + \frac{2IXF}{Aa^2+4I} \fallingdotseq \frac{LF}{2}, \qquad (1)$$

Wieringa, H. (ed), Mechanical Problems in Measuring Force and Mass.

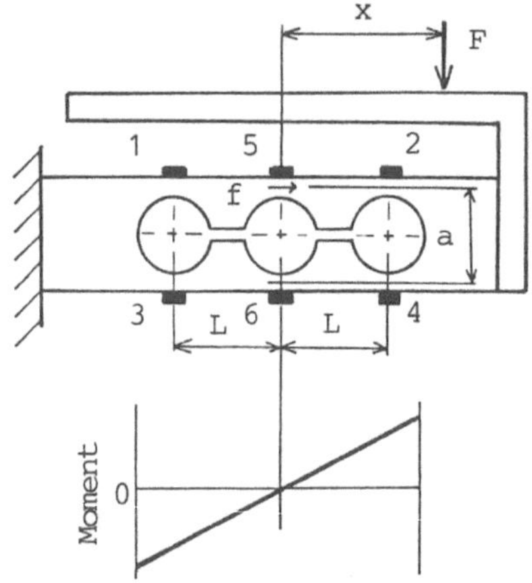

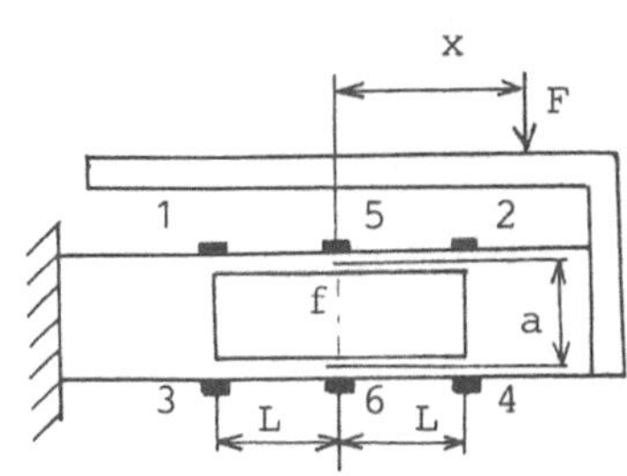

Fig.2 Parallel beams to explain the measuring principle

Fig.1 Schematic drawing of the loadcell

$$f = \frac{xF}{a(1+4I/Aa^2)} \doteqdot \frac{xF}{a}, \qquad (2)$$

where A is the area of cross section and I the moment of inertia. These equations imply that the force F can be measured by the strain gages 1 through 4 and the product Fx by the strain gages 5 and 6. The strain gage arrengements in a Wheatston bridge for measuring F and Fx are shown in Figs.2(a) and 2(b), respectively. The distance x can be obtained through the division, i.e., the output in Fig.3(b) divided by the output of in Fig.3(a). As for the loadcell shown in Fig.1, the above relationships hold in the same manner.

Fig.4 shows the 4-component loadcell designed by coupling the loadcells mentioned above. The weight of this loadcell including the contacting element is 130g and that of the loadcell itself 95g. The length, width and height of the loadcell are 110, 30 and 20 mm, respectively.

Fig.5 shows a 5-component loadcell which can detect an axi-directional force.

2-2 Output of the 4-component loadcell

Table 1 shows the values in sensitivity of the loadcells used for the fingers of a mechanical hand. The value of the output in vertical force

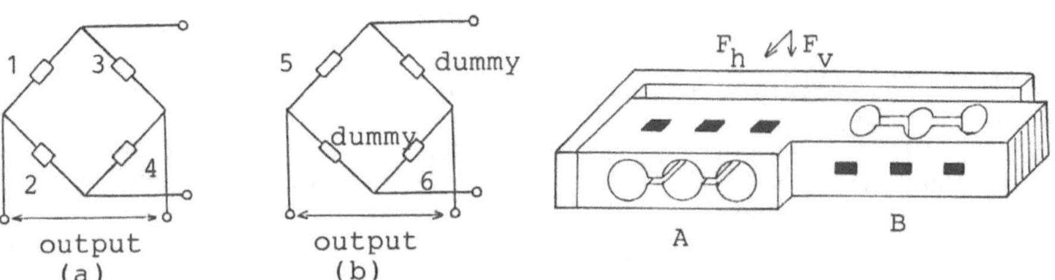

Fig.3 Bridge arrengements for measuring F and Fx

Fig.4 Illustration of the 4-component loadcell

Fig.5 Illustration of a 5-component loadcell

Table.1 Sensitivity of the loadcells

Loadcell (Right)	ε/F_v	0.973μstrain/gf
	ε/F_h	0.594μstrain/gf
	$\varepsilon/F_v x$	0.015μstrain/gfcm
	$\varepsilon/F_h x$	0.031μstrain/gfcm
Loadcell (left)	ε/F_v	0.987μstrain/gf
	ε/F_h	0.575μstrain/gf
	$\varepsilon/F_v x$	0.013μstrain/gfcm
	$\varepsilon/F_h x$	0.031μstrain/gfcm

F_v is about twice the value of the output in horizontal force F_h. The linearity of the output to the increasing or decreasing force is good. In Fig.6 the outputs caused by horizontal force F_h (5N) are plotted against the location of force acting point. Even though the location of force acting point is different, the output in F_h is held constant, which is the characteristic of a double beam-type loadcell. The output in $F_h x$ changes linearly with the change of the force acting point.

Fig.7 shows the stress distribution of the loadcell calculated by means of FEM. A straight line segment G_i shows the place where i-th strain gage is bonded. The gradient of the stress distribution is not flat in the middle of the loadcell(G_5,G_6). Therefore the strain gage must be carefully bonded in the middle to avoid the output error. Fig.8 shows the deflection curve of the loacell calculated by means of FEM. Solid lines A and B show respectively the deflection curves of the parts A and B shown in Fig.4. The deflections are caused by vertical and horizontal force (10N). The amount of the deformation, which agrees well with the measured value,

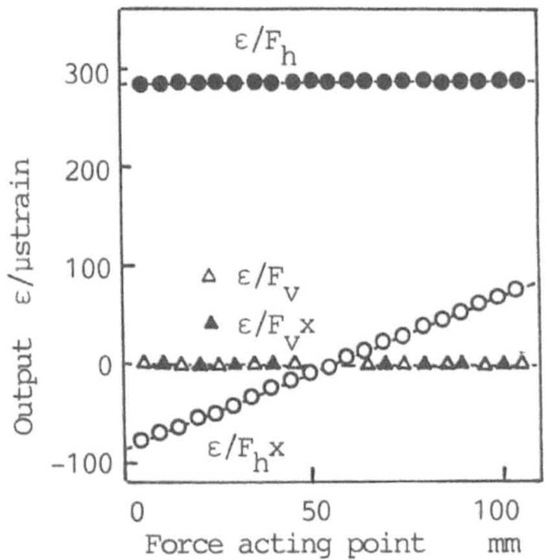

Fig.6 Output caused by horizontal force F_h

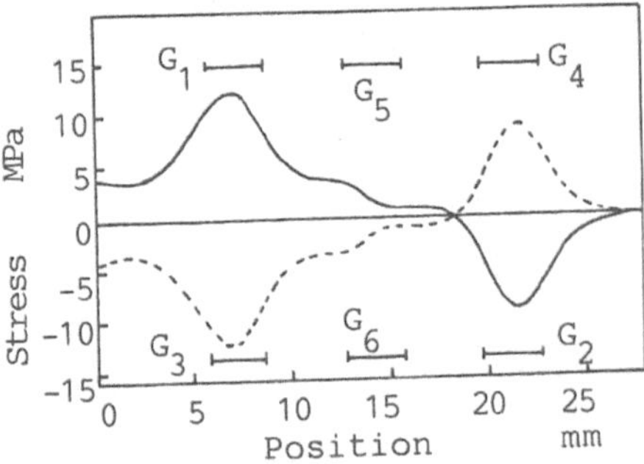

Fig.7 Stress distribution of the loadcell

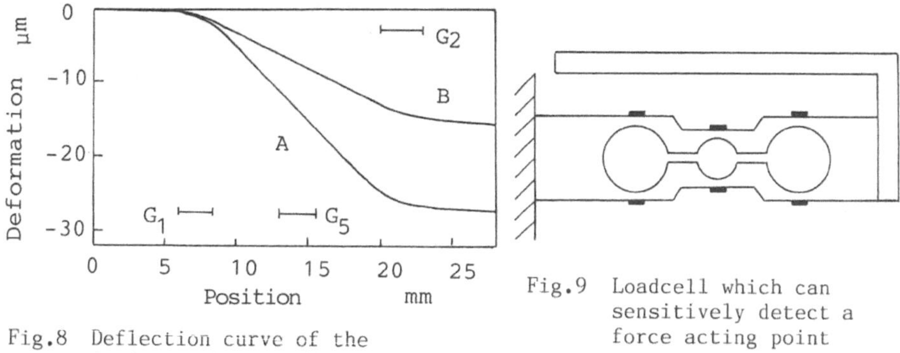

Fig.8 Deflection curve of the loadcell shown in Fig.4

Fig.9 Loadcell which can sensitively detect a force acting point

is small enough to be neglected. In order to obtain more suitable shape, FEM analyses for several shapes are carried out. As the result, the shape shown in Fig.9 is obtained, which can sensitively detect the force acting point.

3 MECHANICAL HAND

3.1 Mechanical hand with two fingers

An Experiment has been carried out by using a mechanical hand having two fingers equipped with the loadcells. The mechanical hand is set to the NC SCARA type robot(see Fig.10). Sponge rubber is pasted on the gripping surface of the fingers. Gripping force can be changed by changing the distance between two fingers. The distance is controlled by a pulse moter. The controlled rate is 0.2 mm per one pulse. The mechanical hand is controlled by a micro-computer and the robot is controlled by a NC controller. The micro-computer and NC controller send signals each other.

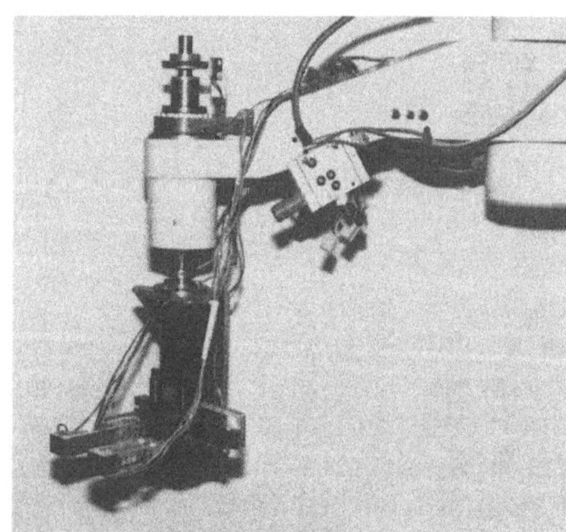

Fig.10 Mechanical hand with two fingers in which the 4-component loadcells are embedded

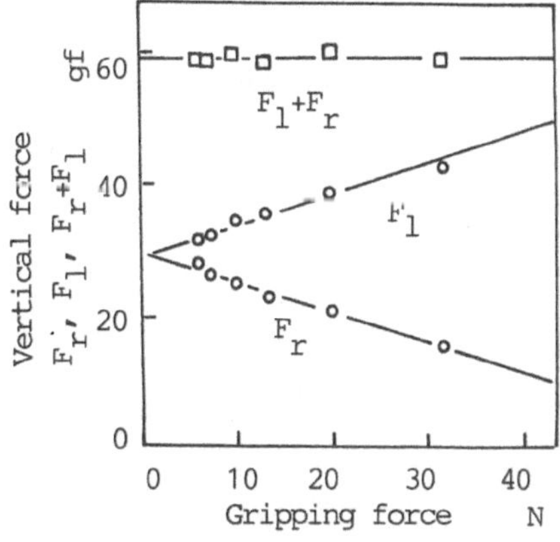

Fig.11 Effect of the gripping force on the vertical forces

The motion of the robot is decided by the signals from the micro-computer.

In Fig.11 the vertical forces acting on the both fingers and the sum of the two forces are plotted against the gripping force. F_l is the force on the left finger and F_r the force on the right finger. Though F_l and F_r change with the value of gripping force, the sum of the two is held constant. This is the experimental proof that the weight of a workpiece can be calculated by using this type of fingers. The reason why the vertical forces depend on the gripping force can be explained as follows. In Fig.12, if the workpiece contacts at a point on each gripping surface, the vertical force, for example, F_{ry} can be written as

$$F_{ry} = mgL_1 - F_{rz}(y_r - y_1) / L_2. \qquad (3)$$

From this equation, we can understand that F_{ry} is affected by both the gripping force F_{rz} and the location of the gripping force acting point. As mentioned before, sponge rubber is pasted on the gripping surfaces of the fingers. If the sponge rubber is not pasted, the standard deviation of the measured values become larger.

3.2 Sensing of the contact position

Fig.13 shows the comparison of the true gripping position (or contact position) with the position calculated from the outputs of the loadcells embedded in the fingers. The workpiece used in this measurement is a cylindrical object. The result shows that the gripping position can be calculated well without the effect of the gripping force. However, if the workpiece has such a shape as the axi-directional force occurs when it is gripped, a 5-component loadcell(see Fig.5) is required to compensate the effect. This method of detecting the gripping position might be applied to the detection of the posture of a workpiece, instead of a visual sensor.

3.3 Mechanical noise

Fig.14 shows the behaviour of the output signal in gripping force. In the state before gripping, i.e., the state that the fingers are approaching towards the workpiece, the output signal is combined with two mechanical noises. One is the noise originated from the forced vibration which occurs at the fingers due to the intermittent drive by a pulse moters. The other

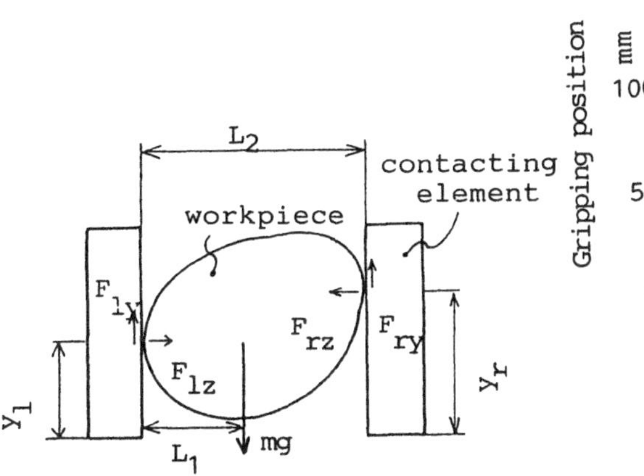

Fig.12 Relation between the gripping force and the vertical force

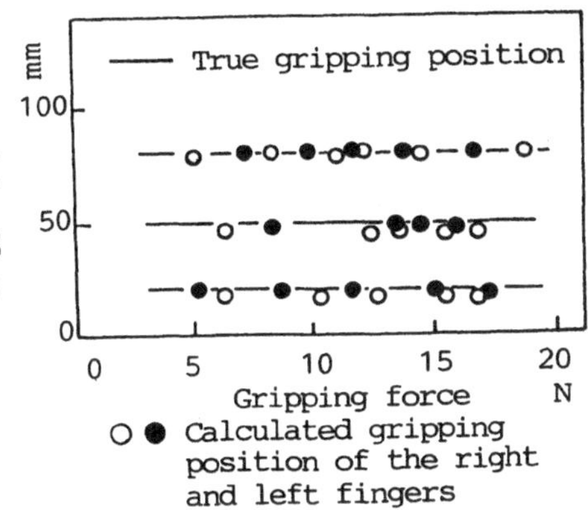

Fig.13 Measurement of a gripping position

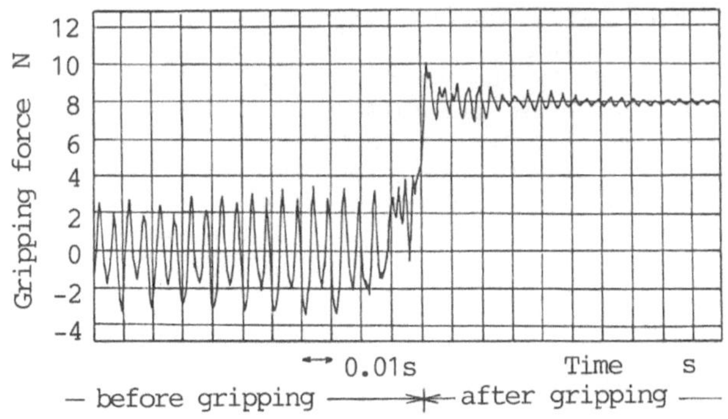

Fig.14 Output signal of the gripping force

is the noise originated from the natural frequency of the fingers induced by the intermittent movement of the fingers. In the state after gripping, only the noise originated from the natural frequency remains, which decreases as the time proceeds. These noises mentioned above can be effectively eliminated by using filtering technics.

4. CONCLUSION

A new loadcell was proposed which can be applied to the gripper of a mechanical hand. The distinctive features of the loadcell are summarized as follows; 1) the loadcell is very small and light sensor to be embedded in the fingers of a mechanical hand, and 2) it has the functions of detecting the weight of a workpiece and detecting the contact force and contact position.

The authors would like to thank Mr. M. Matuno of Ishida Scales Mfg. Co., Ltd. for producing the loadcells and giving helpful advices.

REFERENCES

1. Brady, M.: Robot Motion, MIT Press, 1982.
2. Shimon, Y.Nof.: Handbook of Industrial Robotics, Wiley.
3. Watson, P.C.: Method and Apparatus for Six Degree of Freedom Force Sensing, US Pat. 4,094,192., 1978.
4. Purbrick, J.A.: "A Multi-Axis Force Sensing Finger", 2nd Ann. ASME Conf. Computers, August, 1982.
5. Yamaba, K.: "A Robot Hand with Sensors", Bull. Mech. Eng. Lab., No.37, pp.1-12, 1982.

A NEW BENDING TEST METHOD OF LEAF SPRINGS

Hidetaka IMAI and Kozo IIZUKA

National Research Laboratory of Metrology
1-1-4, Umezono, Sakura-mura, Niihari-gun, Ibaraki, 305, Japan

1. INTRODUCTION

Elasticity of solid materials is effectively utilized as a fundamental characteristic of elastic transducers. Such elastic components as cantilevers, parallel springs, and cross spring pivots are used for transducers or guides of force or mass measuring instruments. Components used in these purposes are made of leaf springs and they are usually performed in bending mode. In order to investigate the performance of such elastic components, precise measuring techniques of elasticity and strength are keenly required.

Spring limit value and Young's modulus are considered to be principal characteristics of spring materials. Spring limit value is determined by cantilever method[1] or three-point bending method[2]. Some kinds of measuring techniques to estimate spring limit value have been reported[3], however, the coincidence among those measuring methods has not been clarified.

In this paper, a new test method which estimates both elasticity and flexural strength is described. The relation between bending moment and curvature of a beam is analyzed theoretically and confirmed experimentally. From the result of analysis it was revealed that the flexural proof stress and Young's modulus in bending mode can be estimated precisely. Estimated spring limit value and Young's modulus are compared with the values obtained by the usual method.

2. THEORETICAL INVESTIGATION

2.1 Relation between bending moment and curvature

In order to investigate the flexural characteristics of solid materials, bending test using a beam is commonly utilized. Stress-strain diagram in flexural deflection of a beam can be represented as the relation between the maximum bending moment M and the maximum curvature c as follows:

$$M/EI=c=(1/r) \tag{1}$$

where, E is Young's modulus, I is the second moment of area, and r the radius of curvature. In this equation, the maximum bending moment M is described as a function of the maximum bending stress S in accordance with the type of a beam. Then, the relationship between the maximum bending stress S and maximum curvature c (this is called S-c diagram hereinafter) can be considered to be similar to the stress-strain diagram in the tensile test.

Fig.1 shows the S-c diagram when it is extended from elastic region to plastic one, and S can be expressed as the following polynomial function.

$$S=\sum_{i=1}^{m} b_i c^{i-1} \tag{2}$$

In this equation, Young's modulus E can be determined from the slope β in the elastic region of the diagram. The bending proof stress can be defined

Wieringa, H. (ed), Mechanical Problems in Measuring Force and Mass.

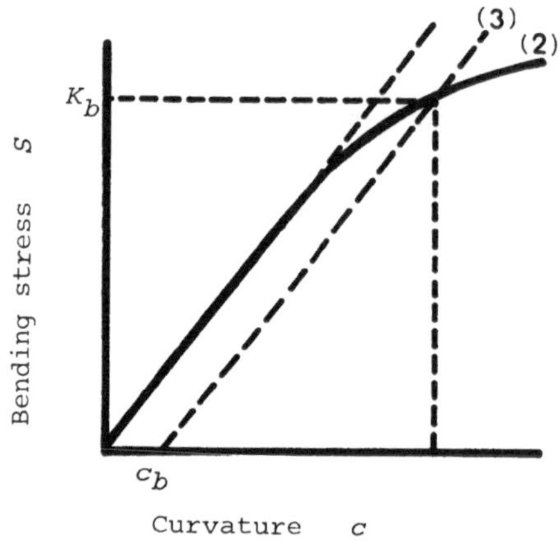

FIGURE 1. Relation between bending stress and curvature

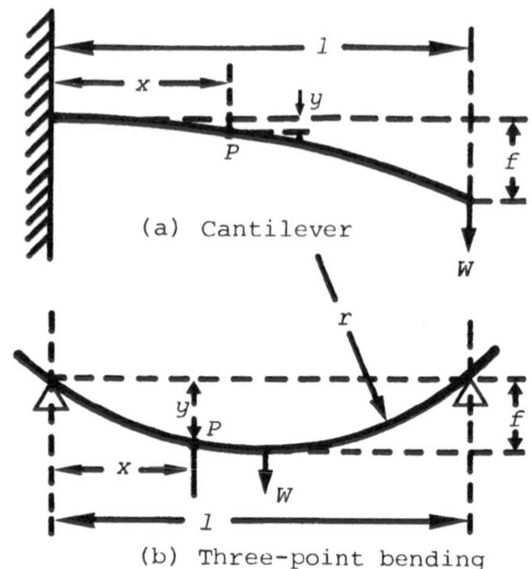

FIGURE 2. Deflection of common beams

as the intersection of equation (2) and the following equation.

$$S=\beta(c-c_b) \tag{3}$$

In equation (3), c_b is the curvature corresponding to the proof strain and it may be defined arbitrarily. In particular, elastic limit can be determined as the stress corresponds to $c_b = 0$.

2.2 Estimation of mean radius of curvature

Generally, bending test is carried out by using a cantilever or simply supported beam as shown in Fig.2 (a) or (b). Bending moment M and curvature c in equation (1) change with the change of the position $P(x,y)$ on the neutral axis of a beam. Therefore, it is required to estimate the mean curvature of a beam precisely. The deflection equation of a beam is described as $y=f(x)$, and this function is determined by solving the following equation.

$$M/EI=\{1+(dy/dx)^2\}^{3/2}=d^2y/dx^2 \tag{4}$$

In the usual small deflection theory, equation (4) is easily solved by assuming that $(dy/dx)^2$ is fairly smaller than 1. On the other hand, in order to extend the effective range of S-c diagram, large deflection theory has to be applied. This will be accomplished by the introduction of power series solution in equation (4)[4].

The mean radius of curvature r can be estimated by fitting the equation of circle to the deflection equation of a beam by means of Deming's method of least squares[5]. Fig.3 and Fig.4 represent the results of calculation in the case of cantilever and centrally loaded simply supported beam (three-point bending method), respectively. In the figures, three kinds of theoretical investigations are compared together with the effect of deflection/span ratio (f/l). Those are small deflection theory (mean radius of curvature is expressed as r_s), large deflection theory (r_m), and the method in which the radius of curvature is determined from only three points (center and both ends) on a beam (r_3). r_0 represents the minimum radius of curvature which yields at the free end (cantilever method) or the center (three-point bending method) of the beam.

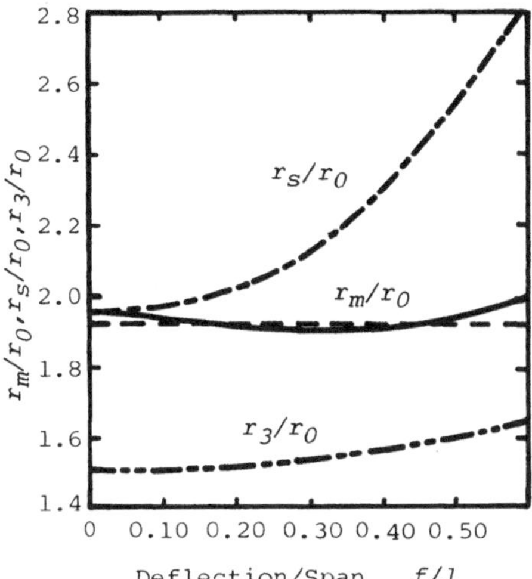

FIGURE 3. Comparison of the theory in cantilever

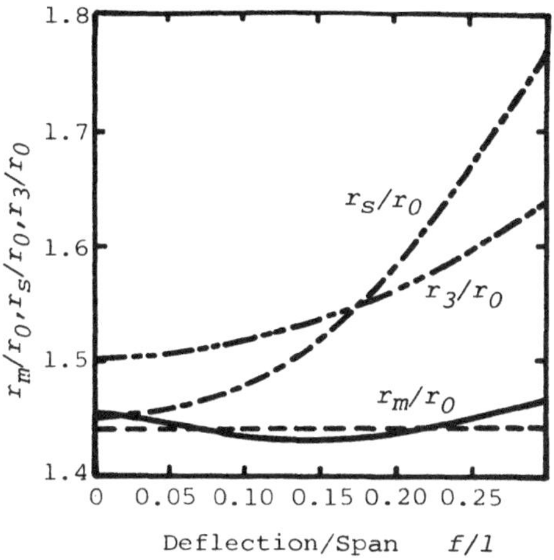

FIGURE 4. Comparison of the theory in simply supported beam

From the result of analysis, it was revealed that r_m/r_0 shows almost constant value for a wide f/l ratio in both cantilever and three-point bending method. The above investigation indicates that deflection equation of a beam can be represented as a circular arc, then mean radius of curvature r_m can be estimated from the measured coordinates on the beam $P(x_i, y_i)$, $(i=1,2,\ldots,n)$. Therefore, the maximum curvature c can be determined as follows:

$$c=1.92/r_m \quad \text{(cantilever method)}$$
$$c=1.44/r_m \quad \text{(three-point bending method)} \qquad (5)$$

2.3 Accurate estimation of Young's modulus

Although Young's modulus may be estimated from elastic region in S-c diagram as described in 2.1, much more accurate estimation can be achieved by using small deflection theory. In the simply supported beam (see Fig.2 (b)), the deflection equation is expressed as follows:

$$F(x,y;E) \equiv 6lEIy - l_1 l_2 W(l_1+2l_2)x + l_2 W x^3 = 0 \qquad (6)$$

where, l is the length of the beam, l_1 is the interval between one of the ends of the beam and loading point, and $l_2=l-l_1$. Then, Young's modulus E can be calculated from the following equation by using Deming's method of least squares.

$$E=E_0-\sum_{i=1}^{n}\frac{F_{Ei}F_{0i}}{L_i}\Big/\sum_{i=1}^{n}\frac{F_{Ei}F_{Ei}}{L_i} \qquad (7)$$

where, $F_{Ei}=\partial F_i/\partial E$, $F_{0i}=F(x_i, y_i; E_0)$, $L_i=(\partial F_i/\partial x)^2+(\partial F_i/\partial y)^2$, and E_0 is the approximate value of E. In this method, as coordinates (x_i, y_i), $(i=1,2,\ldots,n)$ are obtained in a wide range on the neutral axis of a beam, the accuracy of Young's modulus is much higher than that estimated from only the maximum deflection at the loading point (the usual method). Fig.5 shows a comparison between the above two methods as a function of deflecting rate when the number of data is 5. σ_e/r and σ_e/f represent the relative error of estimated Young's moduli by least squares method and the usual method, respectively.

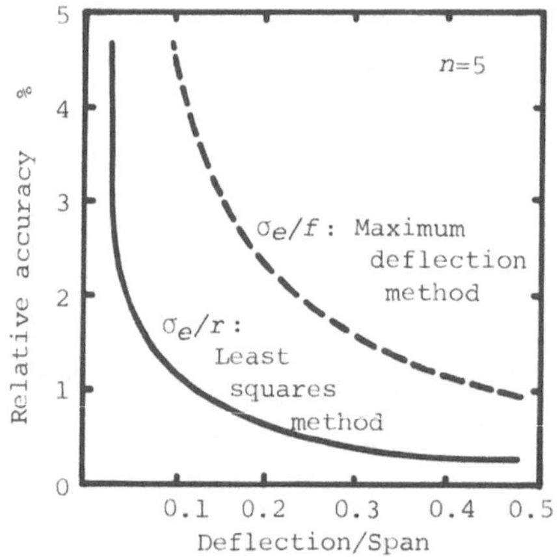

FIGURE 5. Comparison of the accuracy in the estimation of elasticity

FIGURE 6. Testing apparatus by three-point bending method

3. EXPERIMENTAL CONDITIONS

3.1 Testing and measuring apparatus

In the experiment, two types of apparatuses have been utilized. One is the usual cantilever method which applies the stress by bending moment, and the other is the three-point bending method. A new apparatus in which the loading speed and testing temperature are controlled has been constructed for the latter purpose as is shown in Fig.6. In practice, such testing conditions as loading speed, duration of loading, and clamping or supporting conditions of the specimen have to be strictly examined for both apparatuses.

In order to measure the coordinates of the points $P(x_i, y_i)$ on the neutral axis of the beam, photographs of a deflection beam were taken at several loading conditions. Then, measuring microscope or profile projector with x-y coordinate measuring stage was used for the position measurement of the beam in negative films. The minimum reading of the coordinate measuring instrument used in the present experiment was 1 μm.

3.2 Clamping or supporting conditions for the specimen

In the cantilever method, such effects as the shape of clamping pieces and clamping force to the specimen have to be taken into consideration. An experiment was carried out to investigate these effects and it was revealed that there exists remarkable difference among the shape of clamping pieces. As the experimental result for various kinds of clamping pieces, a U-shaped piece with a flat edge of about 1 mm width is found to be satisfactory for both elasticity and strength measurements[6]. In the case of three-point bending method, no significant difference was detected if the sharp edges are used for both supporting and loading points.

3.3 Effective beam length

Not only bending stress but also shearing stress is produced in the deflection of a beam. These effects are investigated theoretically for both cantilever and simply supported beam as are represented in Fig.7. The effect of shearing stress to the deflection of the beam can be neglected theoretically when the span/thickness ratio is larger than 20, however, the experimental result requires much larger span/thickness ratio. Fig.8 shows the experimental result in the case of simply supported beam for several loading conditions. Similar result was obtained for cantilever method and these are summarized as the effective span/thickness ratio (l/t) in Table 1. In the table, effective beam length for strength (spring limit value) measurement is also presented from the experimental result. Effective span/thickness ratio for both elasticity and strength measurement is recommended to be 100~150 in the cantilever method, and 150~200 in the three-point bending method, respectively.

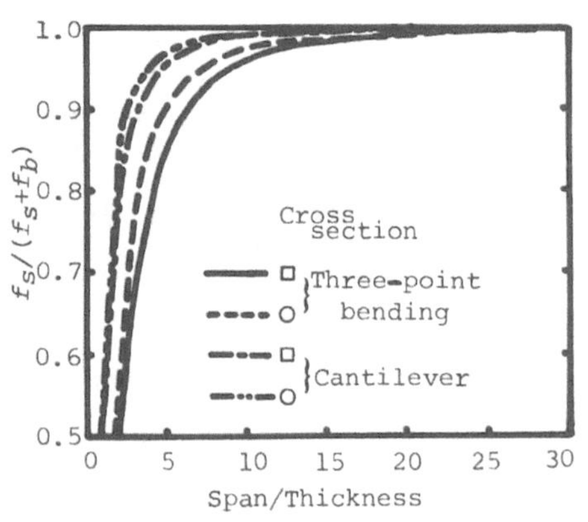

FIGURE 7. Theoretical investigation on the effect of shearing stress
f_s : deflection by shearing
f_b : deflection by bending

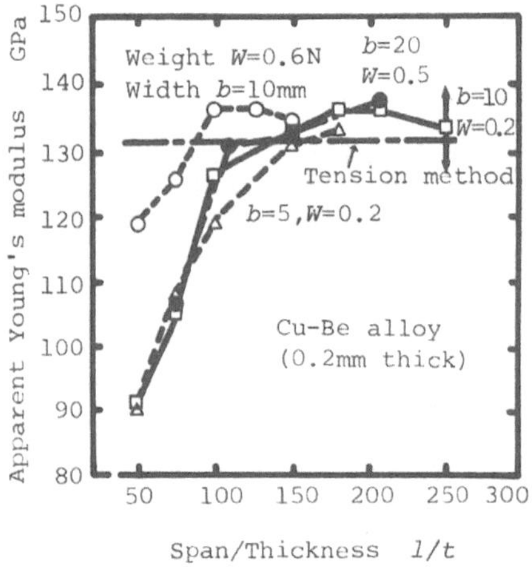

FIGURE 8. Experimental result on the effect of shearing stress (three-point bending method)

TABLE 1. Recommended span/thickness ratio in bending test

Method	Estimation of Young's modulus		Estimation of spring limit value
	Theory	Experiment	(Experiment)
Cantilever	>10	100 - 200	80 - 150
Three-point bending	<25	150 - 250	150 - 200

4. PRACTICAL APPLICATION

From the above theoretical and experimental investigations, the significance of the new bending test method which utilizes the simple relationship between the maximum bending stress and curvature (S-c diagram) was confirmed. In the present application, three-point bending method was introduced for the evaluation of elasticity and strength of spring materials used for measuring instruments. Specimens used in the experiments were commercially available materials and they were tested as rolled condition.

4.1 Measurement of elasticity

Single flexures and cross strips are effectively applied for the pivot of balances because of their fine stability. Cantilevers and parallel strips are used for force and pressure measuring transducers with strain gauges. The performance of those measuring instruments are usually evaluated by their sensitivity, nonlinearity, and hysteresis. These effects can be estimated from the non-elastic behaviour of materials. In practice, time dependence is expressed as the response after the loading. As examples are shown in Fig.9, such effects as loading speed and duration of loading can be evaluated by the curvature measurement of the beam at each testing condition. Time constant of the material for applied stress can also be estimated by the adoption of step response model to the data of time dependence. Fig.10 shows the stress dependence of Young's modulus for some materials. In this experiment, Young's modulus was estimated by the application of Deming's least squares method to the large deflection theory instead of the small deflection theory described in 2.3. Experimental results show that the decrease of elasticity reaches up to 4~6 % even in the elastic region for some materials at higher stress level.

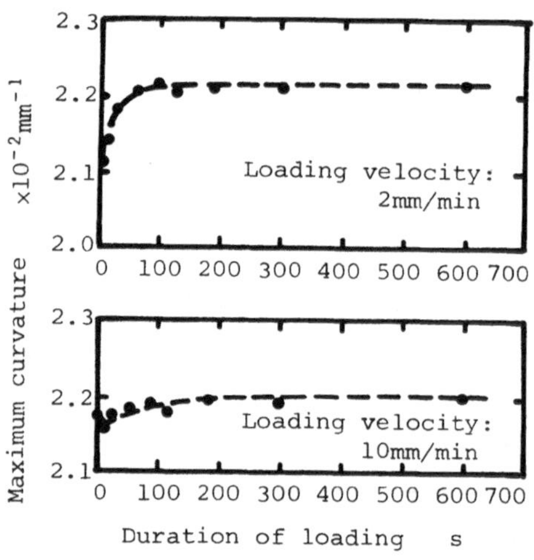

FIGURE 9. Examples of the measurement of elastic after effect

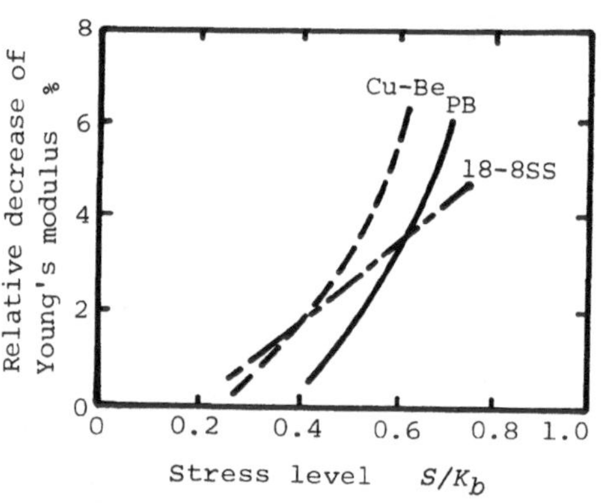

FIGURE 10. Stress dependence of Young's modulus

Linearity and hysteresis of the material against elasticity (stress-strain relation) can be determined more precisely by the *S-c* diagram than by the usual *S-f* diagram which represents the stress-strain relation by means of small deflection theory. Fig.11 and Fig.12 show a comparison of *S-c* and *S-f* diagrams obtained from the same experimental data. In both figures a slight hysteresis error was seen. It is difficult to distinguish the elastic linearity of the material from the non-linearity of *S-f* diagram itself in Fig.11. On the other hand, *S-c* diagram remains linear relation up to high stress levels as is seen in Fig.12.

4.2 Measurement of strength

In selecting a material for elastic elements to be used in bending mode, three parameters should be taken into consideration from the view point of

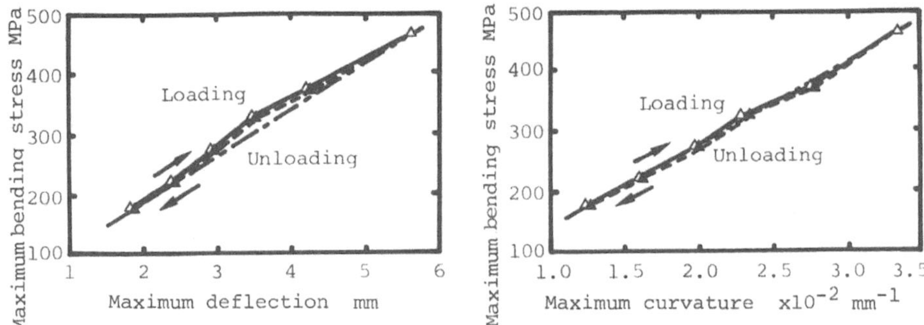

FIGURE 11. An example of S-f diagram FIGURE 12. An example of S-c diagram

strength design. Those are spring limit value K_b, accommodation limit K_L, and elastic limit K_0. K_b has a similar meaning to the 0.2 % proof stress in tensile test and is considered to be an index of the upper limit stress against static loading. K_L may be regarded as an index of the upper limit stress against quasi-dynamic loading and even if the stress less than K_L is applied repeatedly, the permanent deformation never exceeds a constant value.[7] K_0 is the maximum stress which does not admit any sensible permanent deformation. Spring limit value K_b is defined as the maximum bending stress which yields the permanent deflection corresponding to 2.5 % of elastic deformation. Practically, it is expressed as the permanent deflection caused by the stress of 3.675 $E/10^5$ MPa. Not only K_b but also K_0 can be estimated by using S-c diagram as the stress corresponding to $c_b=3/4t\times10^{-4}$ (t: thickness of the specimen in mm) and $c_b=0$, respectively. Fig.13 and Fig.14 represent the S-c diagram for Cu-Be alloy and 18Cr-8Ni stainless

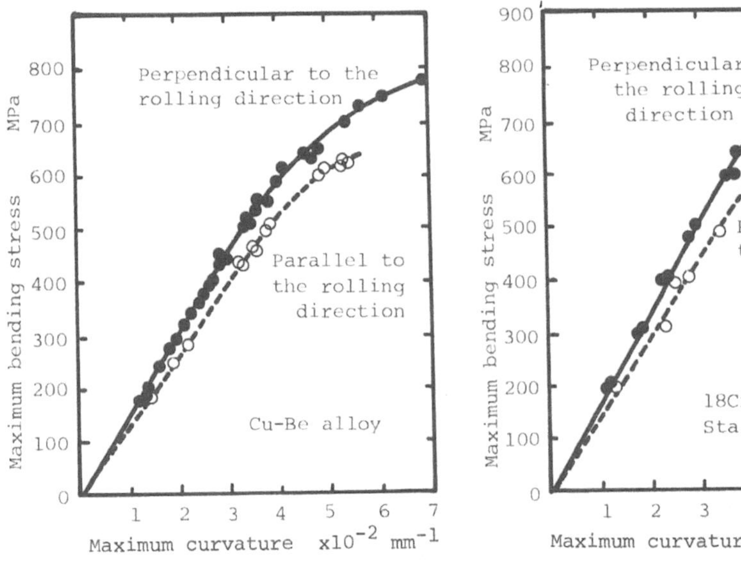

FIGURE 13. S-c diagram for Cu-Be FIGURE 14. S-c diagram for 18-8 SS

steel specimens. These specimens were cut parallel and perpendicularly to the rolling direction and tested as rolled condition.

From the figures, mechanical anisotropy of the materials is clearly seen. Young's moduli and spring limit values obtained from both figures are listed in Table 2. They are compared with the values obtained by the usual cantilever method which uses a flat clamping piece. The difference between the cantilever method and present method may be explained as the effect of the clamping conditions. Because, in the usual cantilever method, elastic deformation of the clamping part affects as the bias to the deflection of a specimen. Therefore, estimated Young's modulus shows smaller value and spring limit shows larger value comparing with the values obtained by the three-point bending method.

TABLE 2. Comparison between cantilever and three-point bending methods

Material	Young's modulus (GPa)		Spring limit value (MPa)	
	Cantilever (corrected)	Three-point	Cantilever	Three-point
Cu-Be alloy	123 (132)	135 ± 4	769	742
Phosphor bronze	111 (126)	123 ± 4	653	597
Stainless steel	177 (209)	205 ± 5	701	606

5. CONCLUSION

A new bending test method of leaf springs which estimates both elasticity and strength from the stress-strain relationship (S-c diagram) has been developed. The significance of the method was analyzed theoretically and confirmed experimentally for some kinds of spring materials. From the result of experiment it was revealed that Young's modulus and the spring limit value can be determined with the accuracy of ±2 % at 95 % confidence interval.

REFERENCES

1) JIS H 3130 (Japanese Industrial Standard).

2) DIN 50151

3) Imai H and Iizuka K: A New Method for Evaluating the Spring Limit by means of Regression Analysis, J. of the Japan Society of Precision Engineering, 42-3(1976), 215-220. (in Japanese)

4) Scott E J and Carver D R: On the Nonlinear Differential Equation for Beam Deflection, J. Appl. Mech., June(1955), 245-248.

5) Iizuka K and Imai H: Form Measurement of Small Spherical Surfaces by means of the Method of Least Squares, Bull. of the Japan Society of Precision Engineering, 4-1(1970), 1-6.

6) Imai H, Takeshita H and Iizuka K: The Effects of the Clamping Conditions on Flexural Deflection of Cantilever Leaf Springs, Report of the National Research Laboratory of Metrology, 24-3(1975), 154-163. (in Japanese)

7) Imai H and Iizuka K: Accommodation Limit in Bending Test, Bull. of the Japan Society of Precision Engineering, 19-1(1985), 24-29.

A CALIBRATION INSTRUMENT FOR ROLLER BRAKE TESTERS

BY

G.J.W.L. Kotte, D. Baksteen, J.A.J. Basten

1. INTRODUCTION

In the Netherlands, since September 1985 a statutory obligation exists for an annual examination of motor verhicles. Resulting from the Act a number of metrological activities with respect to the measuring instruments in use at the test stations are carried out by the Netherlands Metrology Service, i.e.

- the examination and testing of new types of measuring instruments resulting in a "pattern approval";
- the initial verification of newly installed instruments, i.e. individual examination, tests and calibration;
- reverification of particular instruments in use.

An important part of the examination of a vehicle is a brake test with the aid of one of the following instruments:

- a deceleration meter
- a flywheel brake tester
- a plate brake tester
- a roller brake tester.

In particular for the latter type, a test unit was developed by the Netherlands Metrology Service. In this paper we describe:

- the working principles and the way of use of roller brake testers;
- the requirements for a universal test unit;
- the design of the instrument;
- its possibities and its intended use.

2.0 PROBLEMS IN CALIBRATING A ROLLER BRAKE TESTER

2.1 The working principles

A roller brake tester is a measuring instrument intended to determine the braking power of each wheel (or pair of wheels) of a vehicle separately. The instrument consists of two separate measuring units. Each unit contains two steel rollers attached to the floor by a frame. The rollers, which mostly are covered with a rubbing surface, have a diameter of some 190 mm. Between the rollers a movable, free-running contact roller is fitted. The rollers are driven by an electromotor via a transmission of toothed wheels and chains. For the brake test the axle of the vehicle being tested is placed onto the roller bed (see fig. 1 and 2). The arrival of a wheel between the rollers is detected by the contact roller, which also serves as a speedometer. After a few seconds the electromotor is switched on and the rotating rollers drive the vehicle wheels.
As soon as the vehicle's brakes are activated the wheels will exercise a tangential force on the roller surface, opposite to the direction of rotation. This tangential force is measured and displayed as the braking force on the indicator(s). During the test, the electromotor of the tester will keep the circumferential velocity of the rollers approximately constant. Mostly the indicators and the rest of the electronic equipment are housed together in one cabinet (see fig. 2).

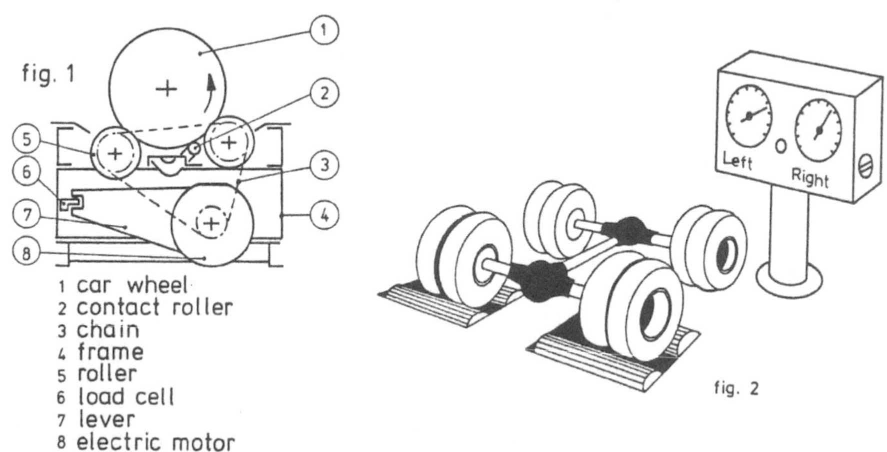

Roller brake testers mostly have a measuring range of 0 - 40000 N. Statutory, the testers must have an accuracy better than ± 5 % of the applied tangential force.

Up to now, 2 measuring principles are applied in roller brake testers in order to measure the tangential force.

A) Measuring the reactive moment of the electromotor by means of a lever and a load cell (see fig. 2). The motor is mounted in such a way that the housing and the stator can rotate freely. To the motor housing a lever is connected supported by a load cell, which impedes the rotation. The reactive moment is exerted and measured by the load cell. From this and from the mechanical transmission ratio of the drive gear the tangential force on the roller surface is calculated and shown on the indicator(s).

B) Determination of the reactive moment of the motor from the absorbed electric power.
A Wattmeter is used to measure the electric power used by the motor to keep the circumferential velocity of the rotors constant. The couple/power curve of the motor is used to compute the moment supplied by the motor. Then the tangential force on the roller surface is calculated from the mechanical transmission ratio of the drive gear and the moment found, and displayed on the indicator(s).

2.2 The calibration of a roller brake tester

Till now, the calibration procedures for roller brake testers have been rather primitive. For testers with working principle A, a set of levers and weights is used to exert a force F_k on the load cell. Using the mechanical dimensions of the roller brake tester the corresponding tangential force on the roller surface (F_r) is calculated.

At the calibration, the indicating device should show the force F_r when the force F_k is exerted on the load cell.

The above calibration method is a static one; moreover, only part of the measuring system is investigated, i.e. the load cell and the indicator. Neither the effects of wear on the roller diameter nor changes in the bearings and chain gear are considered. Although this method provides a fair indication for the adjustment of the roller brake tester it is insufficient for our Metrology Service, as a statutory verification authority. The calibration of testers with working principle B is even less ideal. The only part examined is the indicator/watt meter. Effects of ear of rollers, bearings and chain gear and individual deviations in motor charasteristics are not taken into account. For our purpose, only a calibration method very similar to the functioning of the roller brake tester during an actual brake test, i.e. in which a tangential force is exerted on the roller surfaces is acceptable. Taking into account the statutory tolerances for roller brake testers, the calibration instrument must be able to determine the applied tangential force with an accuracy of $\pm$ 2 %. In the present calibration instrument this tangential force is exerted by a special wheel. To determine the tangential force the braking moment to the wheel is measured from which the tangential force may be calculated with the formula:

$$Mr = Ft * H == Ft = \frac{M_r}{H}$$ see fig. 3.

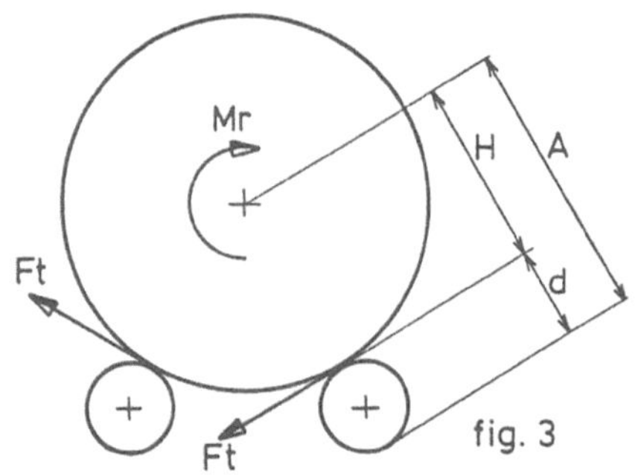

fig. 3

Mr = brake moment
Ft = tangential force
H = wheel radius
d = roller diameter
F_n = normal force
F_{nr} = resulting normal force
F_s = sliding force
F_{sr} = resulting sliding force

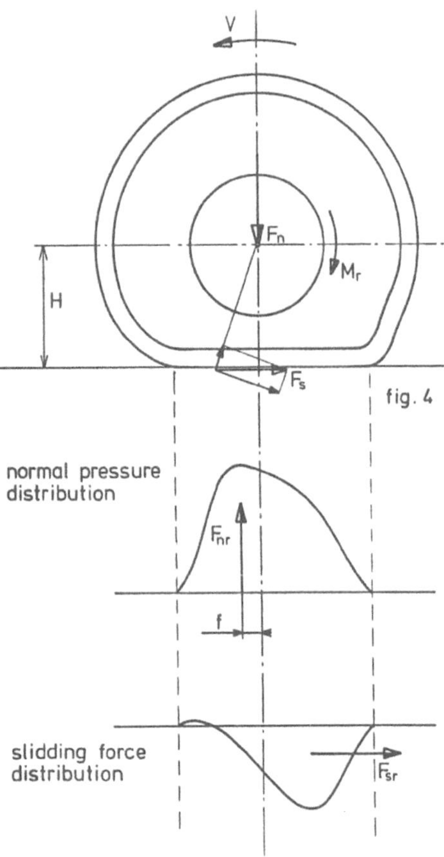

fig. 4

2.3 Disturbing influence factors on measurements taken with a wheel.

In practice the relation between F_r, M_r and H, has shown not to be constant. The effective value of H (H_{eff}) is not simply equal to the distance A -d. A close examination of the behaviour of pneumatic tires whilst braking on a road surface has shown that the following effects affect the value of H_{eff}.

1. The decrease of the distance A.
   Direct measurements of the distance A show that, A decreases 1 to 2 % with increasing braking force. This decrease is the result of deformations in the rotating surface.

2. The phenomenon of a non-symmetric normal pressure distribution over the contact area (see fig. 4).
   Measurements at a flat surface of the normal pressure distribution over the contact area between tire and ground, show that the resulting normal force does not point through the axis of the wheel but at a distance f of the wheel axis in the driving direction. This is caused by hysteresis in the elastic material. Such a normal pressure distribution causes a braking moment on the rotation of the wheel and results in a resistance of the wheel to rolling. The distance f appears to increase with an increasing braking moment of the wheel.

3. The slipping forces in the front and rear of the contact area (see fig. 4).
   Whilst giving a relative equal contribution to the total horizontal sliding force, these forces contribute relatively little to the total braking moment. It is also apparent that the distribution of sliding forces over the contact area is not symmetric and changes as the partial slip increases.

When measuring with a roller brake tester the above effects probably have a larger effect on H_{eff} as at flat surfaces because of the more severe tire deformations.

3.0 THE DESIGN OF A CALIBRATION INSTRUMENT FOR ROLLER BRAKE TESTERS.

With the help of fig. 5 the most important aspects of the construction of the calibration instrument will be explained.

3.1 The measuring wheel

In view of the effects described in 2.3 it is clear that the plasticity of the wheel must be as little as possible to obtain a sufficient measuring accuracy.
The ideal of a completely rigid body should be approximated as much as possible. Contrary to this goes that the required tangential force must be transmitted by friction. In practice this has shown to be only possible with special elastic materials. An acceptable compromise appears to be the "solid wheel". Nave and rim are made of steel. The rim is coated with a synthetic friction surface of 50 mm thickness. Other data of the measuring wheel are:
diameter: 1100 mm
width : 600 mm
mass : 800 kg.

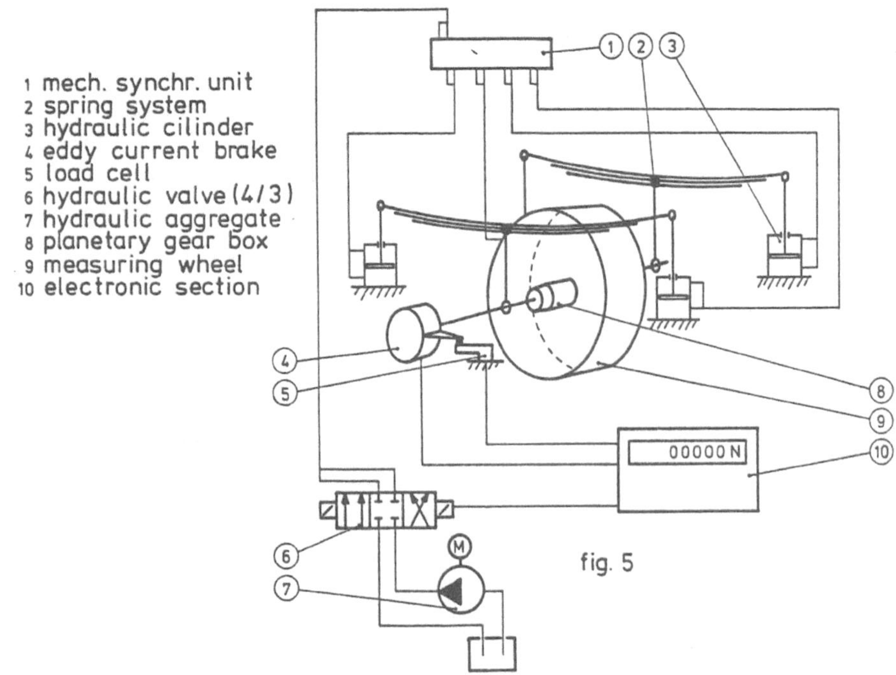

fig. 5

The following requirements apply to the synthetic material:

- The surface must be wear resistant.
- Continuous slip produces heat in the contact surface. The synthetic material will have to be sufficiently heat resistant.
- A high coefficient of friction is desirable, since a certain tangential force then may be transmitted with a lower normal force.

Our choice for the contact surface material was Vulkolan, a wear resistant polyurethane of acceptable rigidity (Shore)/friction coefficient (u) combination:

70° Shore A: u = 0,3 - 0,4 (polished steel)
u = 0,8 - 1,2 (concrete).

However, its mechanical properties deteriorate very quickly at temperature exceeding 65 °C. Since the heat production at the contact surface during braking hardly can be estimated, it had to be determined experimentally. A test wheel was placed on a common roller brake tester and braked for several minutes. Thermo-couples were fitted just under the contact area of the test wheel. The wheel was kept under the load which is expected in the instrument in use. It turned out that during normal tests the temperature would not increase over 45 °C. The minimum observed friction coefficient was 0.55 (on epoxy coated rollers) and the effect of wear on the Vulkolan surface appeared to be acceptable.

3.2 Hydraulic installation

To obtain a tangential force of 40 kN on a roller pair with a friction coefficient of 0.55, the measuring wheel should press on the rollers with a normal force of 80 kN. In the calibration instrument this pressure on the measuring wheel is produced by 4 hydraulic cylinders connected to an aggregate via a mechanical synchronization system. The reactive force needed to apply the normal force is supplied partly by the mass of the calibration instrument itself, i.e. 4500 kg. For the remainder the following possibilities exist:

- additional mass pieces are placed on the instrument;
- using the mass of the vehicle and the loading hoist used for transportation of the instrument;
- clamping the instrument onto the roller bed;
- fastening the calibration instrument to the floor by means of suction.

Practical experience will prove which is the most suitable method.

3.3 The spring system

Practice has shown that the rollers of a roller bed are not always straight. In order to prevent excessive compressing forces between the wheel and roller bed, a progressive spring system was designed through which the normal force is applied to the wheel. The spring system also acts as a mechanical accumulator for the hydraulic system. Leak losses are compensated for by slight back springing so that the normal force is largely remained.

3.4 The brake system

There are several possibilities for braking the measuring wheel. A water brake and a hydro-motor are ruled out, as they both need a liquid storage system and cooling unit to cool away the dissipated heat. Also it is doubtful that the braking power in continuous operation can be regulated properly. Disc and drum brakes are not feasible since warping and vitrification of the friction surfaces make it impossible to maintain a constant high braking force over a longer period.
We have chosen for an eddy current brake because of its low price and compact construction. Moreover the elimination of the dissipated heat is easily performed by air cooling.

3.5 Transmission between measuring wheel and eddy current brake

The selected eddy current brake in operation makes between 700 and 3000 revolutions per minute, where as the measuring wheel rotates at about 12 revolutions per minute. This requires a gear box with an approximate ratio of 1:70. This high ratio also ensures a sufficient increase in the eddy current brake couple. In view of this high ratio, and the requirement for a compact construction, a planetary reduction gear was selected, as normally used for the built-in gear of a ship's winch drum. The bearings of this gear box are strong enough to act as a bearing for the measuring wheel.

3.6 The measuring system

The moment exerted by the brake system on the measuring wheel is determined by measuring the reactive moment of the eddy current brake and the gear box. Other measuring principles appear to be either very fragile and very costly or are not yet completely developed. Determining the active moment by measuring a reactive force at a lever is a very secure method. As sensors we use 2 HBM Z7A shearbeam load cells with a capacity of 5000 kg.

3.7 The frame (not shown in fig. 5)
The main function of the frame is to ensure an adequate strength, stability and rigidity of the test rig. Furthermore the frame should offer the possibility for vertical motions of the measuring wheel which is in a fixed horizontal position.
Finally, it must be suitable for the transportation of the test rig.

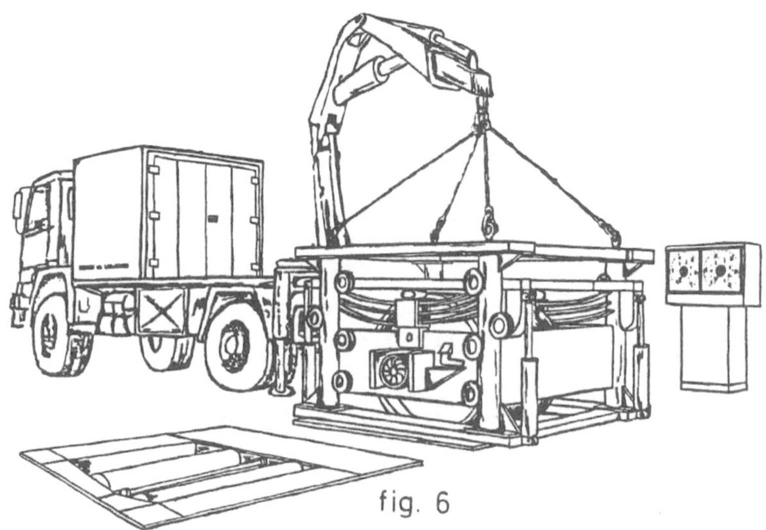

fig. 6

3.8 The electronics
The electronic section is composed of a mini-computer, the power supply for the eddy current brake, indicators for the load cells, an indicator for the pressure gauge in the hydraulic system, connections for the electrically operated valve, pulse counters to determine the rotating frequencies of the measuring wheel and the rollers of the roller brake tester.

3.9 The transport vehicle
The transport vehicle, see fig. 6, is a lorrie with a loading hoist. The lorry has a two-part coach work, of which one part can be moved along the lenght axis of the lorry.

4.0 TEST PROCEDURES

The examination of a roller brake tester proceeds as follows:
The calibration instrument is placed on one roller bed by means of the lorry's hoist and the necessary connections are made. The wheel is lowered hydraulically onto the rollers. The contact roller will switch on the roller brake tester. Using the supply, the braking couple of the eddy current brake can be adjusted and regulate the tangential force on the roller surface. The tangential force is calculated by the computer from the signal produced by the load cells. The system is protected against excessive slip. During the test the monitor of the computer presents the tangential force applied, the pressure in the hydraulic system, the

circumferential speeds of the measuring wheel and the rollers, as well as the percentage of slip. The following properties of a roller brake tester will be checked:

- The brake force indication
  The computer regulates the eddy current brake so as to maintain a constant applied tangential force. This enables the operator to read the indicators of the roller brake tester. The values of the constant tangential forces are stored in computer memory. The operator can automatically or manually increase or decrease the brake force to predetermined values. After introduction of the values from the indicators of the roller brake tester the errors are computed and presented graphically or in tabular form.

- The circumferential speed of the rollers
  Statutory the circumferential speed of the rollers must be at least 2.5 km/h. A pulse recorder and a reflecting strip on the roller surface are used to measure the circumferential speed during the test. If required, this value may be shown simultaneously with the other measuring results.

- The slip prevention on the roller brake tester
  An increase in the tangential force under the same normal pressure results in an increased slip. A roller brake tester normally determines the slip from the speed signal of the electromotor. When the slip percentage becomes too high the roller brake tester stops in order to prevent excessive tire wear. The slip percentage is also calculated by the computer of the calibration unit from the difference in circumferential speed of the measuring wheel and of the rollers. This may serve to judge the slip adjustment of the roller brake tester.

5.0 ACCURACY OF THE CALIBRATION UNIT

At the time of writing this paper, the calibration unit was not yet ready for use. So, it is not yet certain whether the expected accuracy of ± 2 % of the tangential force applied, will be realised. The errors of the calibration instrument are estimated from the following values:

- frictional moments.
  In this respect only the frictional moments of the two end bearings of the measuring wheel are of interest. The estimated error is 0.1 % at maximum with a brake force/normal force ratio higher than 0.55.

- force measurement
  Within the temperature range of 0 - 25 °C the total error of the load cells and the indicator is estimated to be + 0.1 %.

- determination of the length of the force lever
  Estimated error + 0.05 %.

- wheel deformation
  Depending on various assumptions in the calculation, this error is estimated to lie between 1.7 % and 3 %. This error appears to be caused

mainly by the elastic contact surface of the measuring wheel. The following measures were taken to reduce the deformation of the contact surface as much as possible:

A sufficient rigidity of the frame should prevent that during braking the measuring wheel climbs up against the rear roller of the roller bed. Thus it will be avoided that most of the normal force is supplied by only one roller which would cause an extraordinary deformation of the elastic layer. The progressive spring system is designed in such a way that the normal force increases up to 15 %, when the measuring wheel is moved vertically by a roller. In practice it appears that with a slip of approximately 20 % the ratio brake force/normal force is optimal (see fig. 7). Therefore the normal force is adapted to the slip measured by the computer.

The systematic error of the calibration instrument due to deformation of the measuring wheel could be determined by carrying out measurements on a specially designed high precision roller bed under laboratory conditions. From this the value of H_{eff} could be determined so that the accuracy of the calibration unit would be much better than $\pm$ 2 %.

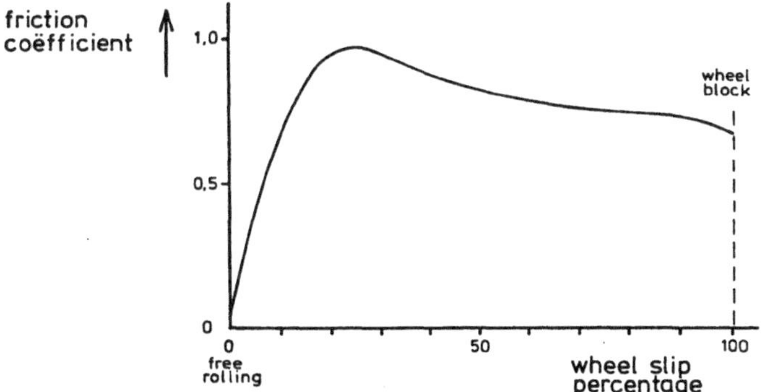

fig. 7 Connection between wheel slip and friction coëfficient

6.0 CONCLUSIONS

As far as we know, the calibration unit described here is the first instrument capable of measuring and testing a roller brake tester in a metrologically acceptable manner. It will be possible to apply a tangential force up to 40000 N on the rollers with an accuracy of better than 2 %.

The use of the calibration unit is not limited to the examination of roller brake testers. After some adaptation, for instance by fitting a hydro-motor, the behaviour of pneumatic tires at well defined high brake or drive couples and with a varying normal force at low circumferential speeds, can be studied.

Moreover, we expect that also other institutes, industry and our sister organisations in Western Europe will be interested in the use of this apparatus.

CHARACTERISTIC DATA OF FORCE TRANSDUCERS TERMS AND DEFINITIONS

W. Weiler

PHYSIKALISCH-TECHNISCHE BUNDESANSTALT, FRG

1. INTRODUCTION

In practice, the terms and definitions for force transducers are used quite differently. The consequence is very often a different interpretation of the same terms, resulting in confusion. Two years ago, therefore, in the Federal Republic, a board of VDI/VDE GMR started on work to standardize these terms and definitions. This work is now completed and will be published as a draft VDI/VDE 2638 (green print) this year. Renewed discussions will take place a year later on the basis of comments received.

2. SCOPE AND FIELD OF APPLICATION

The draft is concerned with the terms and definitions for transducers used in static force measurement. It is restricted to force transducers with strain gauges and related to the VDI/VDE 2637 for load cells used in weighing machines.

3. TERMS FOR THE TESTING OF FORCE TRANSDUCERS

The first section deals with terms for the testing of force transducers. It covers international and national standards in this field. The most important points are:

- repeatability excluding and including rotation
- hysteresis
- temperature coefficient of the rated output and zero signal

Some of these terms are used in standards for classifying force transducers.

4. NOMINAL DATA

A great deal of nominal data is necessary for the use of force transducers. The definition of various terms is necessary for:

- force
- signal
- rated output

- deflection
- resistance
- frequency

With the rated output, for example, it is necessary to distinguish between the rated output, the nominal rated output, the reference rated output and the range of the rated output.

The reference straight line should be very carefully defined. The definition is as follows: Straight line describing the connection between force and output signal. All differences of the output signal are related to this straight line. It can be defined by two different methods (the method chosen must be specified). The reference straight line is calculated from the output signals for increasing and decreasing force.

a) Its ordinate segment is equal to the zero signal. Its increase is so chosen that the sum of squares of all relative signal differences from the straight line become minimal in the range between 20 % and 100 % of nominal force.

b) Its ordinate segment is equal to the zero signal. Its increase is so chosen that the positive and negative relative signal differences are equal in the range between 20 % and 100 % of nominal force.

An illustration is given in figure 1.

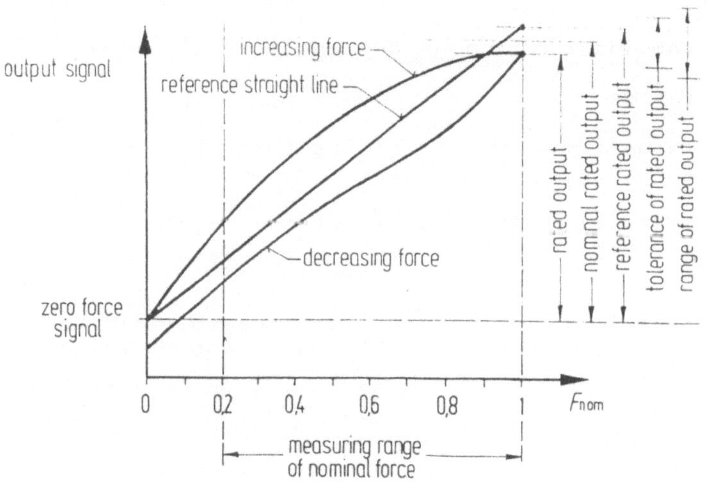

Figure 1. Reference straight line

5. OTHER TERMS AND DEFINITIONS

Terms are defined for the following main spheres

- technical quality of the force transducer (8 terms)
- reference conditions
  temperature, voltage, pressure, humidity
- effects of influence
  temperature, pressure, parasitic components, time, voltage

One of the most difficult definitions is that of creep.

Creep is understood as the variation in time of the output signal under constant force after changing force. A force transducer is evaluated with the help of the times presented in figure 2.

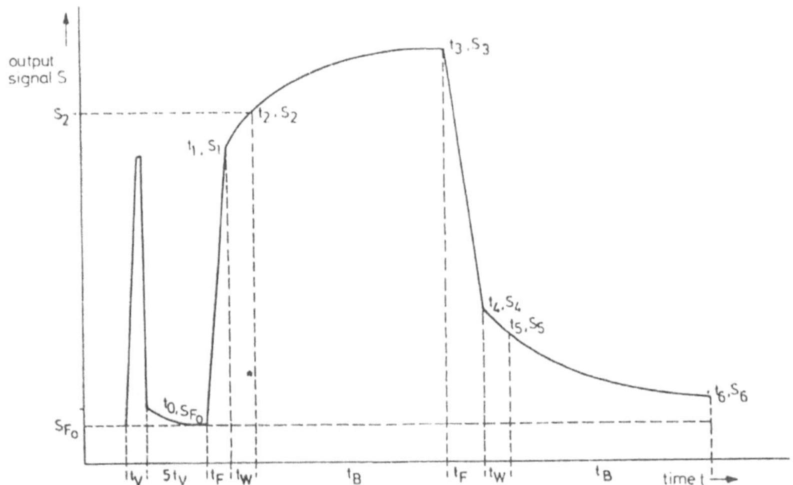

Figure 2. Creep of force transducer

A distinction must therefore be made between the time of the change of force t_F, the waiting time t_W and the evaluation time t_B. Creep d_{cr} is the maximum of the change of the output signal during the evaluation time t_B. The relative creep is e.g. the quotient of the creep $s_3 - s_2$ and the change of the output signal $s_2 - s_{F_0}$ as a result of a change of force.

For a comparison of relative creep values, the change of force must be made to the amount of the nominal load and the times chosen for t_F, t_W and t_B must be indicated. The following times are recommended:

$$t_F = 5\ s,\ t_W = 10\ s \text{ and } t_3 = 30\ s$$

4 different types of creep dishinguished

- increasing creep

$$d_{cr,F} = \frac{s_3 - s_2}{s_2 - s_{F_o}}$$

- decreasing creep

$$d_{cr,E} = \frac{s_5 - s_6}{s_2 - s_{F_o}}$$

- zero drift of force

$$d_{cr,N} = \frac{s_5 - s_{Fo}}{s_2 - s_{F_o}}$$

- rest of zero signal

$$d_{cr,R} = \frac{s_6 - s_{F_o}}{s_2 - s_{F_o}}$$

- permitted limits
  mechanical, thermal and electrical

- reliability established by pattern approval

6. CONCLUSIONS

The draft on terms and definitions is the first paper of this kind: it is something of a pilot project. The working board hopes that it can, also initiate an international draft for force transducers in the English language and papers on other measuring instruments.

A COMPENSATING LEVER AND ITS CONTROL SYSTEM FOR COMPENSATING THE FORCE OF THE FRAME OF A DEADWEIGHT FORCE STANDARD MACHINE

W.W. WEILER AND A. SCHUSTER

PHYSIKALISCH-TECHNISCHE BUNDESANSTALT, FRG

ABSTRACT

When forces are realized by deadweight force standard machines normally the effect of the loading frame is the smallest force realizable. For smaller forces the effect of the loading frame must be compensated. A compensating lever and its control systems are described which render the compensation of the effect of the loading frame possible with an uncertainty of $1 \cdot 10^{-7}$. The advantages are described by the excample of a 5000 N force standard machine.

1. INTRODUCTION

In the field of metrology very high standards are applied to the realization of forces /1/, /2/. The method affected by the smallest uncertainty of measurement is the direct utilization of the effect of a body of the mass m in the earth's gravitational field g_{loc}, corrected by the buoyancy of the air $(1 - \rho_l/\rho_m)$, where ρ_l is the density of the air and ρ_m the density of the body /3/. The principle is shown in Fig. 1.

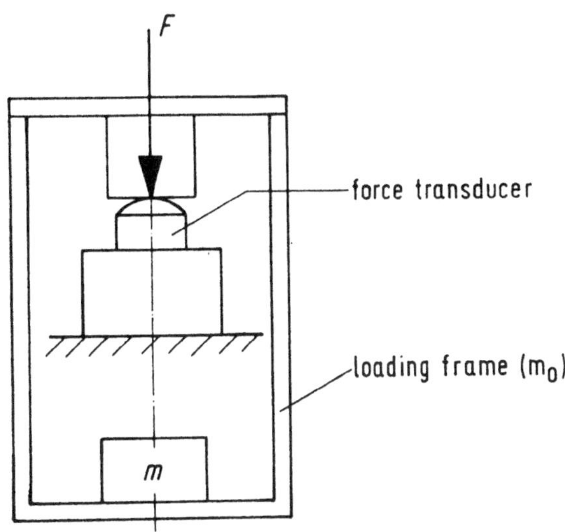

FIGURE 1. Principle of a deadweight force standard machine

The generated force is calculated with the formula:

$$F = (m_o + m)g_{loc} \cdot (1 - \frac{\rho_l}{\rho_m}) \quad (1)$$

The smallest realizable force is determinded by the mass of the loading frame m_o.

$$F_{min} = F_o = m_o \cdot g_{loc} \cdot (1 - \frac{\rho_l}{\rho_m}) \quad (2)$$

For metrological and economic reasons, it is often useful to generate forces $< F_o$. In this case, the effect of the loading frame must be compensated by means of a compensating lever, so that the force $F_o = 0$; the force F_{min} can then be generated by any small mass m_{min} of the first deadweight.

The additional uncertainty of measurement resulting from the compensating lever must then, however, be negligibly small compared with the uncertainty of the force standard machine ifself. Besides, it must be possible to check whether this condition is met. The designs known to date /1/, /2/ do not fulfill these conditions satisfactorily. A compensating lever and its controll system, which meet the required conditions, are described in the following.

2. THE COMPENSATING LEVER AND ITS INFLUENCE ON THE UNCERTAINTY OF MEASUREMENT OF THE FORCE STANDARD MACHINE

Fig. 2 shows a diagrammatic view of the compensating lever. Due to the stable equilibrium the centre of gravity S of the compensating lever lies at the distance p below the knife edge. The deflection of the lever is constant; moreover, it can be kept very small by dimensioning and shaping the lever appropriately. In the concrete case it can also be calculated by simple, mathematical means. The length a of the lever between the two knives is also constant, as long as there is no wear of the knives. This wear can be avoided by an appropriate selection of the knives and the material from which they are made, and it can be checked in accordance with the description in section 3.

The uncertainty of measurement thus results from the uncertainty of the adjustment of the zero position.

$$F_o \cdot a = G \cdot b \quad (3)$$

and the uncertainty due to the deflection of the lever at the angle φ caused by the deformation Δ_s of the force transducer under a force F (Fig.3)

$$(F_o + F)a' = G \cdot b' \quad (4)$$

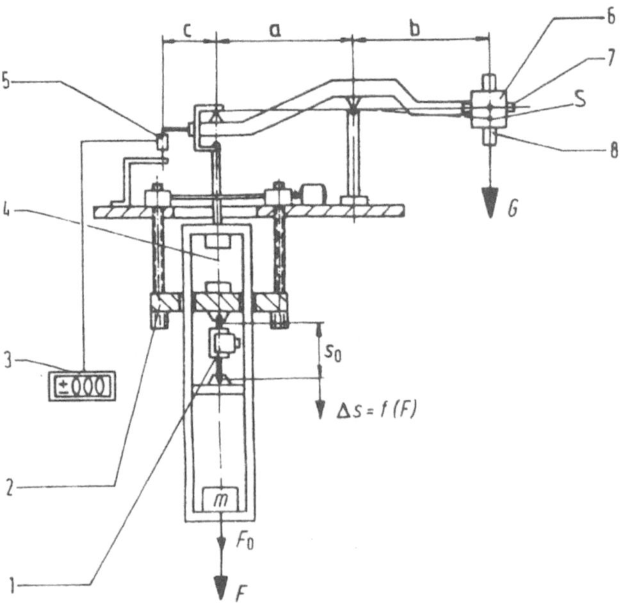

FIGURE 2. Schematic representation of the compensating lever

1 force transducer (force in tension)
2 crosshead, adjustable in height
3 zero indicator
4 space for mounting the force transducer
5 displacement sensing device
6 rough counterbalancing
7 fine counterbalancing
8 adjustment of the centre of gravity
S centre of gravity
G resulting deadweight (weight of lever and counterbalance weight)
F_o force produced by the deadweight of the loading frame
F force produced by the body of mass m
s_o mounting length
Δs deformation of the force transducer

According to the rules of plane trigonometry, the following is valid:

$$\sin \sigma = \frac{p}{s} \tag{5}$$

$$\cos \sigma = \frac{b}{s} \tag{6}$$

$$\cos(\sigma - \varphi) = \frac{b'}{s} \tag{7}$$

$$\cos \varphi = \frac{a'}{a} \tag{8}$$

$$\cos(\sigma - \varphi) = \cos\sigma \cos\varphi - \sin\sigma \cdot \sin\varphi \tag{9}$$

From this it follows, if one multiplies at the same time by s:

$$b' = b \cos\varphi - p \sin\varphi \tag{10}$$

When equation (4) is at the same time divided by $\cos\varphi$, it can be written as follows with

$$\tan\varphi = \frac{\sin\varphi}{\cos\varphi}$$

being inserted:

$$F_o \cdot a + F \cdot a = G \cdot b - G \cdot p \tan\varphi \tag{11}$$

As according to equation (3), $F_o \cdot a = G \cdot b$, the following results if one divides at the same time by a:

$$F = - G \cdot \frac{p}{a} \cdot \tan\varphi \tag{12}$$

This means that the inclination of the lever at the angle φ causes a negative error which is proportional to $\tan\varphi$.

The relative error referred to the maximum force thus is:

$$f_{rel} = \frac{F}{F_{max}} = \frac{p}{a} \cdot \tan\varphi \cdot \frac{G}{F_{max}} \tag{13}$$

or, expressed as absolute value using equation (3)

$$|f_{rel}| = \frac{p}{b} \cdot \frac{F_o}{F_{max}} \cdot \tan\varphi \tag{14}$$

The following assumptions, which definitely go beyond the real conditions, are made for the quantitative evaluation. The effect of the loading frame is assumed to be $F_o = 0{,}01\ F_{max}$.

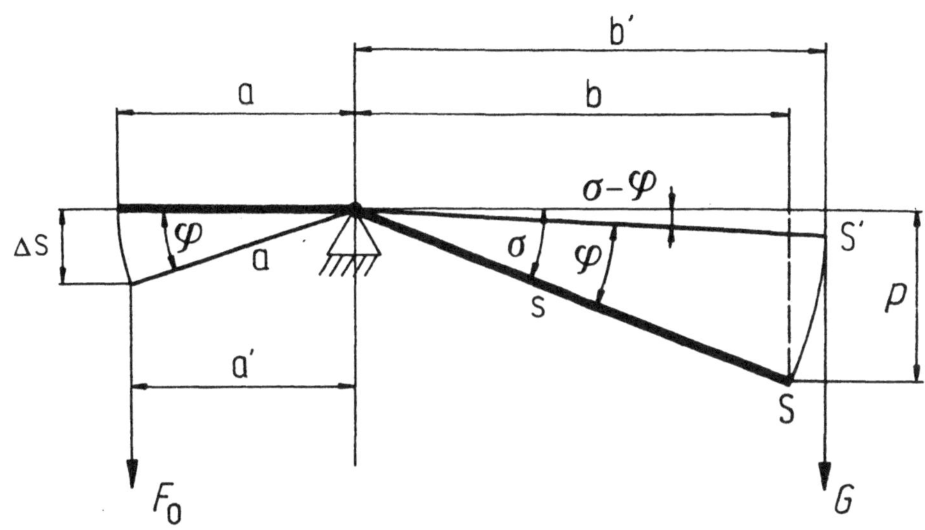

FIGURE 3. Geometry at the lever

The maximum deformation of the force transducer s_{max} is assumed to amount to 0.1 cm, and the distance p of the centre of gravity from the knife edge to between 1 cm and 2 cm.

As b = a holds for the equal-armed lever, for this case, equation (14) can be written as follows:

$$|f_{rel}| = 0.01 \cdot \frac{p}{a} \cdot \tan\varphi \qquad (15)$$

Depending on the length a of the lever, $|f_{rel}|$ can be calculated for each p. The results are shown in Fig. 4 for p = 1 cm and p = 2 cm. The maximum error amounts to $2 \cdot 10^{-6}$.

Dissymmetrical levers are so designed that b > a. The error curves in Fig. 4 change towards decreasing errors by the factor b/a.

If due to reasons of design the relation $F = 0.01\ F_{max}$ cannot be realized, an equally small error can nevertheless be achieved by appropriately changing the position of the centre of gravity (p) by means of the device provided for the adjustment of the position of the centre gravity (No. 8, Fig. 2).

Fig. 5 shows a compensating lever as used in practice for a 1 MN deadweight force standard machine.

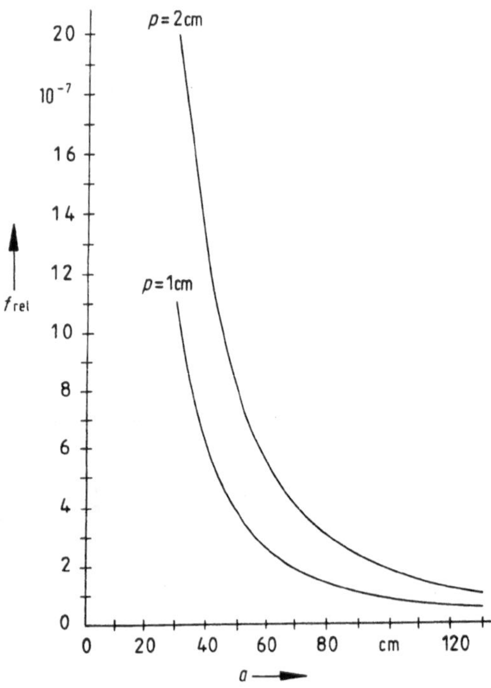

FIGURE 4. Maximum error for an equal-armed compensating lever, as a function of the lever length for the distances p = 1 and p = 2 of the centres of gravity

3. TEST METHOD FOR DETERMINING THE INCLINATION ERROR

The loading frame is counterbalanced with the aid of the zero indicator (Fig.5a,1). The force transducer is mounted and, due to its deadweight, causes a movement of the frame and thus an inclination of the compensating lever. This inclination is compensated by moving the crosshead until the zero indicator again shows zero. The indication at the force transducer which is still unloaded is also zero (Fig. 5a,2). If the force transducer is loaded by a defined control force (F_K) - produced by a standard weight N - then the indication A_1 is displayed at the force transducer as the sum of the indications A_{F_K} as a result of the control force, and A_N in consequence of the inclination of the compensating lever (Fig. 5b).

$$A_1 = A_{F_K} - A_N \tag{16}$$

If, with the force transducer loaded in this way, the crosshead is moved towards the top until the zero indicator displays again zero, the inclination of the compensating lever has been compensated, i.e. the indication A_N as a result of the lever inclination is zero. The indication A_2 at the force transducer now corresponds only to the indication resulting from the control force A_{F_K}.

$$A_2 = A_{F_K} \tag{17}$$

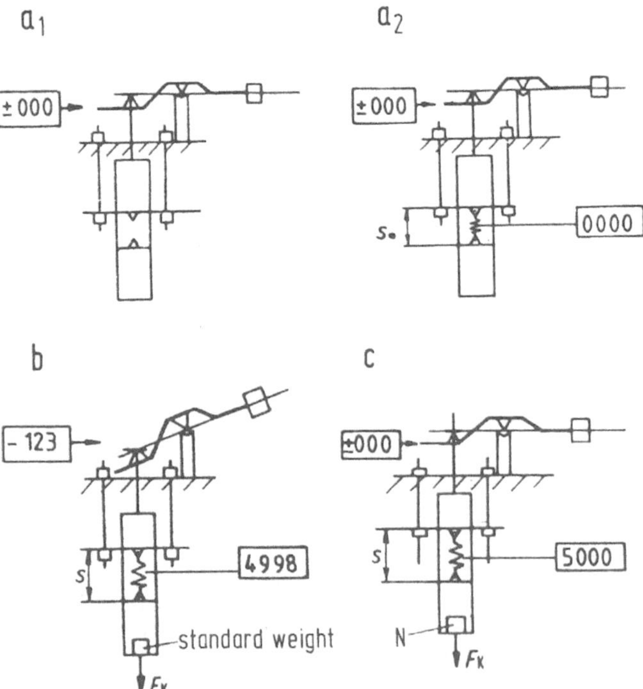

FIGURE 5. Test method for determinig the inclination error

a_1 starting position without test piece loading frame counterbalanced

a_2 test piece mounted indication of position = 0
force indication = 0

b loading with N
indication of position ≠ 0
force indication $A_1 = A_{F_K} - A_N$

c crosshead moved until indication of position = 0
force indication $A_2 = A_{F_K}$

The error of the indication due to the inclination error corresponds to the difference of the indications A_2 and A_1.

$$A_2 - A_1 = A_{F_K} - (A_{F_K} - A_N) = A_N \qquad (18)$$

Thus the relative error can be calculated to be:

$$f_{A_{rel}} = A_N/A_{F_K} \qquad (19)$$

If the relative inclination error is to be kept very small - e.g. $1 \cdot 10^{-7}$ - and is to be established then the force transducer to be tested will usually be unsuitable for this control. For example, the resolution of the indicating device of a force transducer laid out for the nominal force of 5 000 N, would have to be reliable even at 0.0005 N. As the error of inclination depends only on the inclination of the compensating lever, not, however, on the test force, it is possible to use for this measurement a test force transducer with approximately the same displacement but with a control test force, which is substantial smaller to the nominal force.

A force transducer for the nominal force of 5 N connected with an indicating device with 200 000 digit steps (e.g. HBM DK 38) has a resolution of 0.000025 N per digit step. The deviation to be established is then 20 digit steps. The inclination error so determined is valid for the displacement of the test transducer and is smaller for all smaller displacements. The length of the displacement - for example 0.32 mm - can be controlled at the zero indicator (see Fig. 2). A displacement pick-up with a nominal displacement of 2 mm and an error of 0,2 % referred to the final value (e.g. HBM W2 TK), which indicates 1 000 digit steps for the nominal displacement, is suitable for this task. Due to the distance c of the pick-up from the axis of force application, the displacement at the zero indicator is:

$$\Delta s_W = \Delta s \cdot \frac{a + c}{a} \qquad (20)$$

On the assumption that

$$\frac{a + c}{a} = 1.4,$$

$\Delta s_W = 0.45$ mm for $\Delta s = 0.32$ mm.

As 0.002 mm yield one digit step, 240 ± 2 digit steps are obtained in this case. Should this value be exceeded, i.e. should the inclination of the compensating lever become greater than permissible, then this can be indicated by a limiting value contact.

The additional uncertainty of measurement resulting from the lever inclination can be checked again by a control measurement in accordance with the described procedure.

In addition, the test methods described allow the knife-edge bearing and the mobility of the lever system to be controlled. If the knife-edge bearing changed by Δa, then equation (3) for the unloaded lever system would become the inequation

$$F_o(a \pm \Delta a) \neq G \cdot b \qquad (21)$$

The zero indicator displays this as zero error.

The mobility of the lever system can be checked with high sensitivity by insignificantly changing the control test force (standard weight).

4. EXAMPLE

The advantages of the compensating lever can best be demonstrated by a comparison of 5000 N force standard machines of equal design, however, with and without compensating lever. For the sake of clarity, this comparison is made in the form of tabled data:

Maximum force:	50 000 N
Type:	deadweight effect with substitution of forces
Control range:	1 : 100
Force steps:	binary gradation (500 N, 1000 N, 2000 N...)
Force from the movable frame:	5000 N (mass approx. 40 kg)
Possible supplementary force steps:	e.g. 100 N, 200 N, 200 n (outside the control range)

These data allow potential applications for the calibration of force transducers to be derived, the table below making a distinction between calibrations from 10 % and 20 % of the force transducer's nominal force. A calibration from 10 % of the nominal force is to be considered the normal case.

The survey clearly shows the superiority of the force standard machine equipped with a compensating lever. Its range of application is five times larger than that of a machine without lever. Even in the case of limited calibration of a force transducer from 20 % of the nominal force, the range of application is still more than twice as large.

Survey of possible calibrations

Nominal force of force transducer N	Calibration possible without compensating lever		Calibration possible with compensating lever	
	from 10 %	from 20 %	from 10 %	from 20 %
50 000	yes	yes	yes	yes
40 000	yes	yes	yes	yes
30 000	no	yes	yes	yes
20 000	no	yes	yes	yes
10 000	no	no	yes	yes
8 000	no	no	yes	yes
5 000	no	no	yes	yes
2 500	no	no	yes	yes
2 000	no	no	yes	yes
1 000	no	no	yes	yes

At least Figure 6 shows the practical design of the compensating lever.

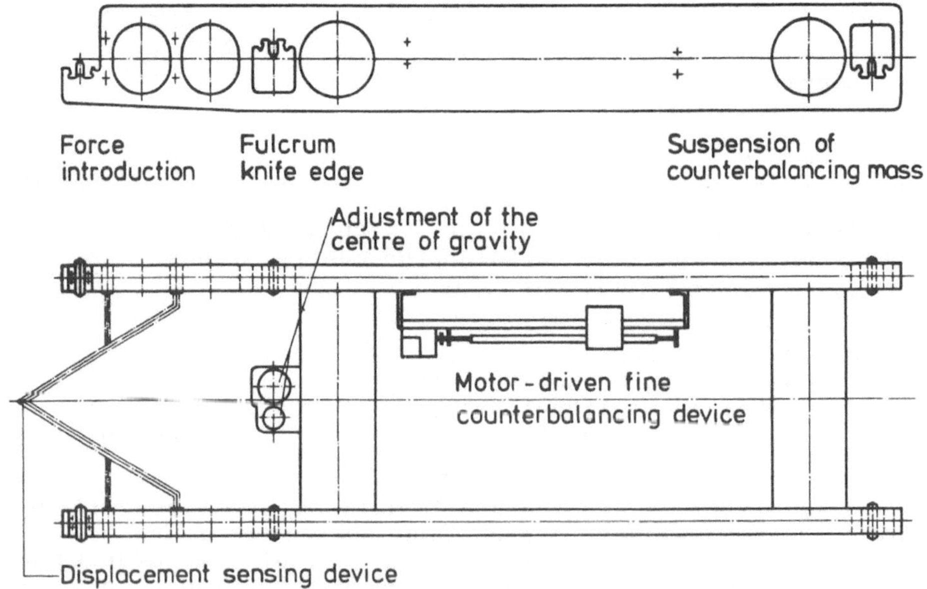

FIGURE 6. Practical design of the compensating lever

5. CONCLUSIONS

In the case of deadweight force standard machines equipped with a lever compensating the effect of the loading frame the application of the above-described method allows the additional unvertainty of measurement resulting from the lever inclination to be limited to a maximum of 10^{-7}. This additional uncertainty of measurement is thus negligibly small compared with a total uncertainty of measurement of $\leq 1.10^{-5}$ aimed at for such devices. By this method it is possible to realize force standard machines of smallest uncertainty, which cover a considerably larger range of force and which consequently make the procurement of several machines unnecessary in many cases.

REFERENCES

1. Weiler, W. and Sawla, A.: Force Standard Machines of the National Insitutes for Metrology. PTB-Me-22, Aug. 1978.

2. Weiler, W. and Sawla, A.: Force Standard Machines of the National Institutes for Metrology, Volume 2, PTB-Me-62, Juni 1984.

3. Weiler, W., Peters, M., Gaßmann, H., Fricke, H., Ackerschott, W.: Die 1-MN-Kraft-Normalmeßeinrichtung der Physikalisch-Technischen Bundesanstalt Braunschweig. VDI-Z. 120 (1978), S.1-6.

SOME CALIBRATION AND TEST FACILITIES AT VSL

G.H. ENGLER
VAN SWINDEN LABORATORY OF DUTCH METROLOGY SERVICE
PO BOX 654, 2600 AR DELFT, THE NETHERLANDS

1 INTRODUCTION

The department "Weighing Machines" of the Dutch Van Swinden Laboratory has two tasks: to carry out the type tests of weighing machines to be used for trade and to calibrate force measuring equipment and torque wrenches.
This paper describes the equipment in use for the type tests of load cells and the calibration of load cells and torque wrenches, as well as some procedures and practical points of view.

2 EQUIPMENT

2-1 Force standard machines

The two dead weight machines of the Van Swinden Laboratory both have been calibrated in kg (mass) rather than in Newton, as the department deals mainly with weighing machines and their components. For calibrations in Newton the acceleration of gravity and the effect of air buoyancy have to be taken into account.

2-1-1 The 550 kg dead weight machine

In the 550 kg dead weight machine of VSL the device under test is mounted under tension or compression between a semi-fixed "table" and a "cradle" in which the weights are laid.
The load is applied to the load cell or proving ring by means of a loading pad or an arrangement of rods, bearings etc., depending on the type of device under test. The construction of the machine is such, that the weight of the cradle (app. 70 kg) plus the loading pad etc. would cause a dead load on the device under test. To nullify this dead load, there is a lever with an adjustable counterweight at the top of the machine. Of course, this lever has to be level at the moment the reading is taken. This is achieved by a feed-back system, consisting of an extension to the lever with inductive sensors and a motor which drives four spindles which in their turn move the semi-fixed

point (table) up or down until the lever is horizontal.
The weights are always applied incrementally and in pairs, starting with the lower ones, in order to keep the centre of gravity in the middle and as low as possible.
Above as well as under the table there is plenty of room, not only for the load cell or proving ring under test, but also for a temperature cabinet which will be described in paragraph 2-2-1.

Currently the machine has two modes of operation:

1. "Normal"
By means of switches we choose the maximum load, the size of the increments and the interval time between the steps in which these increments are applied. If the final load is not a multiple of the chosen increment, the last increment is a smaller one in order to complete the selected maximum load. Between steps, first the lever is levelled (if necessary) and just before the next step a printer can be activated.

2. "Cyclical"
In this mode the machine runs just as "normal", but after the maximum load is reached, the load is removed in the way it was applied (same load increment and time interval) in order to measure hysteresis. After the cycle is completed, a new cycle starts etc. until the machine is stopped by hand. This facility is mainly used to carry out temperature tests.

At the moment, two modifications are under construction:

1. A new, microprocessor-based control panel, which will provide the above-mentioned options in a more flexible way and will make it possible to carry out temperature tests on load cells according to IR 60 automatically.

2. A hydraulic ram will be added, which can lift or drop the cradle with the weights smoothly and makes it possible to apply and remove the maximum load in one load step instead of incrementally. This will enable us to comply with IR-60 in the execution of zero return and creep tests. At present, especially with larger loads, the time needed for the incremental application and removal of the masses and the levelling of the lever between load steps far exceeds the time allowed in paragraph 7 of IR-60 for the application and removal of the load (7,5 s for loads of 10-100 kg, 10 s for loads of 100 to 1000 kg).

2-1-2 <u>The 2,5 / 25 t dead weight / semi dead weight machine</u>

The primary (dead weight) part of the 2,5/25 t machine is similar to the 550 kg machine as described in 2-1-1.
This machine has a lever system too, which does more than cancel out the unwanted dead load on the device under test. The lever has a ratio of 10 : 1, which adds the possibility of measuring up to 25 t in the secondary part of the machine. In the latter case, the device under test is pressed from the lower side against the "table" (semi-fixed point) above it.

In the 2,5 t part of the machine there is sufficient space at both load cell positions to mount a temperature cabinet. This is also the case at the position for a compression type of load cell in the 25 t part of the machine.

The levelling system of the lever makes use of an auxiliary lever which amplifies the deflection of the main lever by a factor of 8. This levelling system is an important contribution to the specified accuracy which was confirmed during the 1980 intercomparison sponsored by the EEC office BCR.

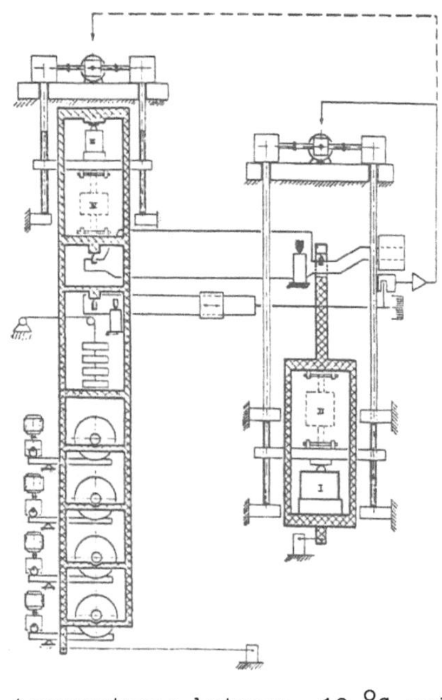

The auxiliary lever has a sliding weight which enables the application of a dead load of up to 60 kg in the primary part or 600 kg in the secondary part.
This machine has the disadvantage of being comparatively slow: it can take up to 2.5 minutes to apply 25 t on a load cell; far too slow therefore to apply loads in the time specified in IR-60 (15 s for loads of 1 - 10 t and 25 s for loads of 10 - 100 t) for the creep and zero return tests.

The microprocessor-based control-panel provides two running modes of the machine:

1. "Normal"
Very similar to the "normal" mode of the 550 kg machine as described in 2-1-1.

2. "Temperature test"
In this case the control-panel is also connected to a cryostat and the machine can carry out a temperature

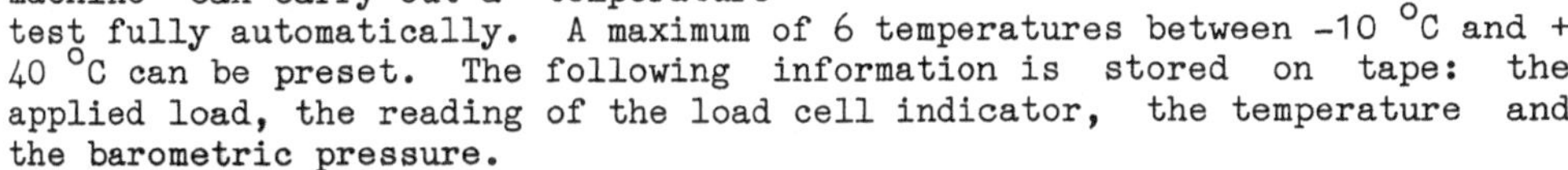

test fully automatically. A maximum of 6 temperatures between -10 °C and + 40 °C can be preset. The following information is stored on tape: the applied load, the reading of the load cell indicator, the temperature and the barometric pressure.

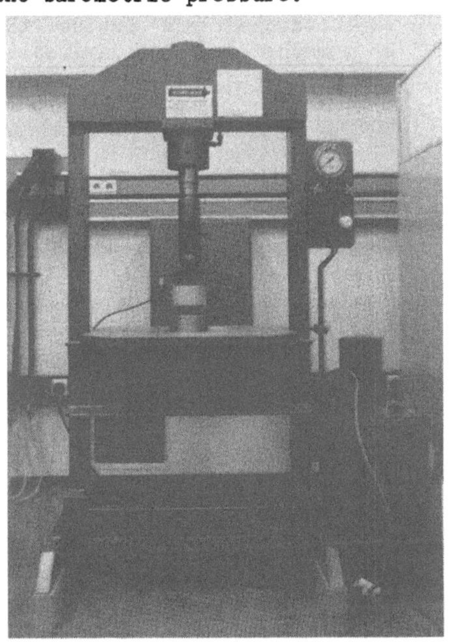

2-1-3 <u>25 t press</u>

As it is impossible to carry out creep or zero return tests on the 2,5 / 25 t dead weight machine and a modification for this purpose was not possible at reasonable costs, a simple hydraulic press has been installed. Basically this is a standard workshop press. As the load is neither exactly defined nor entirely stable, this press is not suited for the creep test but performs well for the zero return test. It is possible to apply and remove a load in accordance with IR-60. Within the frame, there is enough room to mount a temperature cabinet, so it is possible to perform the zero return test at temperatures ranging from -10 °C up to + 40 °C. The press can achieve a maximum load of 25 t in compression and 10 t in tension.

2-1-4 Facilities for small capacity load cells

Load cells too small in capacity to be tested in one of the force standard machines described earlier, are mounted onto a frame and tested by hand. In this case the first load step is performed by one of the platforms, available with calibrated masses of 0,5, 1, 2, 5 and 10 kg.

2-2 Facilities for temperature and humidity tests

2-2-1 Temperature cabinets for the force standard machines

For temperature tests in either of the two force standard machines or for zero return tests under temperature controlled conditions in the hydraulic press, two cylindrical temperature cabinets are available, as well as a cubic one. The basic design of all three is the same:

A metal cabinet is lined with a layer of polyurethane foam. A coiled copper pipe runs along the foam. This pipe is connected by a flexible and thermally insulated tube with a "Colora Ultra Cryostat" which cooles or heats the circulating anti-freeze liquid. This system has some advantages over a fan blowing hot or cold air into the thermally insulated cabinet. Because the cooling unit is installed outside the force standard machine, vibrations in the machine are avoided. Also there is no "wind" in the cabinet causing unwanted forces and thermal gradients are minimized.
Another important precaution in the prevention of thermal gradients is the thermally insulated floor of the cabinet as well as the thermal insulation in the loading pad. The insulating material is "Celoron" (Novotext-ferrozell) which is a tissue reinforced resin, which combines good thermal insulation with high resistance against compression forces.
Both round cabinets have a divisible lid with a hole for the loading pad. The cubic one has a door in the front.
There are two cryostats: an air-cooled one (located outside the laboratory) connected to the 550 kg machine and a water-cooled one inside the laboratory. The latter is used for the 2,5/25 t machine and the hydraulic press. When in use with the 2,5/25 t machine, the temperature is set with

the machine's control-panel. A system of pipes and valves makes it possible to switch cryostats in case one of them is out of order.
The relation between the temperature of the liquid and the final temperature has been estabished empirically: air temperatures of -10 °C and +40 °C require the liquid to be about - 20 °C and + 50 °C respectively. The exact values depend on the cabinet in use and the length of the tube.

Relevant properties of Celoron		
Material	resin	phenol resin
	filler	woven cotton
compression strength		> 15 kN/cm²
thermal conductivity		< 1,1 W/m.K

2-2-2 <u>Temperature cabinet for smaller capacity load cells</u>

For the purpose of temperature tests, low capacity load cells (10 - 50 kg) are mounted by means of a bracket in a modified household-freezer.
This contains a heating facility in the form of breeding lamps, a temperature regulation system, a reinforced floor to prevent bending under the load and a hole in the bottom for a tension-rod (closed by means of a labyrinth filled with silicone oil). Last but not least the whole set-up is placed on a special table. If creep or zero return tests have to be carried out in this set-up, the loaded platform is lifted or dropped via an auxiliary lever with the overhead crane of the laboratory. In order to adjust the time in which the load is moved, springs can be applied between the hook of the crane and the auxiliary lever.

2-2-3 <u>Several temperature cabinets</u>

So far, there has been little demand for component approvals for lower capacity load cells (up to 10 kg), as weighing machines of this range can easily be tested completely. But if necessary tests can be performed by hand in one of the temperature cabinets in use for retail scales.

2-2-4 <u>Climate cabinet</u>

The VSL has a 0,5 m³ climate cabinet made by the Dutch manufacturer of "industrial cold" Grenco. Temperature (-20 °C / +100 °C) as well as humididy (20% / 98% at temperatures from 5 °C to 95 °C) is microprocessor controlled. We use this cabinet for humidity tests on non-hermetically sealed load cells as prescribed in IEC 68-2-30 (test Db: 6 cycles 25 °C / 40 °C at 95 % humidity). It is also very suitable for tests on the temperature behaviour of unloaded load cells. We keep two test control programs permanently stored in the memory of the microprocessor: one for the test according to IEC 68-2-30 and one for a temperature test of -10 °C / 40 °C / -10 °C / etc. with 5 °C intervals. The control system switches to a new temperature setting after a soaking time of 5 hours. Just before a new temperature is set, a relay is activated in order to print the output of the specimen under test.

2-3 <u>Data processing</u>

The data of the temperature and zero return tests are processed by a computer system, consisting of a HP-85 personal computer, a digital tape recorder, a printer and a plotter.

During an automatic temperature test on the 2,5/25 t machine or a zero return test, the data are automatically stored on tape; with other tests the data have to be typed in by hand.

2-4 Pressure cabinet

A functioning prototype of a simple facility to test the influence of changes in barometric pressure on the zero output of load cells consisted of a plastic washing-up bowl, a bucket, a flower-pot and two tubes. Later, a more convenient set-up was made. Blowing air into the inner chamber causes a small change in pressure. Instead of the primitive, but workable "pressure meter" in the drawing, the tube is usually connected to the digital barometer in the control-panel of the 2,5/25 t machine.

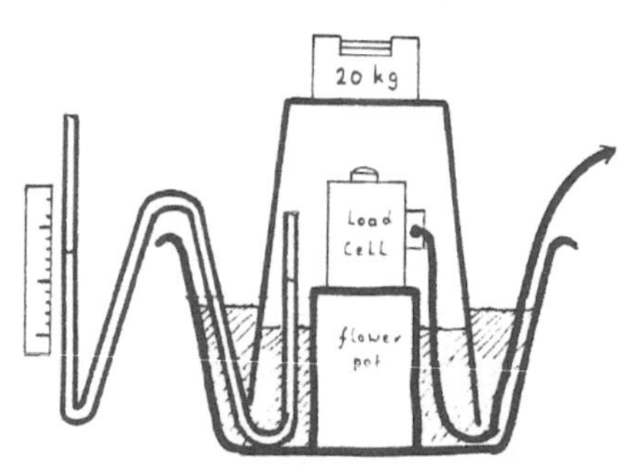

2-6 Torque wrench tester

Due to the continually growing demand for calibration of torque wrenches, a special calibration facility was built in 1985. Basically it consists of a balance and a levelling system. The torque wrench under test is connected to the beam. The handle is connected to a levelling system in the form of an almost vertical lever that can be deflected by a spindle. A load can be applied to either of the two pans, depending on whether the test has to be carried out in clockwise or counter-clockwise direction. After the lever has been levelled by turning the spindle, the moment applied to the torque wrench is known by the weight on the pan and the length of the lever. The length of this lever is 1,019 m, so 1 kg on the pan results in 10 Nm on the wrench.
The measuring range is 10 Nm - 1000 Nm; the accuracy is better than 0,5 %.

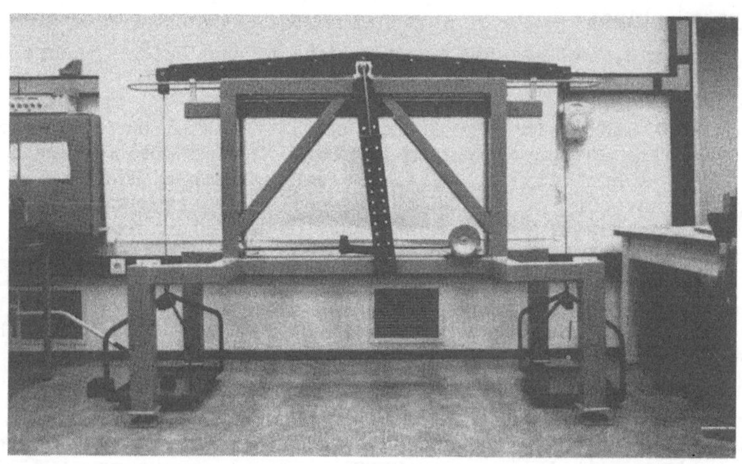

3 CALIBRATION PROCEDURES

Unless otherwise agreed upon with the owner of the device under test, the following measuring procedures are common practice at the VSL:

3-1 Force procedure

The measuring procedure of load cells for force measurement is closely related to DIN 51 301, paragraph 5; the way of calculating the results is slightly different. Recently a BCR sponsored audit took place throughout the EEC. At the time this paper was written, the results were not available yet.

3-2 Torque wrench procedure

For calibrations of torque wrenches, we developed a procedure as shown in the following diagram:

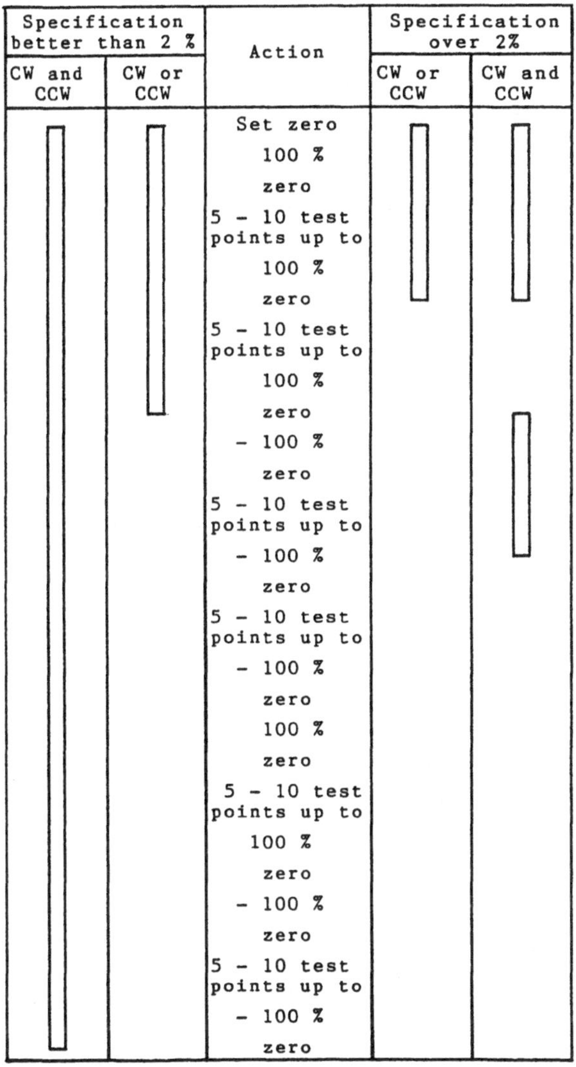

For wrenches with a specified accuracy of 2% or more the shorter procedure can be used. But if:

- the greatest difference of the zero readings is more than half a scale division

  or

- the greatest difference between both 100 % readings in each direction is more than 1% of the full scale

the test has to be continued as for wrenches with specified accuracy of 2 % or less.

CW = Clockwise
CCW = Counterclockwise

100 % = first direction
- 100 % = second direction

Specification better than 2 %		Action	Specification over 2%	
CW and CCW	CW or CCW		CW or CCW	CW and CCW
		Set zero		
		100 %		
		zero		
		5 - 10 test points up to		
		100 %		
		zero		
		5 - 10 test points up to		
		100 %		
		zero		
		- 100 %		
		zero		
		5 - 10 test points up to		
		- 100 %		
		zero		
		5 - 10 test points up to		
		- 100 %		
		zero		
		100 %		
		zero		
		5 - 10 test points up to		
		100 %		
		zero		
		- 100 %		
		zero		
		5 - 10 test points up to		
		- 100 %		
		zero		

4 SOME PRACTICAL ASPECTS OF TYPE TESTING LOAD CELLS

4-1 Deviations from IR-60

Since Jan 1st 1986, IR-60 has been incorporated to a great extent into the national Dutch regulations for weighing machines as far as component-approvals for load cells are concerned. For weighing machines which are tested and approved in their entirety, IR-60 is of no interest.
The most important deviations from IR-60 are:
- The 4 h creep test is not carried out, as in our opinion the zero return test is more severe. Therefore we have not invested in facilities for the creep test on load cells over 550 kg.
- A humidity test as already described in 2-2-4 is carried out on non-hermetically sealed load cells.
- The VSL performs an overload test: if a "safe overload limit" is specified, the load cell is loaded to this value during 5 minutes. After this overload, the zero-output and the characteristic are compared with the values found before. In practical applications however, a load cell should not be overloaded within the measuring range of the weighing machine.
- If the temperature test is carried out on the 2,5/25 t machine, there is a waiting time of 5 min at full load and zero. This is in the program in ROM and cannot be changed easily.

4-2 Precautions during temperature tests

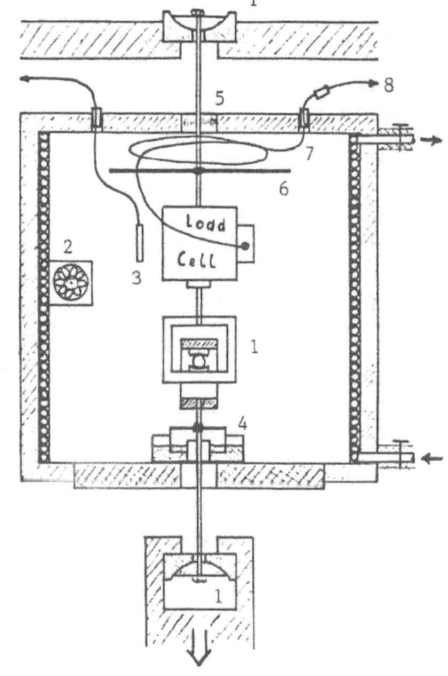

The illustration shows a typical example of a temperature test of a tension type load cell in the qubic temperature cabinet.
In this set-up, the load cell is mounted on bearings (1) to prevent side-loads.
The 220 V fan (2) is running at 110 V in order to keep the dissipation low as well as to minimize unwanted forces due to the "wind".
The ambient temperature is measured (3), rather than the temperature of the specimen itself.
The feed-through of the lower tension rod is a silicon-oil filled labyrinth (4) which prevents cold air from "falling" out of the gap.
The gap (5) in the top is closed simply with a tissue as the influence of side forces at this spot is negligible.
To prevent a temperature gradient caused by heat conduction through the upper tension rod, there is an aluminium or copper disk (6) which acts as a heat-sink, to ensure that the rod has the ambient temperature.The cable (7) of the load cell (usually a four wire system) is in the temperature cabinet. Only a few centimeters and the plug to the extension cable are on the outside. This extension cable (8) to the indicator is a six wire system. The bottom of the cabinet is made of Celoron for the test of compression type load cells.

Table 1: Specifications of the 550 kg and the 2,5/25 t machines

Machine	capacity	550 kg	2,5 t	25t
	accuracy	$2 * 10^{-5}$	$5 * 10^{-5}$	$2 * 10^{-5}$
	repeatability	$2 * 10^{-5}$	$2 * 10^{-5}$	$2 * 10^{-5}$
	built by, in	VSL, 1979	Molenschot, 1968	
	load increments	any multiple of 10 kg up to 90 kg	25/50/100/ 200/500 kg	0,25/0,5/ 1/2/5 t
	time interval	0 - 99 s	0 - 99 s	
Masses	material	stainless steel	mild steel	
	coating	-	black varnish	
	number & weight	110 * 5 kg	48 * 50 kg + 4 * 25 kg	
	accuracy	$1 * 10^{-5}$	$1 * 10^{-5}$	
	cal. interval	4 years	4 years	

Table 2: Specifications of the indicators

make	Servo Balans	Servo Balans	HBM
type	350	380	DMP-39
built in	1976	1978	1980
exitation	10 V dc	10 V dc	1-15 V 225 Hz
resolution	100 000	1 000 000	1 000 000
for	0,5/1/1,5/2/ 2,5/3/3,5/4mV/V	1/2/3/4 mV/V	2/20 mV/V

REFERENCES

1. Organisation Internationale de Metrologie Legale, International Recommendation No 60, Paris, 1985
2. International Electrotechnical Commission, Publication 68-2-30, Geneva, 1968

A GROUP VERIFICATION SYSTEM FOR FORCE STANDARD MACHINES USING HIGH PRECISION LOAD CELLS

Takuro TOJO, Hiroshi MAEJIMA, Junsaku ISHINO and Namiteru HIDA

National Research Laboratory of Metrology
1-1-4, Umezono, Sakura-mura, Niihari-gun,
Ibaraki, 305 Japan

1. INTRODUCTION

In the National Research Laboratory of Metrology (NRLM), a group of the force standard machines of three types, that is dead weight, multiple lever and hydraulic types, are arranged in series corresponding to the magnitude of force and the accuracy of its embodiment, and the standard of force is supplied to industrial and scientific fields.

In general, the accuracy of the magnitude of force generated by standard machines has been evaluated by the determining that of measuring the mass of built-in weights and the ratio of magnification or attenuation of levers and the like.

However, as the various performances of force sensors as well as the resolution have heightened, differences have often been found in the results of the calibration of force sensors according to the standard machines used, even though the fundamental performances of respective standard machines have been sufficiently confirmed. It is considered that this is due to combined effect of force transmitting characteristics of the loading mechanism of the standard machines and detecting characteristics of the force sensors arising when the force is applied to the sensors.

The authors have developed a group verification system, by which the performances in the actual state of use of a group of the force standard machines arranged installed as the national standards up to 15 MN are evaluated. In the system, the intercomparison technique with use of a group of high precision load cells is applied. This paper described details of this group verification system and the results of the calibration of the force standards in Japan.

2. TYPE OF FORCE STANDARD MACHINES AND ABSOLUTE CALIBRATION METHOD

The standard of force can be realized by weights with accurately known mass and the gravitational force acting on them. Accordingly, the force standard machine of any type has the weights, of which the masses are precisely adjusted. The mass of a weight needed for applying a force of definite value F is given by eq.(1).

$$m=\frac{F}{g_l\left(1-\frac{\rho_a}{\rho_m}\right)} \qquad (1)$$

Where, g_l : local acceleration of gravity
ρ_a : density of air
ρ_m : density of weight.

Wieringa, H. (ed), Mechanical Problems in Measuring Force and Mass.

2.1 Dead weight type

The standard machines which can generate the force of precise value only by weights and the gravitational force acting on them are dead weight force standard machines (D.W.M). As for the standard machines of this type, the absolute calibration can be carried out by measuring accurately the mass of the built-in weights and the acceleration of gravity at the place of installation. The accuracy is the highest among the force standard machines, and they are called primary standard machines. However, when their capacity becomes large, the machines becomes also large, and there is such defect that the measurable range of force is dependent on the steps of the built-in weights.

In NRLM, there is a series of the D.W.M with the capacity of 20, 54 and 540 kN, which have the similar mechanisms as shown in Fig. 1. For loading, the oil jacks at the top of the standard machine, which support the built-in weights, are moved up and down, and thus the desired weights are successively added and removed. The specifications of the respective standard machines are shown in Table 1 in the lump together with those of the standard machines of other types.

Table 1 Specifications of force standard machines

Type / Specification	Dead weight machine			Multiple lever machine					Hydrauric machine	
Maximum Capacity (kN)	20	54	540	2	20	100	360	1000	5500	20000*
Lower Limit (kN)	1	9.8	29.4	0.1	0.5	5	10	50	98.0	100
Interval (kN)	1	4.9	4.9	0.01	0.05	1	10	10	98.0	100
Member of transmission				2	3	4	5	3	1	1
Transmission ratio (1:a)				10	100	200	500	1000	500	1000

* : Compression only

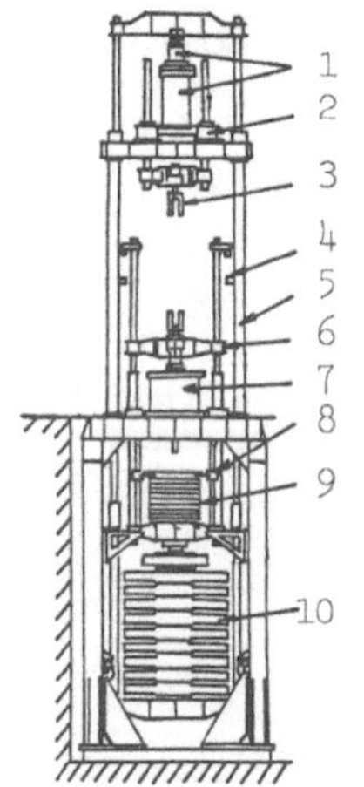

1:Oil-jacks
2:Tension crosshead
3:Tension rig
4:Lifting block
5:Lifting frame
6:Loading frame
7:Compression table
8:Lifting block
9:Small weights
10:Large weights

Fig.1 Structure of dead weight force standard machine

Fig.2 Calibration method of lever ratio using standard weights

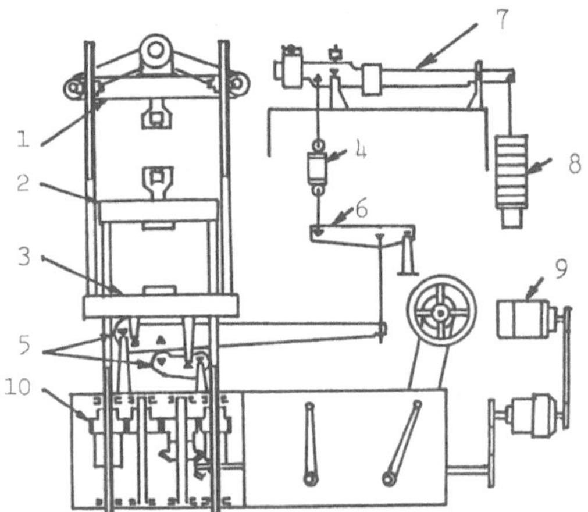

1; Tension crosshead
2; Loading crosshead
3; Compression table
4; Monitoring load cell
5; 1st lever
6; 2nd lever
7; 3rd lever with scale graduation
8; Built-in weights
9; Motor and loading speed controler
10; Transmission gear

Fig.3 Structure of multiple lever type force standard machine

2.2 Multiple lever type

The standard machines which generate large forces precisely by magnifying smaller standard force utilizing the principle of levers are of lever type. Those with a single lever arm are generally used, but when a number of levers are combined, the standard machines with large transmission ratio can be attained. The latters are multiple lever type force standard machines (M.L.M)/1/.

Since the built-in weights are relatively small in the mass as compared with the magnitude of force to be measured and are easy to operate the multiple lever type force standard machines have many features such that conversion of the unit systems, for example, from the gravitational unit to SI unit can be made easily and calibration of force sensors can be done at an arbitrary force up to the weighing capacity.

It is impossible to determine the combined lever ratio of the multiple levers in the built-in state by measuring the length of respective lever arms, that is, the distances between fulcrum, power application point and load application point. In practice, weights are loaded on a table as shown in Fig. 2, and being balanced with the built-in weights of the standard machine, the sensitivity and lever transmission ratio are measured.

In NRLM there are a series of the multiple lever type machines with the maximum capacities of 2, 20, 100, 360 and 1000 kN , and an example of the mechanism is shown in Fig. 3.

2.3 Hydraulic type

For setting up the standard of large force exceeding 1 MN, a pair of large and small ram-cylinders are connected hydraulically, and based on the principle of a hydraulic press of Pascal, the force applied to the small cylinder with a weight is magnified by the ratio of the cross-sectional areas. The large force is thus precisely generated with use of the hydraulic force standard machines (H.M)/2/.

In the standard machines of this type, by producing both the ram-cylinder mechanisms in geometrically similar shape, the elastic deformation of the ram-cylinders accompanying pressurization can be maintained at a certain proportion, and the ratio of the effective cross-sectional areas of the ram-cylinders is kept always constant.

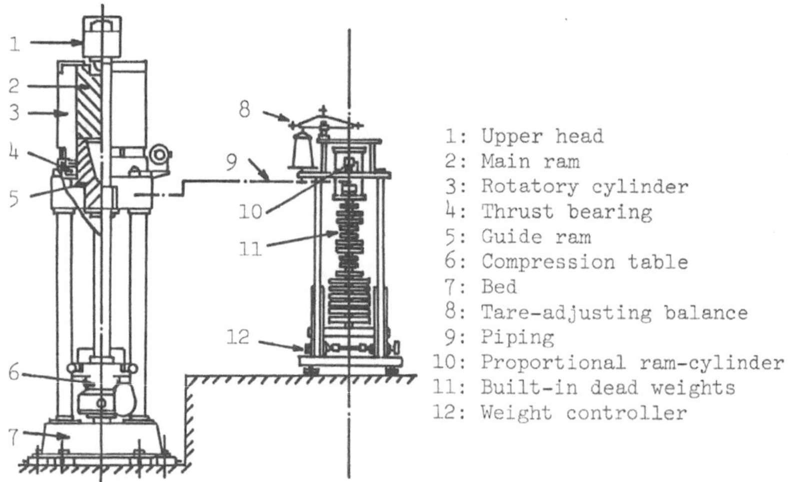

Fig.4 Structure of hydraulic force standard machine

The accurate measurement of the forms of rams and cylinders is the determinant factor for the transmission ratio.

In NRLM there is a series of hydraulic force standard machines with the maximum capacities of 5.5 and 20 MN, and the mechanism is shown in Fig. 4.

3. INTERCOMPARISON METHOD

3.1 Necessity of intercomparison

In the case where the plural number of force standard machines are used, differences are sometimes found in the output of a force sensor calibrated by the standard machines, even if the built-in weights of respective force standard machines and the transmission ratio have been sufficiently calibrated. It had been considered in the past that the dead weight machines were the most accurate among the force standard machines without any doubt, but recently it has been clarified that there are some problems.

Besides the errors due to the mechanism of standard machines themselves, the errors due to the misalignment of the loading axis of the standard machines and the principal axis of force sensors and the initial condition when a force is applied are to be taken in account. These errors are caused by the parasitic components of the force standard machines mixed with the geometrical errors of force sensors and the errors of setting the force sensors on to the standard machines /3/.

Instead of evaluating individually the performances of the standard machines simply from the measured values of the mass of the built-in weights and the transmission ratio, the group verification system has been developed. In the system, the accuracy of setting force standard is determined on the whole by intercomparison methods using a group of high precision load cells and the performance in the actual state of use including the characteristics of the loading mechanism of a series of force standard machines can be evaluated.

3.2 Method of intercomparison

As the method of intercomparison, the force standard machines shown in Table 1 were divided into four groups as shown in Fig. 5 according to their capacities, and the intercomparisons were carried out using 16 load cells.

Group 1 : On the basis of the 20 kN D.W.M, in the loading range up to 20 kN for the 20 and 100 kN M.L.M, the intercomparison was carried out by using the load cells of 10 and 30 kN (10kN×3 parallel built-in type).

Group 2 : On the basis of the 54 and 540 kN D.W.M, in loading range up to 100 kN for the 100 and 360 kN M.L.M, the intercomparison was carried out by using two load cells of 50 kN and the one of 100 kN (30kN×3 parallel built-in type).

Group 3 : First, on the basis of the 540 kN D.W.M, in the loading range up to 300 kN for the 360 and 1000 kN M.L.M, the intercomparison was carried out by using six load cells of 300 kN. Next, in the loading range above 300 kN for the 1000 kN M.L.M, the correction value was obtained by the build-up procedure of load cells described above, because the calibration of lever transmission ratio with use of weights can be made up to about 500 kN at the maximum, and further more, the intercomparison was carried out by using the 5.5 MN H.M and 1 MN load cells /4/.

Group 4 : The 5.5 and 20 MN H.M were compared each other up to 15 MN by the build-up procedure of three load cells of 5 MN.

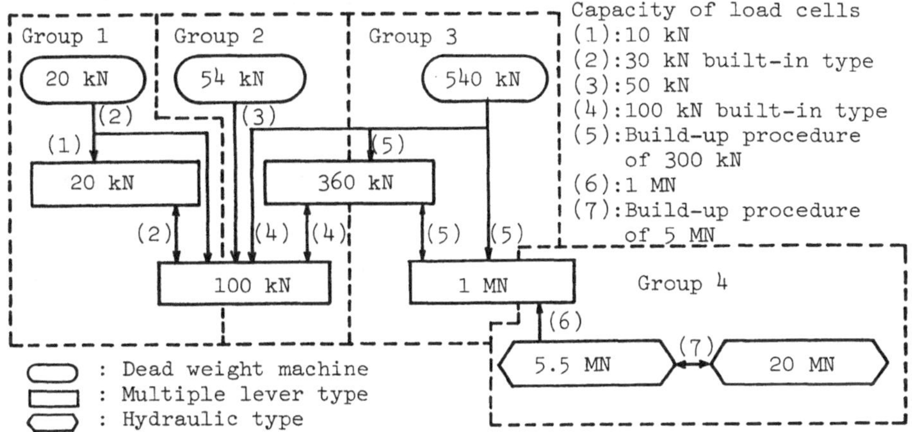

Fig.5 Block-diagram of the group calibration system

When the intercomparison is carried out with load cells, it is necessary to sufficiently grasp fundamental characteristics of the respective load cells. In order to obtain the highly accurate results of intercomparison, it is important to control the time when the respective standard machines are adopted for loading.

In case of the dead weight force standard machines, loading is carried out stepwise as shown in Fig. 6(a), and in case of the lever type and hydraulic force standard machines, loading is carried out at a definite speed as shown in Fig. 6(b), and the target force is thus attained. For the whole measuring system, sharing of time for measurements as shown in Fig. 6(c) was taken, and the time after the attainment of the target force to the measurement was controlled as shown in Figs. 6(a) and (b).

For examining the characteristics of a standard machine and the individual load cells used during loading, the measurement was carried out at the rotational positions of 90° intervals. The preloading for these measurements was carried out once. The comparative measurements were carried out in both directions of increasing and decreasing load and repeated three times, and these measurements were duplicated twice.

Table 2 Specifications of the load cells

Specification	Load cell						
Rated Capacity (kN)	10	30	50	100	300	1000	5000
Rated Output (R.O;mV/V)	2.0	3.0	3.0	3.3	2.5	3.0	2.0
Bridge Resistance (Ω)	350	350	700	700	700	700	700
Non-linearity (%R.O)	<0.022	<0.005	<0.023	<0.010	<0.014	<0.007	<0.102
Hysteresis (%R.O)	<0.026	<0.005	<0.002	<0.006	<0.012	<0.010	<0.017
Repeatability (%R.O)	<0.005	<0.004	<0.003	<0.003	<0.002	<0.003	<0.004
Excitation Voltage (V)	12	12	20	30	30	30	30

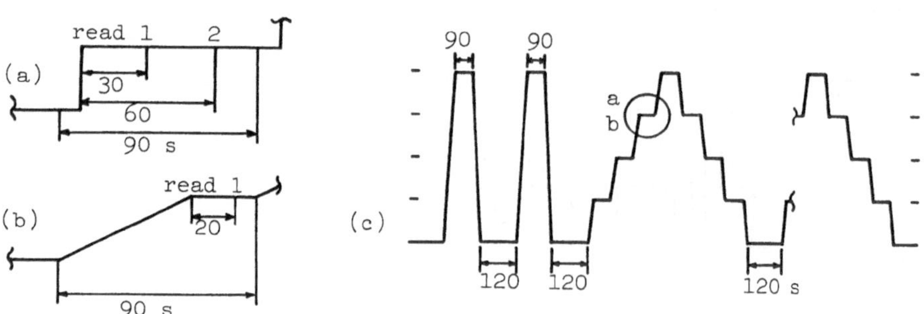

Fig.6 (a) Loading condition of dead weight machine. (b) Loading condition of lever type and hydraulic machine. (c) Time-table for intercomparison method.

Moreover, temperature compensation circuits are built in the load cells but their characteristics have not yet been elucidated. The temperature in the installation room of force standard machines was therefore kept at 23°C. All the load cells used for the intercomparison were specially designed and ordered by NRLM, and their specifications and performances are shown in the lump in Table 2.

4. RESULTS OF INTERCOMPARISON

As shown in Fig. 5, the comparative experiments were carried out so that a closed circuit was formed among the standard machines in the respective groups. To make the understanding easy, the loading characteristics of the respective standard machines were expressed as the difference from referred to the output of the load cells for the dead weight force standard machines or from that for 5.5 MN hydraulic force standard machine.

The results of calibration of the 20 kN M.L.M are shown in Fig. 7. In the figure, the abscissa gives the values for the 20 kN D.W.M, and the relative deviation is given by the percentage in relation to the applied force. The uncertainty of the intercomparison is estimated to be within $\pm 5.0\times 10^{-5}$.

The results for the 100 kN M.L.M are shown in Fig. 8. For the load cells of 30, 50 and 100 kN, the differences from the values respectively for the 20, 54 and 540 kN D.W.Ms were taken. It can be understood that in the smaller load range, the indication becomes more unstable. The uncertainty of the results of intercomparison is $\pm 3.0 \times 10^{-5}$ in the case of the 50 kN load cell, and $\pm 3.5 \times 10^{-5}$ in the case of the 100 kN load cell.

The results for the 360 kN M.L.M are shown in Fig. 9. The intercomparison was carried out by using six load cells of 300 kN capacity. Three of them are the shear spring element type and data are shown by the marks ○ , ◎ and ● , and others are the type combining column and cylinder spring elements being shown by the marks × , + and * . Both types have nearly the same height in their bodies, but the difference is thought to arise from the form of spring elements.

As for all the load cells, the differences from the values for the 540 kN D.W.M are indicated. The uncertainty of the results of intercomparison with use of individual load cells was $\pm 2.5 \times 10^{-5}$, and the uncertainty of the results of intercomparison with use of all six load cells was $\pm 4.5 \times 10^{-5}$.

The results of calibration of the 1 MN M.L.M are shown in Fig. 10. In the range up to 300 kN, six load cells of 300 kN were separately used, similarly as the above, and the mean values of the differences from the 540 kN D.W.M were shown.

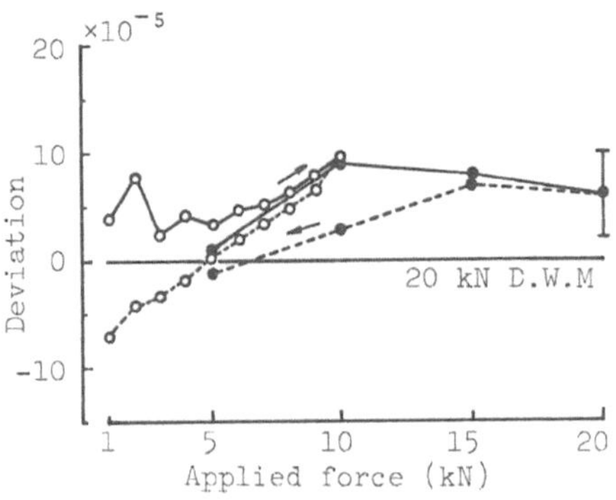

Fig.7 Relative deviations of 20 kN multiple lever machine

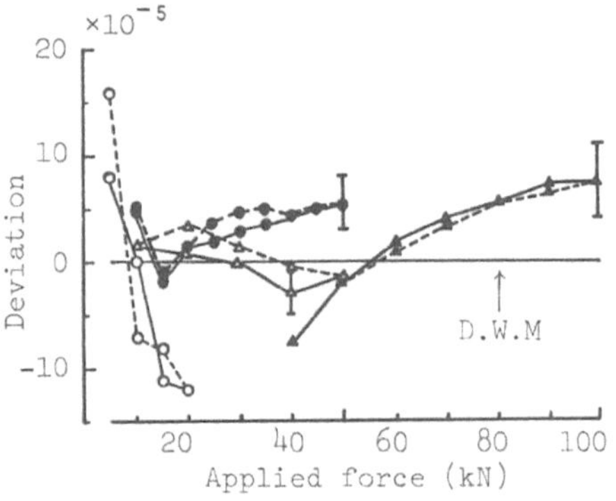

Fig.8 Relative deviations of 100 kN multiple lever machine

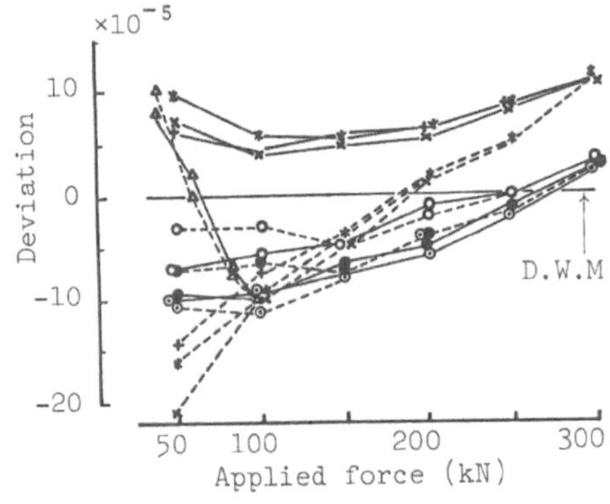

Fig.9 Relative deviations of 360 kN multiple lever machine

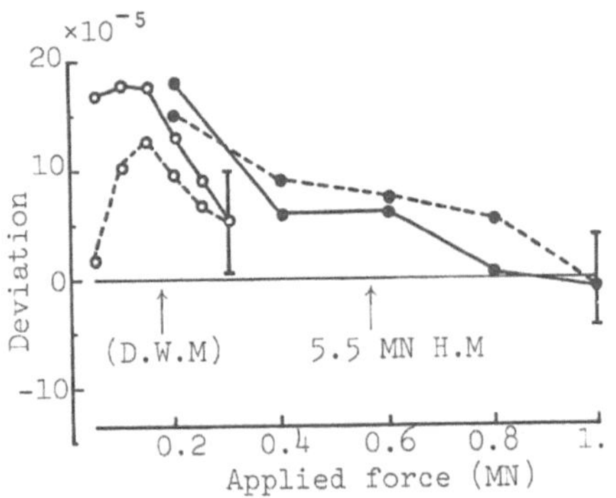

Fig.10 Relative deviations of 1 MN multiple lever machine

In the range exceeding 300 kN and up to 900 kN, the calibration was carried out by the build-up procedure of each type of three load cells, and with reference to the 540 kN D.W.M, and further more, a 1 MN load cell was also used and the difference from the 5.5 MN H.M was shown. This was to avoid lowering of the accuracy of intercomparison by the accumulation of the errors in respective stages when plural stages were piled up. The uncertainty of the results of intercomparison was $\pm 5.0\times 10^{-5}$.

The results for the 20 MN H.M are shown in Fig. 11. For the range up to 5 MN, three load cells of 5 MN were individually used, and in the range from 5 to 15 MN, the intercomparison was carried out by the build-up procedure, and the difference from the 5.5 MN H.M was shown. The uncertainty of the results of intercomparison is $\pm 3.0\times 10^{-5}$.

Since the variations of the load cell output for the different rotational positions of the load cells depended on the standard machines and load cells, a general evaluation can not be derived for this installation effect. For examples, the results for the 540 kN D.W.M and the 100 kN M.L.M using the 100 kN parallel built-in type load cell are shown in Fig. 12.

Finally, the whole of the results of the intercomparison measurements described above are synthetically evaluated, and as seen in Fig. 13, it was proved that the standard of force in every loading range has been established within $\pm 10\times 10^{-5}$.

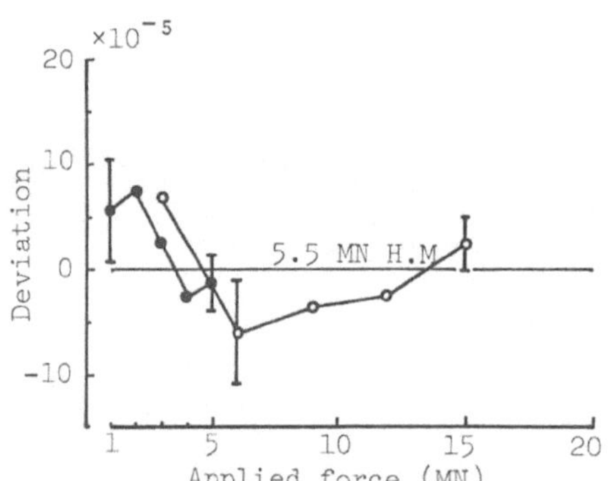

Fig.11 Relative deviations of 20 MN hydraulic machine

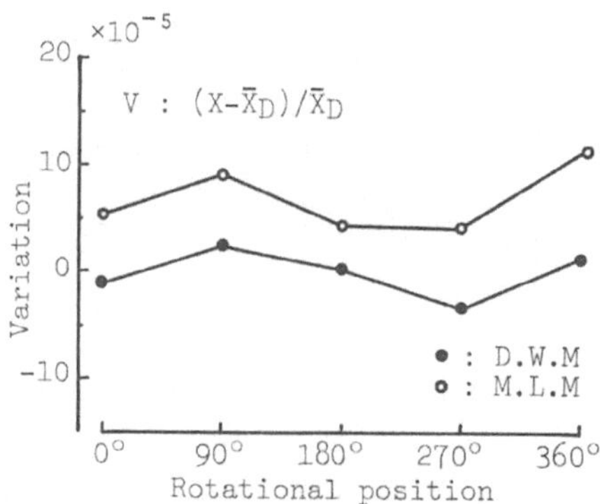

Fig.12 Relative deviations of a load cell caused by the rotational effect with different standard machines

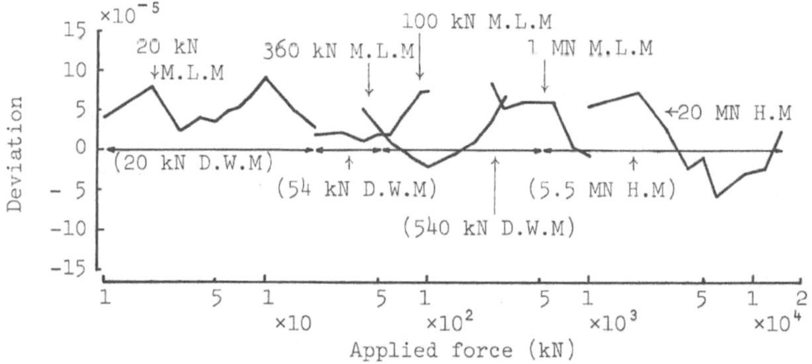

Fig.13 Estimated accuracies of national standards of force up to 15 MN

5. CONCLUSION

When the standard of force is set up by installing a number of the force standard machines with different measuring ranges, the intercomparisons of them are very useful to verify mutually the accuracies of the force standard machines in their running state, as well as the performance of the loading mechanism. These evaluation can not be made by merely measuring the mass of the built-in weights, and by measuring the magnification ratio for the types of having the magnifying mechanism.

The precision of the intercomparison measurements was better than $\pm 5.0\times 10^{-5}$ in every case. It was confirmed that the force standard machines for which the intercomparison has been carried out are capable to the standard of force with the accuracies better than $\pm 10\times 10^{-5}$ in all cases.

To evaluate the performance of force standard machines, many problems remain, such as the elucidation of paracitic components which have been discussed in the 10th IMEKO TC-3 Kobe Conference. In effect, performance of lever type force standard machines is verified by placing a large number of weights on the table to balance with the built-in weights, spending a lot of expense and time, and that of hydraulic force standard machines with large capacity is likely impossible to be confirmed. On the contrary, the verification can be made with high accuracy and in a short time by the intercomparison using high precision load cells.

In Japan, two public organizations besides NRLM who conduct the calibration of force sensors and also the manufacturers of force sensors have their own force standard machines. By utilizing the intercomparison technique developed here, the transfer system for supplying the national standard to these working standard machines in the state of actual use and with high accuracy was completed.

REFERENCES

1. T.Tojo, H.Maejima, N.Hida and M.Murata : Structure and Performance of Newly Developed Compound Lever Type 360 kN Force Standard Machine, Proc. IMEKO TC-3 Kobe Conf. (1984) (to be published)
2. N.Hida, T.Tojo, H.Maejima and M.Murata : Performance of Newly Developed 20 MN Hydraulic Force Standard Machine, Proc. IMEKO TC-3 Kobe Conf. (1984) (to be published)
3. M.Murata, T.Tojo and N.Hida : Error Due to Paracitic Components in Force Standard Machine and Its Detection, Proc. of Round-Table Discussion on IMEKO TC-3 Kobe Conf. (1984) (to be published)
4. T.Tojo, H.Maejima and M.Murata : Calibration Method by a Build-Up Procedure for Large Forces, Proc. Autumn Meeting of Japan Soc. Precision Engg. (1981) 582-584, (1982) 713-715

CALIBRATION AND CLASSIFICATION OF FORCE TRANSDUCERS BY USING A TRACEABLE SET OF HIGH-PERFORMANCE FACILITIES

P.J. BRANDENBURG, J.M. APPIJ, J. LOOMANS AND H. WIERINGA
(TNO-IWECO, DELFT)

SUMMARY
A description is given of the calibration procedure as derived and followed by TNO-IWECO when calibrating force transducers. From the measuring results some parameters are being derived by which the transducer is classified. The classification scheme is also presented. A short description is given of TNO-IWECO's facilities for carrying out the calibrations. Attention is being given to the system which has been set up for securing the traceability of the used facilities. In this respect special reference is made to results of intercomparisons which have been carried out. In particular the results of an intercomparison between PTB and TNO in the case of three transfer standards are given. Two transfer standards which are of the build-up type are used for linking force generating machines to the 550 kN deadweight standardizing machine of TNO-IWECO. They can also be used for calibrating force transducers by positioning them upon the transfer standard in a force generating machine.

1. INTRODUCTION

Force transducers are used in a wide variety of application. In order to use the output results of force transducers with some degree of confidence they need to be calibrated from time to time. The frequency of such calibrations and the facilities to be used depend on the type of application of the transducer. Calibrations are carried out by laboratories which are specialized in this kind of work.
In general calibration work is a kind of work that must not give much room for different interpretations/results or whatsoever. Nevertheless different procedures for calibrating force transducers have been designed and adapted by the various laboratories which are still active in this field.
Some laboratories calibrate the transducers only in an increasing load cycle. Also the number of cycles is different. The influence of the procedure upon the results is now subject of research work within the European Community by BCR (Bureau Communautair de References).
In 1984 TNO-IWECO has adopted a new calibration procedure with a classification schedule. The DIN 51301 was partly used as a basis for the new procedure wich is roughly described in the next paragraphs.

2. CALIBRATION PROCEDURE

Before the calibration is started some general data for defining the transducer and its accessories and the environmental conditions are recorded. It is also necessary that the isolation resistance of the strain gauge force transducer has a value of at least 2000 Mohm. In case the isolation resistance is less the calibration will not be carried out.

2.1 Preload

The transducer will be preloaded three times with its nominal force (F_n). The load will be present during one minute followed by a zero load period of three minutes. The transmission time for loading and unloading may form part of the one and three minutes periods respectively if the transmission times are not longer than 25 s.
The reading will take place at least 30 s after full load or zero load has been applied.

2.2 Load series

The calibration will take place with six load series at four different positions of the transducer in the reference-facility. A load series preferably consists of 10 equally spaced steps of 10 % of the measuring range. Load is applied in an upwards going direction as well as in a downwards going direction. It may happen that a reference facility is not capable of generating 10 equal steps for a certain nominal load. In such cases it is allowed to select another number according to the following criteria:

- 5 for 0,5 and 1 % class transducers
- 8 for 0,25 % class transducers
- 10 for 0,1 % or less class transducers.

A reading will be recorded after the load has been present for at least 30 s. In cases where the loading time will be less than 25 s it is allowed to change the load and to take a reading every minute.
When the transducer has to take another position in the reference facility it is necessary to apply a preload of nominal capacity for a period between 30 and 60 s.
The time span between a preload and a series of measurements or between a series of measurements and a preload must be kept equal.
If the calibration between two series of measurements is interrupted for a period of time longer than 1 hour a preload procedure as described in para 2.1 should be applied before the calibration is to be proceeded. If the interruption time lies between ¼ of an hour and one hour only one preload of nominal capacity should be given. It is not allowed to interrupt the calibration between series 1 and 2 and between 2 and 3.
When applying or removing weights one must pay attention to the fact that each step must be taken in such a way that it coincides with the present direction.

2.2.1 Tension or compression force transmitters. The calibration comprises six series of measurements of rising load: two of them are followed by a series of measurements with falling load. After the first three series of measurements the transducer will be rotated over 90° before each following series of measurements. So the sequence is:

Series			
Series	1.	0° position	rising load
	2.	0° position	rising load
	3.	0° position	rising- and falling load
	4.	90° position	rising load
	5.	180° position	rising- and falling load
	6.	270° position	rising

2.2.2 Tension/compression force transducers, all classes. For this type of transducer the measurement series as described in the preceding paragraph will be split up into two groups of 3 measurement series in turn to be performed during tension and compression. Changing from tension to compression or vice versa two preloads must be applied. The sequence is as follows:

- three preloads (see para 2.2.1)
- series 1, 2 and 3 in compression
- two preloads
- series 1, 2 and 3 in tension
- two preloads
- series 4, 5 and 6 in compression (one preload between each series)
- two preloads
- series 4, 5 and 6 in tension (one preload between each series)

3. DERIVATION OF CALIBRATION PARAMETERS

3.1 Measuring value

From the readings measuring values will be determined for each series of measurements. The measuring value is defined as the difference between the indicator reading at a certain load and the indicator reading at zero load preceding the series of measurements under consideration. The notation for the measuring value will be $x_{n,i}$ for rising load and $x'_{n,i}$ for falling load. (n = series number, i = load)

3.2 Characteristic

From the measuring values of series 1, 4, 5 and 6 an average value $\bar{x}_{v,i}$ will be calculated. These $\bar{x}_{v,i}$ values are then used to derive the transducer characteristic as a first, second or third degree curve running through zero. For transducers of class 0,1 % or less a third degree curve must be used: $F/F_n = A.x + B.x^2 + C.x^3$.

3.3 Relative interpolation error

By using the characteric $F/F_n = f(x)$ a force $F_{a,i}$ can be calculated for each value of $x = \bar{x}_{v,i}$. $F_{a,i}$ differs from the load F_i that resulted in $\bar{x}_{v,i}$. The relative interpolation error is now defined by:

$$f_{r,i} = \frac{F_i - F_{a,i}}{F_{a,i}} \,.\, 100\ \%$$

3.4 Relative spread

In order to have a measure for the spread of the measuring values two different values are introduced, viz. one for the transducer in the same position (b_{rg}) and one for the transducer in different positions (b_{rv}). The repeatability values are calculated from:

$$b_{rg,i} = \frac{x_{n,i}(\max) - x_{n,i}(\min)}{\bar{x}_{g,i}} \,.\, 100\ \%\ (n = 1,\ 2,\ 3)$$

$$b_{rv,i} = \frac{x_{n,i}(\max) - x_{n,i}(\min)}{\bar{x}_{v,i}} \,.\, 100\ \%\ (n = 1, 4,\ 5,\ 6)$$

Where $\bar{x}_{g,i}$ is obtained by averaging the measuring values $x_{n,i}$ for series 1, 2 and 3.

For tension/compression transducers separate values will be calculated for tension and compression.

3.5 Relative hysteresis

The hysteresis of the transducer will be calculated from the differences between the measuring values at falling loads and at corresponding rising loads. For series 3 and 5 a value U_n is calculated for each load according to:

$$U_{n,i} = x'_{n,i} - x_{n,i}$$

An average will then be calculated:

$$\overline{U}_i = \frac{|U_{3,i}| + |U_{5,i}|}{2}$$

The relative hysteresis (u_i) is obtained from:

$$u_i = \frac{\overline{U}_i}{\overline{X}_{v,i}} \cdot 100\ \%$$

3.6 Relative zero shift

The relative zero-shift will be calculated from the indicator readings obtained during the preload tests and obtained during each series of measurements. Values for the zero-shift are calculated as follows:

- during the preload tests the indicator readings before and after the third preload are taken as well as the reading at the third preload

$$Z_o = \frac{\text{reading "before"} - \text{reading "after"}}{\text{reading "full load"} - \text{reading "after"}} \cdot 100\ \%$$

- during the various measurement series the indicator readings at the end of each series and the one at full load are taken. The zero-shift is obtained as:

$$Z_n = \frac{x'_{n,0\ \%}}{x_{n,100\ \%}} \cdot 100\ \%$$

3.7 Number of indicator units

In view of the classification of the transducer after its calibration it is necessary to determine the number of indicator units for the lower limit of the measuring range. This number is obtained as the ratio between the smallest possible change of the reading (resolution). For analogue indicators the resolution is usually 0,2 - 0,1 scale unit. For digital indicators the resolution is the smallest difference between two successive readings of the indicator.

4. CLASSIFICATION

In order for the user to have one single value a classification figure is adopted. For classification use is made of the following parameters:

- minimum number of indicator units for the lower limit of the measuring range
- zero shift, Z
- relative spread at one transducer position, b_{rg}
- relative spread at different transducer positions, b_{rv}
- relative interpolation error, f_r
- relative hysteresis, u

In table 1 the classification scheme is given. For a certain classification it is necessary that the numerical values of all parameters satisfy the requirements. A partial classification is not allowed. The classification at least applies for a part of the measuring range with a certain lower limit. The lower limits depend on the number of load steps that were applied.

5. CALIBRATION FACILITIES

For carrying out calibrations TNO-IWECO has available the following facilities:

- a 2 kN deadweight machine (all classes)
- a 100 kN semi-deadweight machine, TD10 (class $\geqslant$ 0,05 %)
- a 550 kN deadweight machine (all classes)
- a 1 MN hydraulic machine (class $\geqslant$ 0,5 %)
- a 5 MN hydraulic machine (class $\geqslant$ 0,5 %)
- a 1,65 MN and a 4,95 MN transfer standard (class $\geqslant$ 0,1 %)

For specifying the uncertainty of the various facilities extensive measurements have been carried out in the past and are still in progress. Based on the results the facilities have been made traceable to each other. The traceability system is presented in figure 1. From this figure it can be seen that quite some of our facilities are linked to our 550 kN deadweight machine. For a schematic drawing of this machine see figure 2. To determine the uncertainty of this device the weights have originally been calibrated by the Netherlands Metrology Service. The machine has been completed in 1967. Thereafter the machine has taken part in intercomparisons within the EEC. Results of the most recent intercomparison measurements are presented in figure 3. Based on these results the uncertainty of the machine is specified as 2.10^{-5}.

Two more of our facilities will be discussed in more detail, viz.

- two transfer standards with capacities of 1,65 MN and 4,95 MN
- a 100 kN semi-deadweight machine

Transfer standards have been designed and built by TNO-IWECO for two main purposes:

- to make force generating machines with capacities up to 5 MN traceable to the 550 kN deadweight standardizing machine
- to calibrate force transducers by placing the transducer itself on top of the transfer standard. Both units are then placed in the force generating machine.

Both transfer standards are designed according to the build-up principle and consist of 3 or 9 force transducers with a capacity of 0,55 MN each. The designs of the force measuring elements and the associated electronics were made in such a way that the full calibration of both standards was thought to be carried out by calibrating the individual 0,55 MN elements in the 550 kN deadweight-machine and by averaging the results.

For a cross-section of the 4,95 MN unit reference is made to figure 4. After having completed the standards and after having done the calibrations in the 550 kN machine PTB at Braunschweig was so kind as to make available their 16,5 MN hydraulic standard machine. The uncertainty of this machine is given as 3.10^{-4} (1)

The results of the calibrations done at TNO and at PTB are summarized in table 2. From these results the following conclusions can be drawn:

- the differences between the PTB and TNO calibration results are less than 0,01 % (1.10^{-4}) for loads above 1500 kN
- with reference to the uncertainty of the 550 kN deadweight machine (2.10^{-5}) the uncertainty of the 4,95MN was specified as 2.10^{-4}
- the differences between the two calibrations carried out at TNO are indeed very small (2.10^{-5} of end value)
- in view of the observed differences it is justified to have the calibration of the individual elements done on the 550 kN machine.

Another facility of interest is a 100 kN semi-deadweight machine. A drawing of the machine is given in figure 5. Load is applied via a lever (ratio 1:10). All supports are of the knife-edge suspension type. In order to adjust the correct lever ratio the knife-edge suspension of the weight can be shifted. Care has also be taken of the fact that the edges of all three knives on the lever are in the same plane. As a result of this the sensitivity of the machine is independent of the load. At the longer end of the lever there is a vertical rod with dishes. On these dishes 9 disc-shaped weights of each 100 N and 9 weights of 1000 N can be placed. The force transducer is mounted between the two clamps at the short side of the lever. By means of a special cage, that can be mounted between the two clamps, also compression forces can be generated.
For the calibration of the machine two different methods were used, viz.:

* the 9 100 N weights were calibrated by using 5 precision weights of each 2 KN and two precision weights of each 50 N for creating intermediate steps. The uncertainty of the precision weights is 2.10^{-6}
* the 9 1000 N weights were calibrated by using a precision force transducer with a capacity of 100 kN as a transfer standard to the 550 kN deadweight machine.

For the 100 N weights a systematic error was found which could easily be adjusted. Also for the 1000 N weights a systematic error was found that was adjusted.
The results of the calibration carried out with a precision 100 kN force transducer that was made available by the Netherlands Metrology Service are presented in table 3. It can be seen that differences are zero or 1 unit (resolution of the indicator). The results indicate that the uncertainties of this semi-deadweight machine and the 550 kN machine are of the same magnitude, 2.10^{-5}.

REFERENCES

1. Material prüfung 16 (1974), p 165-169
2. Ir. H. Wieringa, "Parallel schakelen van opnemers" april 1983, TNO-IWECO, rapport no 5221002-831 (Dutch)
3. Ir. H. Wieringa, "ontwikkeling 1,65MN en 4,95MN kalibratie standaards", 1983, TNO-IWECO, rapport no. 5221002-83-2 (Dutch)
4. Ir. H. Wieringa, Deadweight standardizing machine 5500 kgf, VDI-Berichte nr. 137, 1970
5. Ir. H. Wieringa, Design of a 1,65 and 4,95 MN transfer standard, based on the "build-up" procedure, Proceedings 10th Int. Conf. of the IMEKO Techn. Comm. TC-3 on Measurement of Force and Mass, Kobe, Japan, Sept. 11-14, 1984, pp 205-208.

Table 1 - classification requirements

Requirements	Class					
	0,025 %	0,05 %	0,1 %	0,25 %	0,5 %	1,0 %
Minimum number of indicator units for the lower limit of the measuring range	10.000	5.000	2.000	1.000	500	500
Zero shift, $z \leqslant$	±0,004 %	±0,01 %	±0,02 %	±0,04 %	±0,01 %	±0,2 %
Relative spread at one position, $b_{rg} \leqslant$	0,01 %	0,02 %	0,05 %	0,10 %	0,2 %	0,4 %
Relative spread at different positions, $b_{rv} \leqslant$	0,02 %	0,05 %	0,10 %	0,20 %	0,3 %	0,4 %
Relative inter-polation error, $f_r \leqslant$	±0,01 %	±0,02 %	±0,06 %	±0,10 %	±0,1 %	±0,4 %
Relative hysterisis, $u \leqslant$	0,03 %	0,05 %	0,15 %	0,30 %	0,3 %	1,0 %

If the requirements for class 1 % are not met, classification will be specified as "class greater than 1 %".

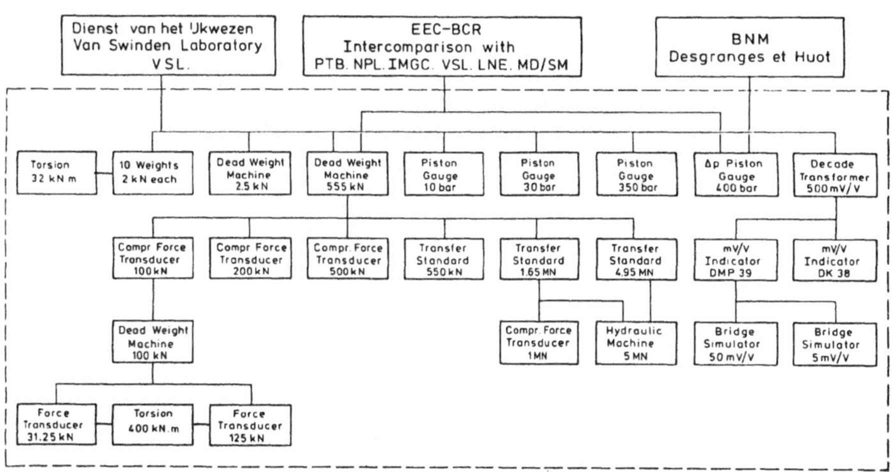

Fig. 1. Traceability system.

Table 2 - Comparison of $\overline{x}_{v,i}$ found at TNO and PTB (4,95 MN unit)

Load	TNO (I)	PTB	TNO (I) - PTB	rel. diff.	TNO (II)	TNO (II) - PTB	rel. diff.
(kN)	nV/V	nV/V	nV/V	%	nV/V	nV/V	%
0	0	0	0	0	0	0	0
500	212150	212188	- 38	- 0,0180	212182	- 6	- 0,0028
1000	424129	424203	- 74	- 0,0170	424146	- 57	- 0,0130
1500	635951	635953	- 2	- 0,0003	635972	19	0,0030
2000	847653	847632	21	0,0025	847665	33	0,0039
2500	1059198	1059177	21	0,0020	1059208	31	0,0029
3000	1270572	1270513	59	0,0046	1270582	69	0,0054
3500	1481734	1481644	90	0,0061	1481751	107	0,0072
4000	1692683	1692573	110	0,0065	1692706	133	0,0079
4500	1903416	1903261	155	0,0081	1903458	197	0,0100
5000	2113945	2113839	106	0,0050	2113985	146	0,0069

Table 3 - Differences of the TD10 and 550 kN machines via a precision force transducer of the Netherlands Metrology Service.

Applied load	Transducer characteristics				Differences	Relative
	$\overline{x}_{v,i}(I)$	$\overline{x}_{v,i}(III)$	$\overline{x}_{v,i}(av)$	$\overline{x}_{v,i}(II)$	$\overline{x}_{v,i}(av) - \overline{x}_{v,i}(II)$	difference
(kN)	(nV/V)	(nV/V)	(nV/V)	(nV/V)	(nV/V)	(%)
0	0	0	0	0	0	0
10	96747	96745	96746	96746	0	0
20	193533	193531	193532	193533	- 1	0,00052
30	290345	290342	290343	290344	- 1	0,00035
40	387183	387175	287179	387179	0	0
50	484023	484017	484020	484021	- 1	0,00021
60	580869	580861	580865	580865	0	0
70	677711	677704	677708	677709	- 1	0,00015
80	774551	774543	774547	774548	- 1	0,00013
90	871388	871380	871384	871384	0	0
100	968230	968223	968227	968226	1	

$\overline{x}_{v,i}$ (I) : characteristic when calibrated in the 550 kN machine.
$\overline{x}_{v,i}$ (II) : characteristic when calibrated in the TD10 machine.
$\overline{x}_{v,i}$(III) : characteristic when recalibrated in the 550 kN machine.
$\overline{x}_{v,i}$(av) : $(\overline{x}_{v,1}(I) + \overline{x}_{v,i}(III))/2$.

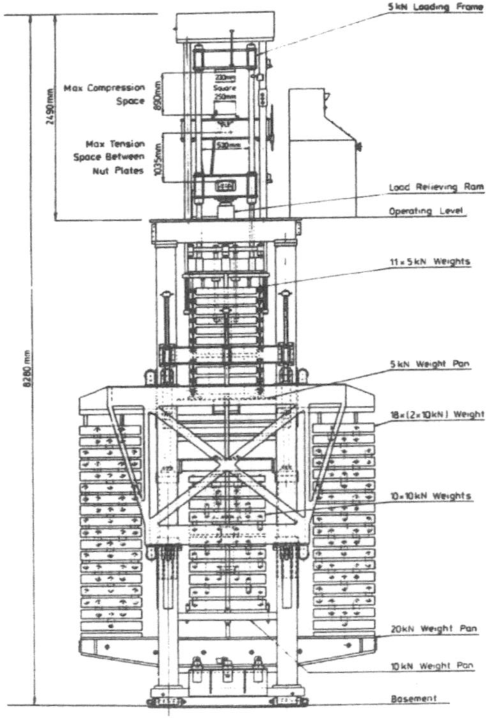

Fig. 2. 550 kN deadweight machine.

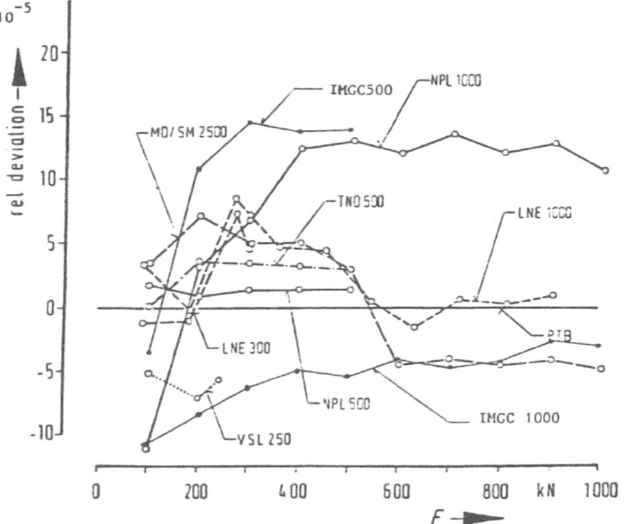

Fig. 3. Results of intercomparison measurements.

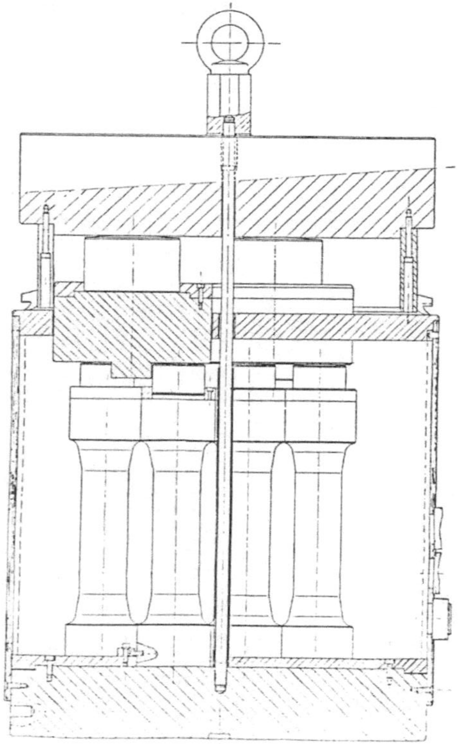

Fig. 4. Design of 4,95 MN transfer standard.

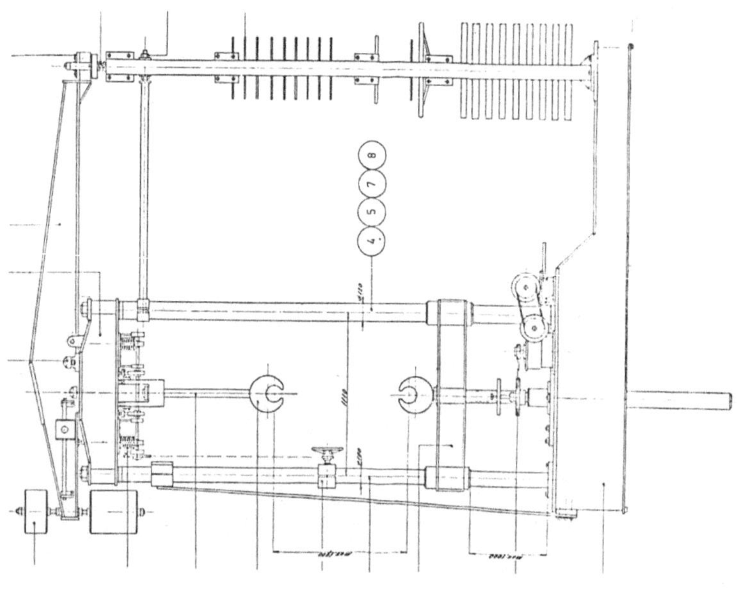

Fig. 5. 100 kN semi-deadweight machine.

ABOUT THE UNCERTAINTY OF THE HIGHER CAPACITY PRIMARY STANDARD AND THE DISSIMINATION OF FORCE IN THE ASMW

M. Dietrich, K. Hasche and D. Peschel

Office for Standardization, Metrology and Quality Control (ASMW), GDR

1. INTRODUCTION

This contribution deals with a qualitative and quantitative discussion of the uncertainty of force generated by the 2 MN dead weight primary force standard (KNME 2 MN) of the ASMW and with the application of tension force transfer standards developed in the ASMW laboratory for force measurements.

The KNME 2 MN was introduced in /1/. Details of construction, an explanation of its function and some information about the uncertainty of forces generated by the KNME 2 MN were given in /2/. One of the main sources which may influence significantly the uncertainty is a magnetic field between the dead weights. The recognition of this influence and taking measures for the reduction of this effect on the uncertainty of generated force are discussed.

Contemporaneous with the KNME 2 MN the force transfer standards on strain gauge basis for forces in the range of 50 kN up to 5 MN have been developed and tested. They are used in connection with the known precision measurement instrument DMP 39. The transfer standards are equipped with three additional measuring bridges for measuring the actual values of torque and two bending moment components in parallel to the tension force interesting in first line. Reports about the transfer standards were given in /3,...,6/. Latest metrological parameters of the transfer standards are explained briefly. The calibration of a lever type 1 MN secondary force standard machine (KNME 1 MN) and the calibration of a hydraulic transmission type 10 MN force standard machine (KNME 20 MN) in the range of force up to 4 MN are explained as an example for their practical use in the ASMW. The latter force standard machines have been described in /1/.

2. THE UNCERTAINTY OF FORCES GENERATED BY THE KNME 2 MN WITH SPECIAL REGARD TO THE FERROMAGNETIC MATERIAL CHARACTERISTICS OF THE DEAD WEIGHTS

The equation

$$F = m.g.(1-\frac{\varrho_L}{\varrho_m}) + \Delta F(MF) + \Delta F(TH) \qquad (1)$$

may serve for a simplified description of forces generated by the KNME 2 MN (see Fig. 1). The symbols in (1) - see also /7/ - have the following meaning: F - generated force, m - mass of dead weights (in some cases mass of scale pan included), g - local gravity, ϱ_L - density of air, ϱ_m - density of dead weight material, $\Delta F(MF)$ - force generated by magnetic field effect between dead weights, $\Delta F(TH)$ - additional force component when scale pan mass compensation by compensating lever is used (esp. in case of generation of forces < 50 kN). When known systematic error components are completely corrected, the random deviation $\Delta F/F_o$ related to the defined value of force

$$F_o = m_o \cdot g_o \cdot (1 - \frac{\varrho_{Lo}}{\varrho_{mo}})$$

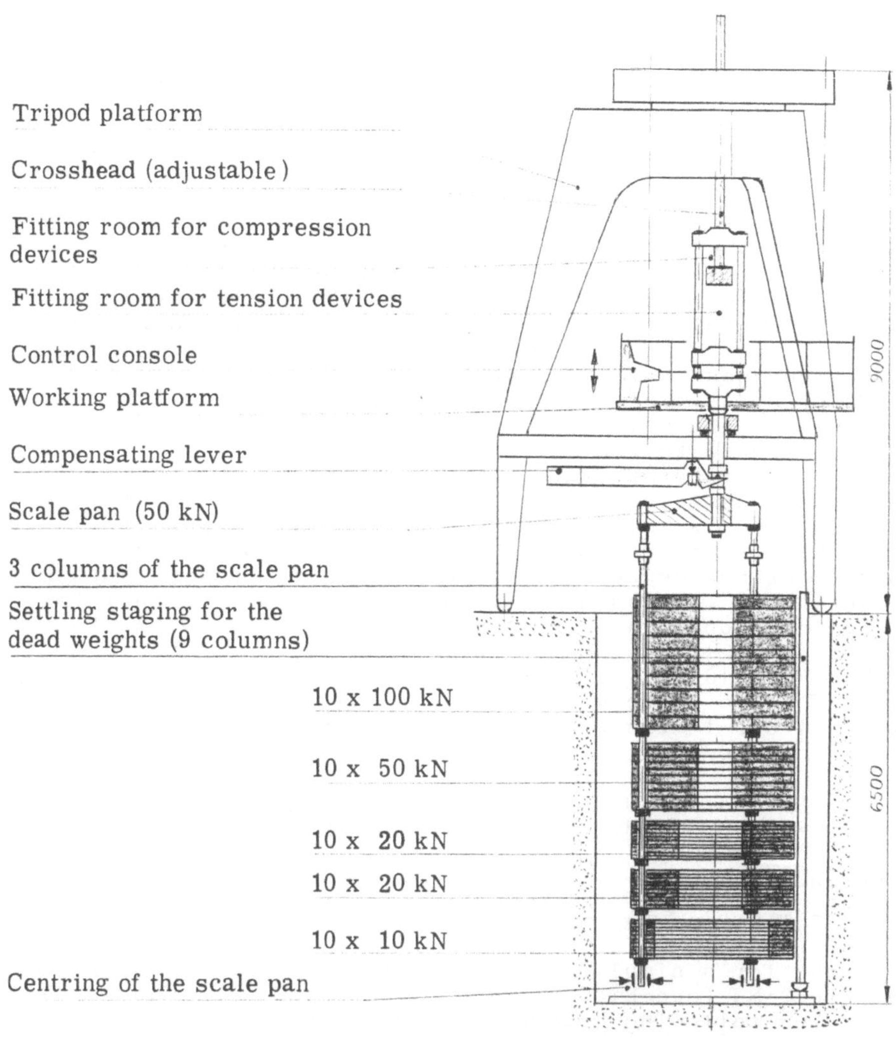

Fig. 1: Deadweight force standard machine KNME 2 MN

follows from the linear component of a Taylor series from (1):

$$\frac{\Delta F}{F_o} = \frac{\Delta m}{m_o} + \frac{\Delta g}{g_o} + \frac{\varrho_{Lo}}{\varrho_{mo}} \cdot \left(\frac{\Delta \varrho_L}{\varrho_{Lo}} + \frac{\Delta \varrho_m}{\varrho_{mo}}\right) + \frac{\Delta F(MF)}{F_o} + \frac{\Delta F(TH)}{F_o} \qquad (2)$$

$\Delta F/F_o$ may be characterized by an estimated variance $\sigma^2(\Delta F/F_o)$. In view of uncertainty of force the worst case of force generation by the KNME 2 MN is that of F_o = 10 kN. The variance σ^2 $(\Delta F/F_o)$ results according to the known mathematical rules for this case from the following values of standard deviations, $\sigma(\Delta F(MF)/F_o)$ excluded:

$$\sigma\left(\frac{\Delta m}{m_o}\right) = 3.10^{-6},\ \sigma\left(\frac{\Delta g}{g_o}\right) \leq 1.10^{-6},\ \sigma\left(\frac{\Delta \varrho_L}{\varrho_m}\right) \leq 2.10^{-6},$$
$$\sigma\left(\frac{\Delta \varrho_m}{\varrho_m}\right) < 1.10^{-6},\ \sigma\left(\frac{\Delta F(TH)}{F_o}\right) \leq 5.10^{-6}$$

Now there will be discussed the magnetic field effect and $\sigma(\Delta F(MF)/F_o)$. The extremely high expenses for dead weights made from austenitic steel and the technological difficulties of production for a high capacity force standard machine are known. Therefore we decided to use for the dead weights a rolled steel BST 3 sp acc. to GOST (made in Soviet Union) instead of austenitic steel.

The conceivable difficulties related to durability and long time drift of the calibrated value of mass have been solved by special corrosion-resistant lacquer coatings and repeated measurements of mass. Tests were also made before the production of the dead weights. But there is no quantitative information about the increase of the uncertainty of force caused by a magnetic field interaction between dead weights settled on scale pan and dead weights settled on settling stagings.

The material of the above mentioned dead weights shows soft magnetic characteristics.

Measurements with test pieces showed a relatively magnetic permeability $\mu_{rel} \approx 75$ and a polarization coercitivity force $I^{Hc} \approx 13$ A.cm^{-1}. Therefore a magnetization resulting from mechanical processing is not surprising. When the magnetic field of the earth is homogeneous a force interaction between the dead weights made from soft magnetic material is not to be expected. Supposing, tangential magnetic field components are neglected - which seems to be permitted - the force interaction resulting from magnetic flux density B is defined by

$$\Delta F(MF) = \oint \frac{1}{2\mu_o} \cdot B^2 \cdot dA.$$

According to this formula a force interaction may appear only in the case when the magnetic flux density integrated over the intervace between dead weight and air in direction to the force F_o is unequal zero.

For the reason of a representative number of dead weight combinations on the upper and lower surface of the dead weights used for the actual generation of force, the actual magnetic flux density had been measured by a Hall probe and a point by point procedure. Measurement results are shown in Fig. 2 for

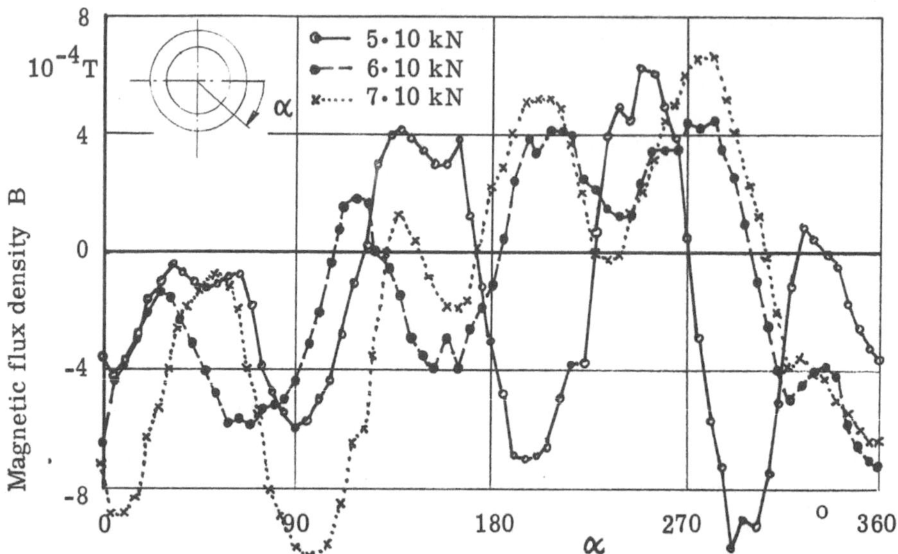

FIGURE 2. Measured magnetic flux density distribution for 3 force steps generated by dead weights of the lowest stack

the case where the forces of 50, 60 and 70 kN are generated by the lowest stack (10 x 10 kN dead weights). The distance between the dead weights is 15 mm.

The diagram shows the mean values of magnetic flux density B (mean in radius direction) as function of the angle α in circumferential direction. The variations of flux density are explainable by borings in the dead weights. But general quantitative conclusions cannot be drawn from measurements on a limited number of dead weight combinations. All interesting combinations must be examined individually. The mean magnetic flux density on the lower surface of a dead weight stack which was distant $\approx$ 200 mm from the next stack proved approximately independent from the number of the dead weights in the stack. The difference of force interaction on the upper and the lower surface of a complete dead weight stack had been estimated by $\Delta F/F_{\,} \leq 1 \cdot 10^{-8}$. Directed to the lowest stack (10 x 10 kN dead weights) the values of systematic force components $\Delta F(MF)$ caused by the magnetic field effect have been analysed by experiment. The minimum distance between dead weights settled on the scale pan and the other dead weights is 15 mm for the lowest dead weight and 39 mm for the 9th dead weight. This is conditioned by construction. Depending on the value of the generated force - e.g. when the other stacks are settled on the scale pan - there is a maximum distance increase of 20 mm caused by the extension of scale pan and deformation of the tested measuring device. The dependence of the force $\Delta F(MF)$ from the distances of the lowest dead weight to the other dead weights of the 10 kN dead weight stack is given in Fig. 3.

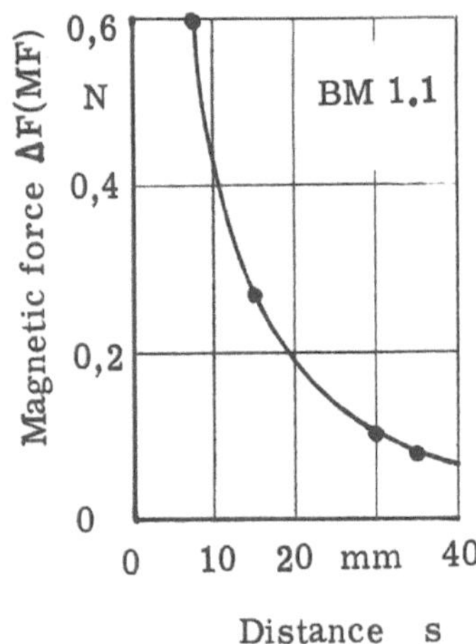

FIGURE 3. Example for the dependence of force generated by magnetic field from the distance between two dead weights

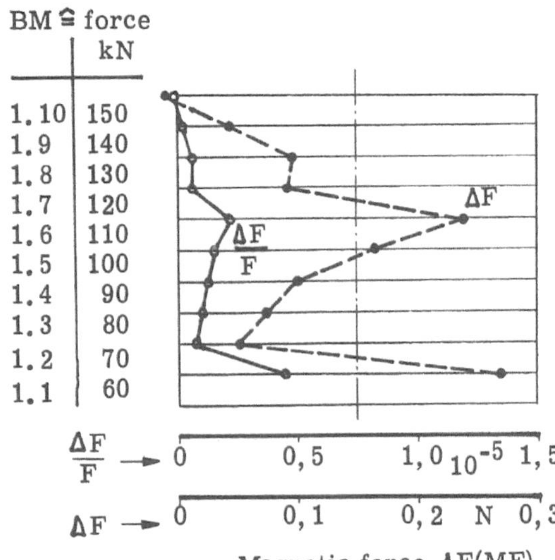

FIGURE 4. Maximum relative systematic error of force generated by magnetic field between 10 kN dead weights

Random variations of the experimental results are nearly neglegible small.

Having made this preliminary experiment the value $\Delta F(MF)$ was measured for all force steps with the lowest stack and for the minimum distance between the dead weights. The results - see Fig. 4 - demonstrate that under the most unfavourable conditions the systematic contributions $\Delta F(MF)/F_o$ are not greater than 5.10^{-4} and mostly smaller than 2.10^{-6}, taking into consideration the force of 50 kN generated by the scale pan mass.

The magnetic field effects only for F_o = 10 kN a component $\Delta F(MF)/F_o > 4.10^{-6}$. In this case a mathematical correction of the measured systematic deviation or a correction by changing the value of dead weight would be possible.

The above mentioned experiments were repeated in a two years time interval. The measurement results related to $\Delta F(MF)/F_o$ corresponded within an interval of $\pm 5.10^{-2}$. From that fact we draw the conclusion that the temporal stability of the magnetic field characteristics seems to be evident. The results have been checked by the measurement of a force F_o generated by different combinations of dead weights. The deviations of the mean for the force F_o = 250 kN are represented in Fig. 5. In the lower part of Fig. 5 is indicated which dead weight combinations have been used. The forces generated by the two complete dead weight stacks (10 x 20 kN) and the scale pan (50 kN) agree within 1.10^{-6}. The two other forces generated by parts of dead weight stacks agree only within 1.10^{-5}. The magnetic field effect reduces the force generated by the dead

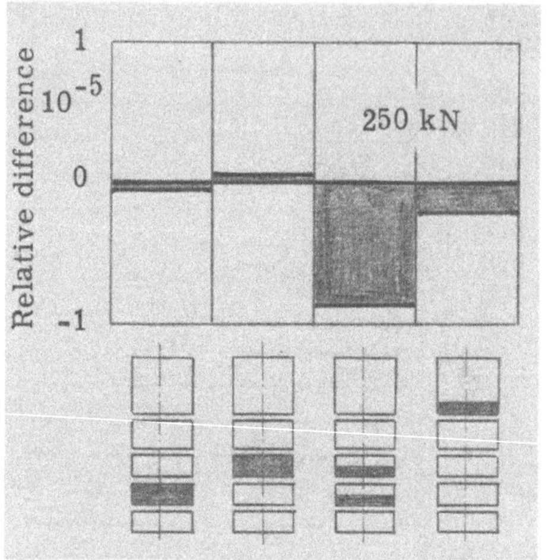

FIGURE 5. Systematic differences in generation of a force of 250 kN by different combination of dead weights

weights. This is in correspondence with the conclusions drawn by the measured magnetic flux density. Using the compensating lever the deviations in the lowest dead weight stack (10, ..., 100 kN) have been checked similarly.

The above mentioned discussion is based on extensive measuring results. It is recommended to draw the conclusion that the use of austenitic steel for dead weights of high precision force standards is no imperative compulsion when the distance between the dead weights is greater than 30 mm. When the distance is smaller than 30 mm under certain circumstances the force component generated by the magnetic field can be corrected. In this case a very extensive experimental work has to be done. In about 10 per-cents of the 200 possible dead weight combinations the force components caused by the ferromagnetic material characteristics of the dead weights of the KNME 2 MN are not neglegible. For the reason of the construction of the KNME 2 MN with 5 stacks of 10 dead weights for 10, 20, 20, 50 and 100 kN (see Fig. 1) 20 steps of force generated by complete stacks are free of magnetic field effects.

Summarizing one can state the relative uncertainty of force generated by the KNME 2 MN (defined as $2\,\sigma(\Delta F/F_0)$) does not exceed the value 2.10^{-5} in the range up to 40 kN and 1.10^{-5} in the range 50 kN ... 2.05 MN.

3. FORCE TRANSFER STANDARDS

Tension force transfer standards on strain gauge basis (example see Fig. 6) have been build-up for nominal forces 200 kN, 500 kN, 1 MN, 2 MN and 5 MN. A transfer standard especially for the calibration of the hydraulic transmission force standard machine KNME 10 MN has been constructed for tension and compression forces. Measures were taken for the reason of reducing the measuring uncertainty, the reduction of bending and torque moment sensivity and for the correction of creep and temperature influence.

So far the dependence of output signal from bending moments (originally relative variations up to 2.5×10^{-4} depending on the angular position) has been reduced by additional resistance in parallel circuits (changing the effective strain gauge factor) that the influence of bending moments of the KNME 2 MN changes the output signal not more than 1.10^{-5} related to the actual value (see Fig. 7). The influence of the

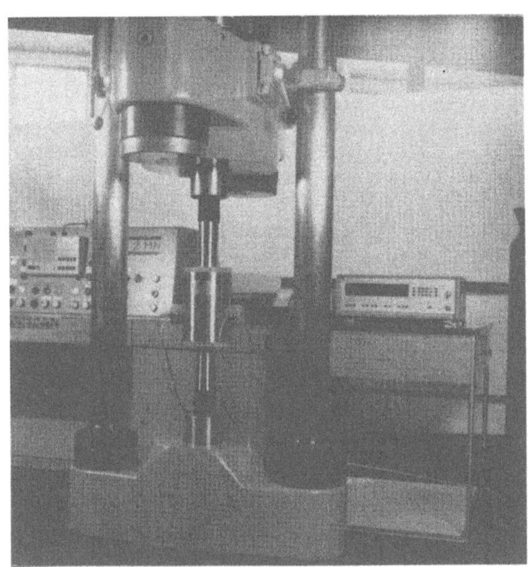

FIGURE 6. Tension force transfer standard TN 1 MN in the force standard machine 2 MN

torque moment has been reduced from originally 2.10^{-5} to practically zero.

A temperature sensor is fixed on the surface of the deformation body near the strain gauge circuit. Calibration provided temperatures can be measured with the DMP 39 with a resolution of almost 0.01 K inertialess.

This option is of advantage because of the fact that a remaining temperature dependence can be considered without difficulties after adjustment. Beside transfer standards where a creep effect is nearly absent there are specimen with creep effects in the range of 1.10^{-4} related to the output signal. When the realization of identical conditions for comparisons e.g. force-time-function is not possible corrections have to be made. Representative experiments have confirmed that the creep function of transfer standards can be simulated by superpositions of two exponential functions with different time constants. By this way measurement results made for different force-time-functions can be transformed into a comparable reference situation. The correction allows to reduce the influence of a creep error on smaller than 1.10^{-5} under worst case conditions.

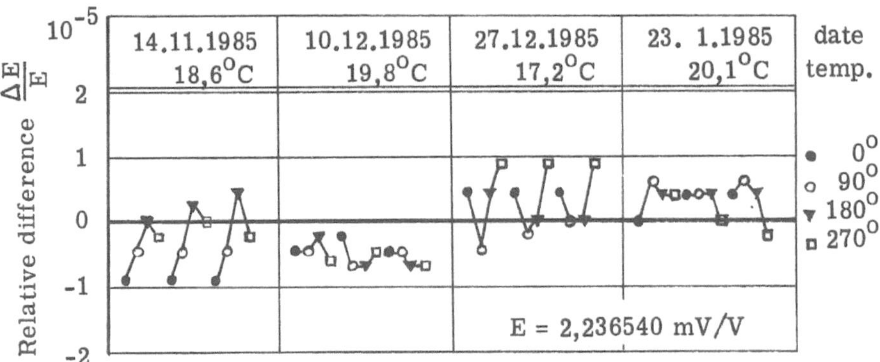

FIGURE 7. Relative differences $\Delta E/E$, output signals in 4 different angular positions for observation of the time dependence of the sensivity E of a transfer standard TN 500 kN

The decisive characteristic of transfer standards is the stability of sensivity in time. Fig. 7 shows measurement results of sensivity of the 500 kN transfer standard over a period of 2 months in preparation of a comparison between force standards. The mean values computed from values which were measured at one and the same time are distributed within $\pm$ 5.10^{-6} around the mean of all values. This is a variation which is near the stability characteristics of the DMP 39 used for the measurements. With reference to the experiences made up to now one can draw the conclusion that the above described transfer standards are suitable for high precision comparisons. Uncertainties can be reached which seem to be comparable with those in /9/.

4. CALIBRATION OF THE KNME 1 MN

The uncertainty of force generated by the lever type transmission of the KNME 1 MN had changed. Therefore a comparison of force was made with the primary standard KNME 2 MN using the transfer standards TN 500 kN and TN 1 MN. As a result of the comparison an adjustment of the KNME 1 MN was made by defined changing the transmission ratio and correction of mass of the dead weights (compensation of force-dependence of the transmission ratio). The deviations between the force values of the KNME 1 MN and the KNME 2 MN before and after adjustment are shown in Fig. 8. The comparison was made in the range of force up to 500 kN with both transfer standards. The differences of the measuring results of both instruments are smaller than 1.10^{-5}.

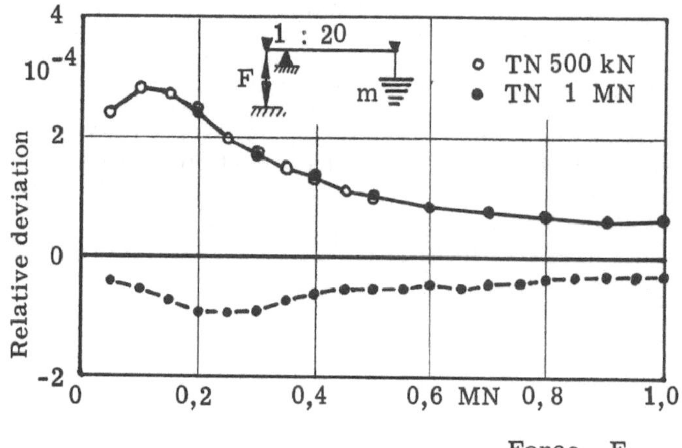

FIGURE 8. Differences between force generated by the secondary standard KNME 1 MN and the primary standard KNME 2 MN. Full line before correction, dotted line after correction

5. CALIBRATION OF THE KNME 10 MN IN THE RANGE UP TO 4 MN

The hydraulic transmission type force standard machine 10 MN (KNME 10 MN) of the ASMW had been calibrated in 1968 by a step by step procedure using a 40 t dead weight /8/. The uncertainty of force generated by the KNME 10 MN has been estimated with 5.10^{-4} as a result of this calibration. A new adjustment in the range of force up to 4 MN was made in 1985 by using the advantage of construction which allows the parallel build-in of a tension and a compression force transfer standard in the machine. When 2 MN transfer standards are calibrated in the KNME 2 MN in the first step the calibration of the KNME 10 MN up to 4 MN is possible in a second step. The upper part of Fig. 9 illustrates the force which is transmitted by the tension force and by the compression force transfer standard.

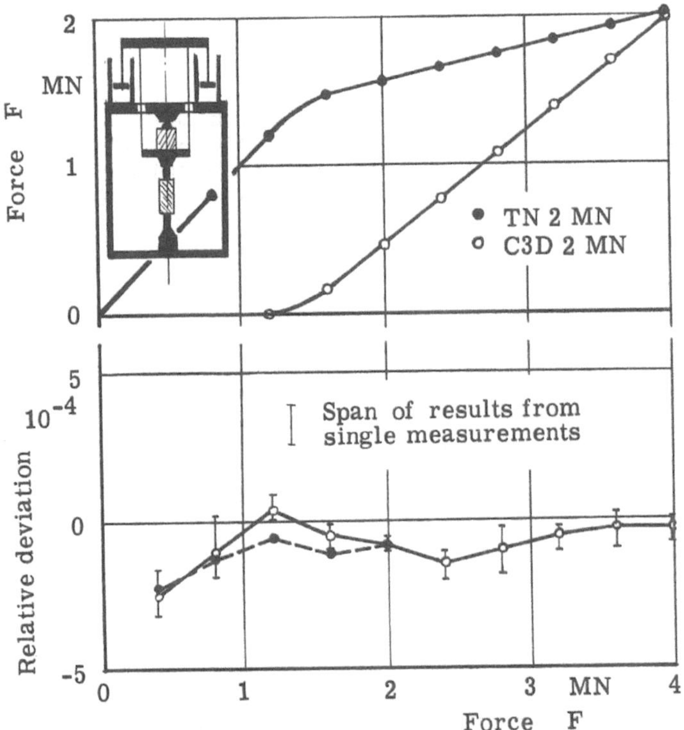

FIGURE 9. Result of comparison between KNME 2 MN and hydraulic transmission type secondary standard KNME 10 MN in the range of force up to 4 MN
(• Comparison by tension force transfer standard up to 2 MN,
o Comparison by loading of a tension and a compression force transfer standard 2 MN)

The lower part of Fig. 9 shows the deviation of the values of force generated by the KNME 10 MN from the values of the transfer standards. From Fig. 9 we draw the conclusion that the uncertainty of force generated by the KNME 10 MN in the range from 1 MN to 4 MN does not exceed 2.10^{-4} related to the actual force value. In near future the calibration of the KNME 10 MN in the range 4 ... 10 MN is planned with 5 MN transfer standards calibrated on the KNME 2 MN in the range up to 2.05 MN and on the KNME 10 MN in the range > 2.05 MN. The uncertainty of force 5.10^{-4} in the range up to 10 MN shall be ensured in this way.

6. FINAL REMARK

The analysis of characteristics of the KNME 2 MN, the construction and test of special transfer standards, the improvement of methods for comparison measurements and calibrations are part of preparations for comparison measurements between the highest force standards of European CMEA-countries.

REFERENCES

1. Weiler, W., Sawla, A.: Force standard machine of the National Institute for Metrology. PTB Bericht Me-22, 1978.
2. Adolf, K., Peschel, D.: Normalmeßeinrichtung zur Darstellung von Kräften bis 2 MN (KNME 2 MN). Metrologische Abhandlungen des ASMW, 3 (1983) 2.
3. Dietrich, M.: Development of Strain Gauge Precision Load Cells for the Comparison of Force Standard Measuring Devices. Proceedings of the 8th Conference of the IMEKO-TC "Measurement of Force and Mass", Krakow, 1980.
4. Dietrich, M., Steinhäuser, E.: Stand der Entwicklung und Anwendung von Transfernormalen der Kraft. Metrologische Abhandlungen des ASMW, Berlin 3 (1983) 3.
5. Dietrich, M.: Störkraftkompensation an Präzisions-Kraftmeßwandlern. Metrologische Abhandlungen des ASMW, Berlin 4 (1984) 4.
6. Dietrich, M.: Einfluß der Nichtlinearität auf die Meßunsicherheit der Transfernormale der Kraft des ASMW. Metrologische Abhandlungen des ASMW, Berlin (in print).
7. Weiler, W. et al.: Die 1 MN-Kraft-Normalmeßeinrichtung der Physikalisch-Technischen Bundesanstalt in Braunschweig. VDI-Zeitschrift 120 (1978) Nr. 1/2 - Jan. (I/II). Düsseldorf.
8. Bauschke, H.: Parameter der Normalbelastungsmaschine NBM 1000 Mp. Feingerätetechnik 19 (1970) 9.
9. Peters, M. u.a.: Final report-EC-comparison measurements for forces up to 1 MN. PTB-ME-54, 1985.

HIGH ACCURACY CALIBRATION METHODS FOR FORCE TRANSDUCERS

M. PETERS

PHYSIKALISCH-TECHNISCHE BUNDESANSTALT, FRG

ABSTRACT

This paper describes methods for the calibration of force transducers. The influence of the loading process on the measured value and other effects are taken into account. In addition, advantages and disadvantages of various adjustment methods are compared.

1. INTRODUCTION

In the following, various methods for the calibration of precise force transducers will be explained and discussed. These methods aim essentially at the transfer of forces with the highest possible accuracy. Based on various existing standards and on the results of comparison measurements carried out during the past years on an international level /1,2/, extensive investigations were carried out which have substantially improved the use of force transducers as transfer standards /3,4/. This improvement is due to the fact that, in addition to rotation, time and overlapping /5/ effects, the influence of the loading process on the measured value was taken into account. Even the creep behaviour of the transducers has been taken into consideration with a view to increasing the accuracy of measurement /6,7/. In addition, reference is made to new evaluation methods which are of great importance for the use of force transducers. Advantages and disadvantages of the various adjustment methods are compared in order to achieve optimum solutions for the different applications.

2. QUALITATIVE DIFFERENCES BETWEEN CALIBRATION METHODS

Force transducers can be calibrated by various methods which are of different quality and which should be chosen in accordance with the intended practical application of the transducer. Some of these procedures have been compiled in various directives and regulations, e.g. DIN 51 301, BSI, ASIM etc., as they have proved to be of practical importance for a large range of application. No regulations or directives exist, however, when the requirements to be met by the calibration method are high. For this reason, the criteria necessary for this purpose will be discussed in more detail, taking the latest findings in the field of measuring technique into account.

In the simplest case a force transducer is calibrated by once applying increasing forces in steps of 10% of the nominal force. This calibration is very inaccurate as no statement on the reproducibility and hysteresis can be made. For investigations of the hysteresis to be carried out later on certain materials, the transducer's hysteresis must, however, be known as it would otherwise result in a falsification of the measurement results. Moreover, a calibration only with increasing forces would presuppose that also during application the transducer is solely loaded with increasing forces.

For this reason the transducer should be loaded at least twice with increasing forces and once with decreasing forces. As due to interaction of force transducer and force standard machine, the value indicated under load depends in addition on the mounting positions, it is definitely useful to repeat the calibration in another mounting position rotated about the transducer axis. The mean value of both measurements, together with the respective relative deviation, furnish an additional point of reference for the quality of the transducer and that of the calibration carried out. This way of proceeding will be sufficient for many applications.

In order to derive from the calibration the respective fitted function of the indicated values, it must be known whether in practical application, the transducer will only be loaded with increasing forces or whether decreasing forces will also be applied. In the first case the fitted function - which will be discussed later - will only be computed from the values indicated under increasing load, whereas in the second case, the values displayed at decreasing load will also be included. As there is always a hysteresis, the uncertainty will necessarily be greater when the decreasing forces are taken into account.

If the quality of the calibration is to be further improved, the force transducer must be tested in at least four mounting positions rotated about its axis by 90°. In comparison measurements which require utmost precision of the calibration, the measurements are even carried out in six mounting positions, each rotated by 60°. Here a reasonable statement on the influence of the interaction between force transducer and force standard machine can be made, which leads to a reduction of the uncertainty of measurement during calibration. These influences - also referred to as rotation and overlapping effects - have already been discussed very thoroughly in the past years /5/ and will not, therefore, be the subject of the present paper. The following discussion is based on the assumption that the interaction effects of force transducer and force standard machine are taken into account.

3. INFLUENCE DUE TO THE ZERO FORCE SIGNAL

The force signals of the transducer which, during calibration, are assigned to the various forces are composed of the differences between the indications in the loaded and unloaded state. The signal of the unloaded state is referred to as zero signal, and it strongly depends on the time creep of the transducer.

Due to the creep effect the value indicated upon loading also depends on the time and on the history of the transducer. With the technical progress achieved, the creep has been compensated for to a large extent, however, it may still lead to errors in precise calibrations.

Fig. 1 shows the creep due to loading of a force transducer as a function of time, with differing transducer histories. Curve 1 was recorded after the force measuring device had remained unloaded for more than 15 hours; curves 2 and 3 show the situation after a 30-minute unloaded state. The relative difference between curves 2 and 3 amounts to $5 \cdot 10^{-6}$, whereas curve 1 differs from the two other curves only by about $2 \cdot 10^{-5}$ to $3 \cdot 10^{-5}$. This supports the point of view that

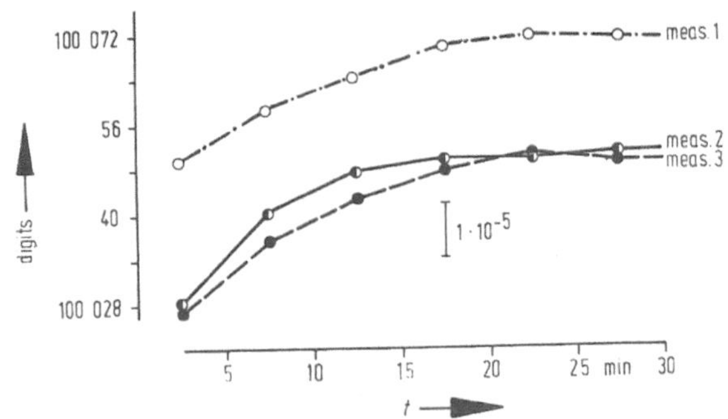

FIGURE 1. Creep due to loading, with the zero signal taken into account prior to loading

prior to the calibration proper, a force measuring device must be loaded at least once with nominal load; otherwise the first measurement series must not be included in the evaluation. Fig. 1 furthermore shows that a stable calibration signal is obtained only after about 20 to 30 min.

For a calibration, the expenditure of time would, however, be disproportionately high. A minimum period must therefore be determined for each measurement during which the uncertainty of measurement attainable is still within tolerable limits. A rhythm of between 3 min and 5 min has proved to be optimal. Within this period, the zero point - which is allowed for in the load signal - is on a relatively stable level, i.e. the indicated value remains constant. The level is reproducible and is found in a similar form for the zero creep of almost all types of force transducers. Only this value should therefore be used as zero point. As a consequence, a waiting period of at least 3 min after unloading must be observed before a new load is applied. In the case of shorter periods, unnecessary reproducibility errors will result from the very beginning.

Another characteristic behaviour of the zero signal is shown in Fig. 2. Represented is the development of the zero signal after various loading processes which are indicated by the upper curve. A rise of the zero signal with the number of loadings increasing can be recognized. If the loads are applied in the same way, the development of the zero signal as a function of time can be very well calculated as a approximative function $y = f(t)$. For the determination of the force signal, the respective fitted value of the zero signal must be employed; as a consequence, short-term electric interference effects acting on the zero signal will affect the uncertainty of measurement to a lesser degree. The normally decaying zero signal which appears after the hysteresis has been measured, can be corrected as well.

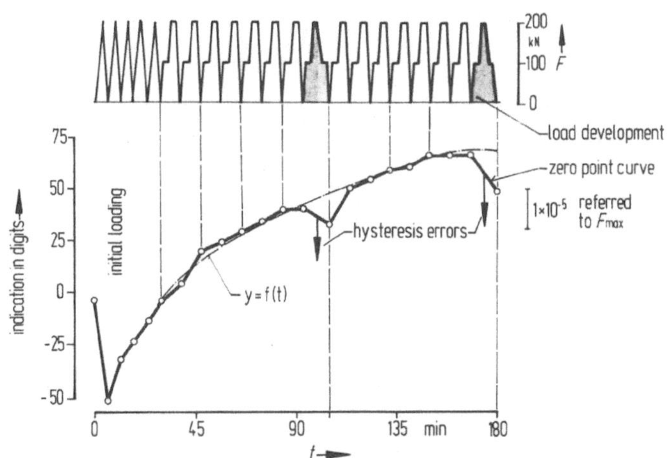

FIGURE 2. Influence of the load on the zero point

Independent of the zero point signal, other influences exist which must be taken into account with the load signal.

4. INFLUENCE OF THE LOADING PROCESS ON THE MEASUREMENT VALUE

The loading process, i.e. the number and graduation of the selected forces, also exerts an important influence on the measurement signal obtained. The individual forces can be directly adjusted by starting from zero, or in steps by taking other force steps as a basis. Due to the different creep processes the two measurement signals are not, however, identical. This can be inferred from Fig. 3.

Represented is the development of the measurement signal of a 200 kN force transducer at the nominal load of 200 kN, measured during various loading processes which have been outlined in the upper curve. The first three values and the last six ones were determined by directly adjusting the nominal force

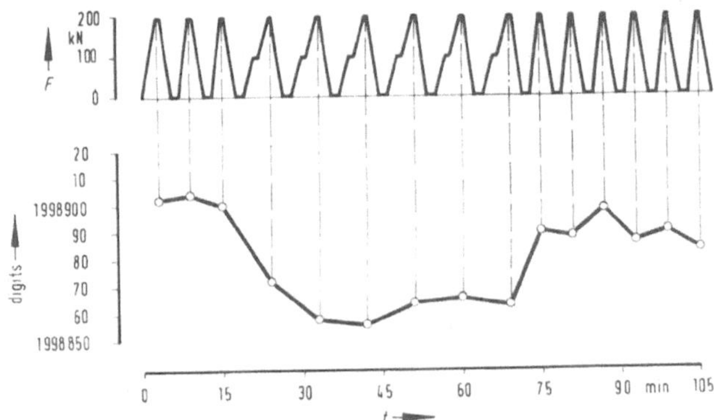

FIGURE 3. Influence of the loading process on the measurement value shown by an example of a 200 kN force transducer

starting from zero. Measurements were made in a 3-minute cycle. The five other lower values lying in between were determined by loading after an intermediate step of 100 kN, i.e. they were measured after 6 min. The difference between the measurement values is about $2 \cdot 10^{-5}$. Fig. 4 shows in a similar way the dependence of the measurement reading on the loading process measured with a 1 MN force transducer. The curve represents the relative differences of the force signals at different forces obtained from direct or stepwise adjustment of

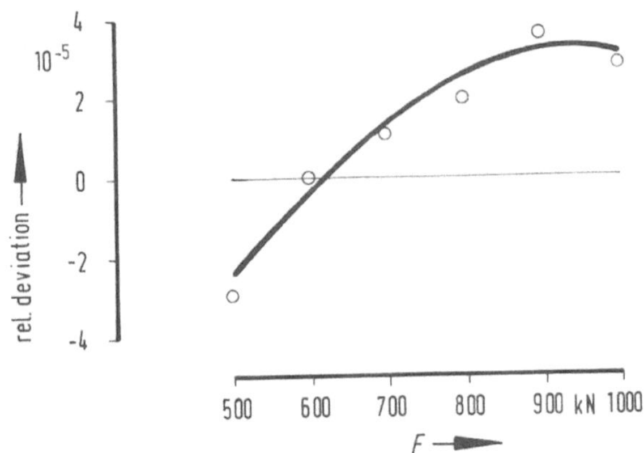

FIGURE 4. Relative differences in the force readings obtained with direct adjustment (curve) compared with stepwise adjustment (horinzontal line)

the respective force. If, for example, the force of 800 kN is directly adjusted by starting from zero, the measured value is by approx. $2 \cdot 10^{-5}$ higher than in the case of an adjustment in steps of 10%. The reasons for this behaviour are different time- and force-dependent creep influences of the transducer. For a very precise calibration, it must therefore be exactly known how the force transducer will be loaded in the future.

At the force steps where the transducer exhibits an increased creep influence, this effect will, of course, also have an influence on the rotational behaviour. This becomes evident in Fig. 5. After the transducer has been rotated about its axis, i.e. in the $0^o/360^o$ position, the relative deviations from the mean value are no longer identical. Here quite definitely a time-dependent creep is concerned as in the meantime the transducer had also been repeatedly loaded in the other mounting positions. The creep influence can, however, be corrected by raising the curve from the 360° value to the 0° value. The values obtained in the other mounting positions must be corrected in the same way, however, with appropriate weighing, as can be seen in the figure. The mean value of the indicated force can then be determined from the new rotation curve which is independent of the creep behaviour.

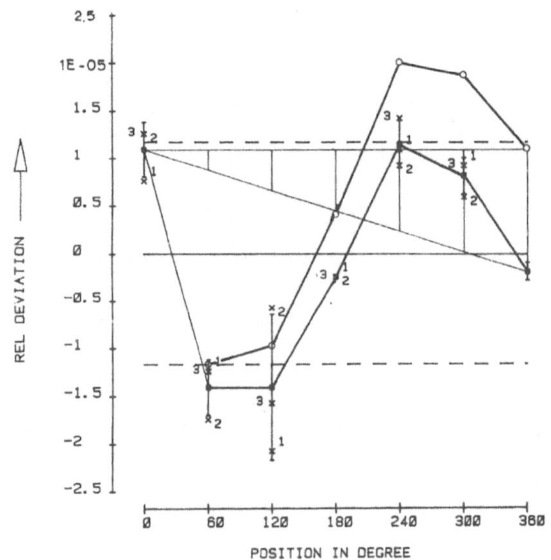

FIGURE 5. Taking into account of the creep behaviour in the rotation effect

5. SELECTION OF THE SUITABLE FITTED FUNCTION

After the calibration of a force transducer has been completed which was carried out for strict accuracy requirements and with all the above-mentioned criteria being taken into account, a calibration curve can be computed by the adjustment calculus. In most cases linear interpolation is applied. As in force measurement the relative uncertainties of measurement are always related to the measured value, errors of up to the order of magnitude of 10^{-3} can occur, in particular in the low force steps. A higher-order adjustment calculus is therefore reasonable. A function of the third degree with a constant is frequently used, however, with the restriction that interpolation may only be made as from 10% of the nominal force. Mathematical studies of a large number of different calibrations have, however, shown that the cubic parabola is not always the optimal fitted function. This is explained in closer detail in Fig. 6. Represented are in each case the relative deviations of the interpolated function from the force signals actually measured, as a function of the different force steps. The relative deviations have here been related to the measured values.

The degree of the fitted function is given by n. It can be clearly seen that with increasing degree, the relative deviations from the measured value become increasingly smaller. They amount to $< 1 \cdot 10^{-4}$ for functions of the third degree and higher. For the function of the seventh degree which cannot

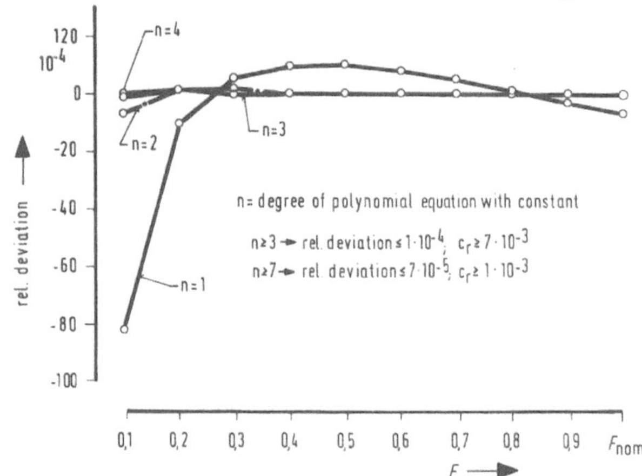

FIGURE 6. Relative deviations of interpolated functions with constant of degree n from the measured values

be represented in the figure, the relative deviations amount to $< 2 \cdot 10^{-5}$. The disadvantage of the fitted function with constant which is chosen and applied in almost all cases, is the constant itself. In low force steps it affects the relative deviations very strongly. If the constant is related to the 10% measurement value, relative deviations C_r between 10^{-4} and 10^{-3} result near the zero force step. With the degree of the fitted function rising, the constants first become smaller and then assume higher values again after having reached the maximum which lies between the functions of the fourth and sixth degree. Fitted functions with a constant can therefore definitely be used only for force steps of 10% of the nominal force and more. This is, however, inappropriate for practical application.

It is the fitted function without constant which approximates the actual shape of the curve very closely. This is shown in Fig. 7. It is true that the deviations become smaller only at a higher-degree fitted function; however, even with the curve represented in Fig. 7 which was determined with a poor force

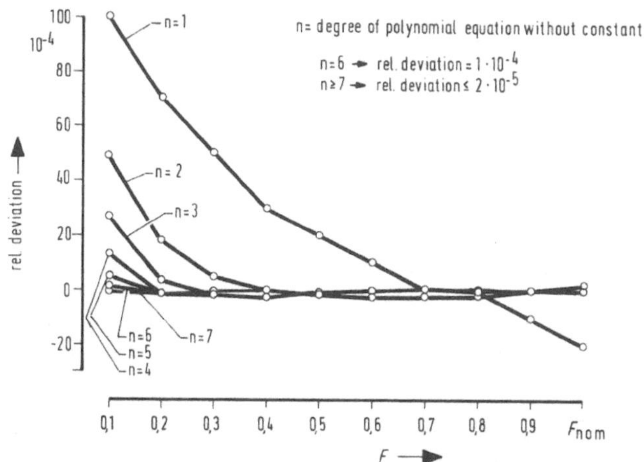

FIGURE 7. Relative deviations of interpolated functions without constant of degree n from the measured values

transducer, the relative deviations for a function of the sixth degree amount to $\leq 1 \cdot 10^{-4}$. As the function has no constant, interpolation can be made up to the zero point without important errors being introduced.

Figs. 8 and 9 show the corresponding fitted functions obtained with a good force transducer. The tendency is the same. With a

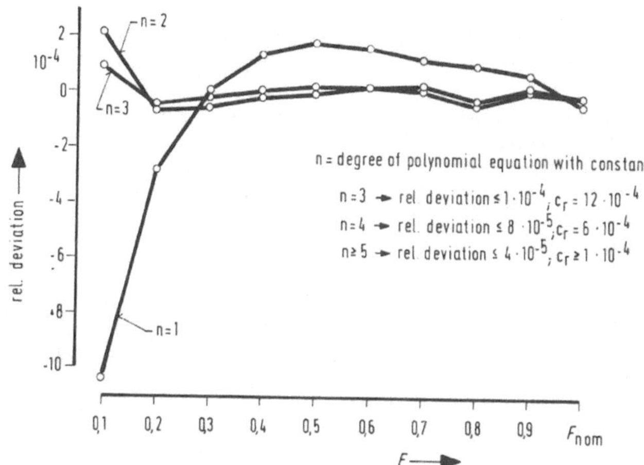

FIGURE 8. Relative deviations of interpolated functions with constant of degree n from the measured values

fitted function of the fourth degree without constant, the relative deviations amount to $\leq 1 \cdot 10^{-4}$ (Fig. 9). The corresponding function with constant (Fig. 8) reaches these

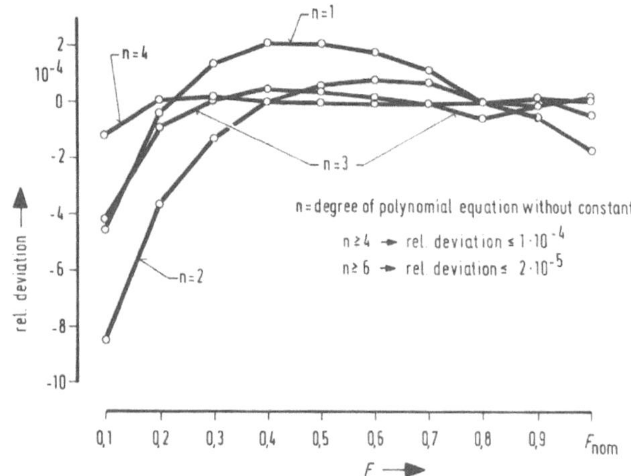

FIGURE 9. Relative deviations of interpolated functions without constant of degree n from the measured values

small values already at the 3rd degree, however, due to the relatively great constant, close to the zero point, relative errors of approx. $12 \cdot 10^{-4}$ still occur. This is why no interpolation may be made below 10%. The constant leads to the smallest value of $1 \cdot 10^{-4}$ with a function of the fifth degree. With an even higher degree, it again increases considerably.

It can thus clearly be stated that, if exact calibration is desired, the choice of a fitted function of a higher degree without constant is imperative for interpolation.

If a suitable mathematical statement is made, on the basis of the desired limit of the relative uncertainty, the degree of the fitted function can be determined automatically with the aid of modern computers at present available.

5. SUMMARY

Various possibilities for the calibration of force transducers have been explained and discussed. It turned out that a precise calibration of force transducers can only be realized if sophisticated evaluation procedures are applied. Otherwise, the uncertainty of measurement can be increased due to too many influences and effects. It is, however, very time-consuming to take all criteria mentioned in this paper into account; this will certainly be necessary only for the calibration of a force transducer intended for use as a transfer standard of highest accuracy.

REFERENCES

1. Peters M.: Problems of high-precision force measurement. Proc. R.T. Disc. 8th IMEKO TC-3. Krakow, 12-15, 1981.
2. Peters M., Sawla A., Wilkening G.: EC comparison measurement for forces up to 1 MN. PTB-Me-54,1-124, 1983.
3. Peters M.: The realization of forces up to 1 MN on an international level. Proc. 10th IMEKO TC-3, Japan, 1-6, 1984.
4. Sawla A.: Problems of the high accuracy transfer of large forces of up to 8 MN. Proc. Weightech 83, London, 1-8, 1983.
5. Peters M.: Reasons for and consequences of the rotation and overlapping effect. Proc. R.T. Disc. 10th IMEKO TC-3, Japan, 1-5, 1984.
6. Peters M., Wilkening G.: Präzisionsmessung in der Kraftmeßtechnik. Meßtechnische Briefe, Vol. 16, 61-65, 1980.
7. Peters M.: Limits to the uncertainty achievable in force transfer. Proc. Weightech 83, London, 1-9, 1983.

EVALUATION OF PARASITIC COMPONENTS ON NATIONAL DEADWEIGHT FORCE STANDARD MACHINES

C. FERRERO, C. MARINARI, E. MARTINO

Istituto di Metrologia "G. Colonnetti" (IMGC) - Torino (Italy)

1. INTRODUCTION

The results of several intercomparisons of force standard machines show uncertainties higher than can be expected on the basis of the uncertainties with which mass and gravity acceleration values are known (1, 2, 3).

The main sources of the differences are parasitic effects caused by undesirable components (transverse forces and moments) generated by asymmetry in machine structure, non-symmetric deformations of the loaded machine, faulty load cell positioning on the machine, and machine -load cell interaction (4, 5, 6, 7).
Significant progress can be made as regards force standards if these components are measured as functions of various parameters, e.g., load level, load transmission system, type of machine - dynamometer interface.

Only in this way and by reducing defects and bringing parasitic effects under control, can the performances of force standard machines be improved.

The present paper describes the main metrological characteristics of 3 national force standards: the IMGC 105-kN deadweight machine (Italy), the 500 kN deadweight machine of the National Physical Laboratory (NPL, Teddington UK), and the 550 kN deadweight machine of the TNO Institute (Delft, Netherlands) as determined by a 6-component dynamometer expecially designed and constructed by IMGC.

Several tests were carried out, to measure the influence of different parameters on parasitic components, namely,
a) eccentricity of the load application point to the dynamometer axis; b) load application rate; c) load transfer system (LTS); d) different weight-piece combinations, whenever possible; e) different vertical (height) positions of the adjustable crossbeam to the main machine frame.

2. THE IMGC SIX-COMPONENT DYNAMOMETER AND EQUIPMENT

The IMGC dynamometer is a prototype standard to measure vertical force (up to 100 kN) and parasitic components: side forces (X, Y = up to $\pm$ 2 kN) bending moments (L, M = up to $\pm$ 400 N m) and twisting moment (N = up to $\pm$ 100 N m) introduced in calibration systems by defects in force transmission and support deformation.

This dynamometer is an assembled balance. Fig. 1 shows the dynamometer mounted on the IMGC-machine with the multicomponent calibration system and the ancillary instrumentation.

The dynamometer and its main metrological characteristics were described in detail in previous papers (8, 9).

The composite configuration chosen has several advantages: a) high sensitivity to transverse components (detection of small inclination angles) and to twisting moment because of the low interaction between the axial component and the transverse components; b) the possibility of varying the range measurement and, consequently, improving the sensitivity, by replacing secondary cells by others of different capacity; c) lower dependence on interface conditions and, consequently, on parasitic components during the load transmission phase; these components may cause permanent changes in other cell-machine systems (10); d) low stiffness, so that it is possible to provide a blocking protection against overloads and obtain better adaptation of the dynamometer to the force axis and make the interaction effect less significant (11).

The flexibility characteristics of the dynamometer and the measurement method adopted make it possible to study dynamic phenomena of the machine/dynamometer system, understood both as load-application transients and as free oscillations of the system under constant load.

Loading conditions were made uniform for the three machines, with a cross-knife joint (CK) serving as a load transfer system (LTS).

For special tests and to detect possible effects of the type of interface on parasitic components, another kind of LTS was used, namely, a ball (dia. = 50.8 mm, hardness = 60 HRC) with a flat spacer (PB). A special device, using the same ball, was constructed, to apply loads with a known ± 1 mm eccentricity.

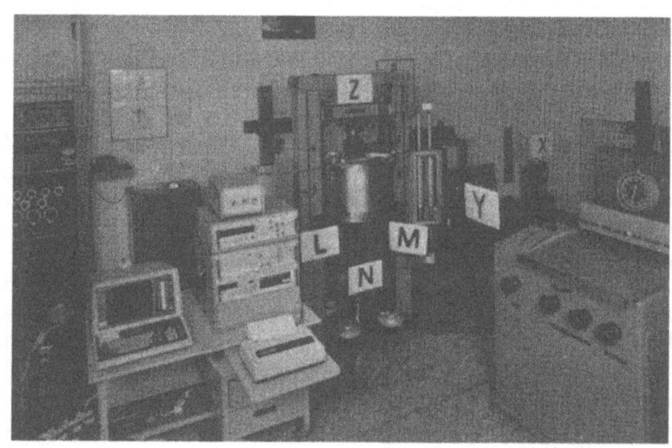

FIGURE 1. IMGC-105 kN machine with 6-component dynamometer and ancillary equipment.

The measurement chain (Fig. 1) includes: a) a digital indicator (HBM-DK37A) sensitivity 2 mV/V/100000 digit and 2 mV/V/200000 digit; b) an automatic scanning device (HBM-UPH3200) for signal acquisition; c) a minicomputer with a printer and hard copy equipment; d) a HBM calibrator BN100.

3. MEASUREMENT CONDITIONS

To reduce errors caused by creep and creep recovery, a carefully timed loading schedule was followed in tests on each machine. Readings were taken three minutes after load application.

Measurements were repeated with the dynamometer at different positions to machine axes and were made at a temperature of 20 ± 2 °C. The dynamometer was kept in the laboratory at a constant temperature for more than 24 hours, before starting load application with the load cells connected to the relevant instrumentation.

The six-component dynamometer was preloaded three times after each rotation when it had been at rest for more than one hour, and only once for shorter rest times.

4. EXPERIMENTAL RESULTS

4.1. The IMGC 105 kN standard deadweight machine

Fig. 2a shows the graphs of side components (X, Y) vs axial load: average values and standard deviations (S_x, S_y) were calculated for all the tests regarding dynamometer rotation to machine axis according to the methods applied in intercomparison with high accuracy one-component load cells (12, 13). In this way the

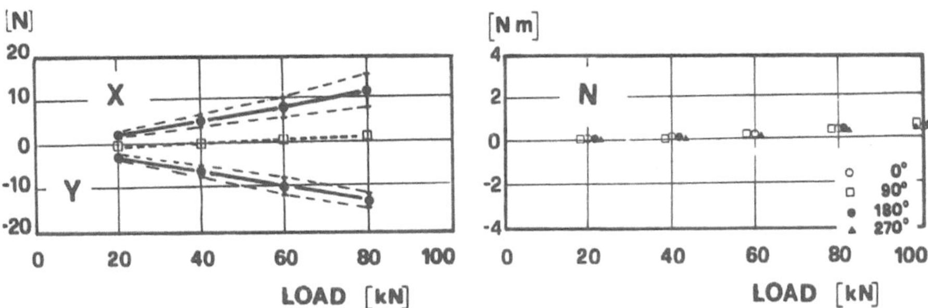

FIGURE 2. a) Side components vs axial load: average values and standard deviations at several angular dynamometer positions (□ final value obtained with machine inclination β). b) Twisting moment vs axial load at several angular positions.

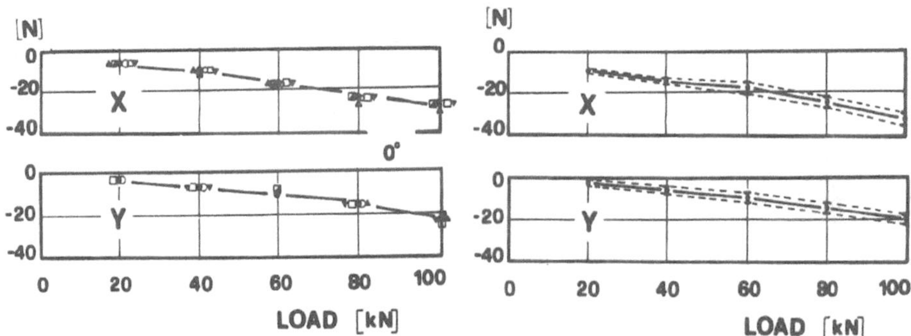

FIGURE 3. Side components vs axial load on NPL machine: a) at 0° angular position and usual operation method, (□) with different position of adjustable crossbeam, and (●) with different weight-piece combinations; b) average values calculated for several angular dynamometer positions;

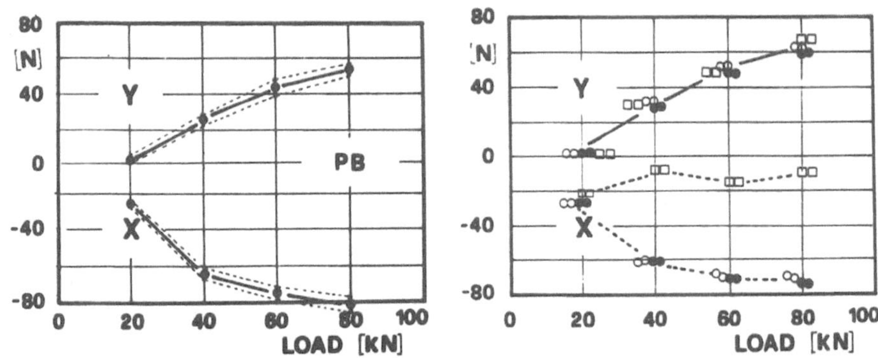

FIGURE 4. Side components vs axial load on TNO machine: a) average values calculated for several angular positions. b) with three different weight-piece combinations (A (●), B (○), C (□)).

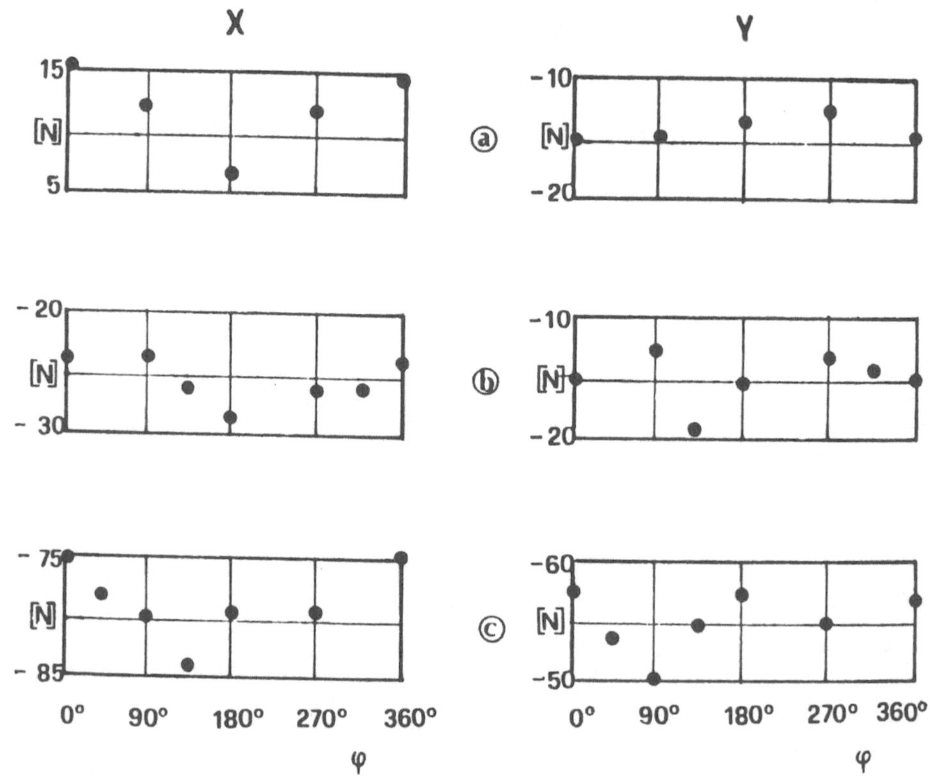

FIGURE 5. Side components at different angular positions on IMGC (a), NPL (b) and TNO (c) standard deadweight machines (Z = 80 kN).

TABLE 1. General metrological characteristics of the IMGC, NPL, TNO standard deadweight machines at Z = 80 kN.

Machine	X/Z 10^{-4}	Y/Z 10^{-4}	N N m	$\Delta y = L/Z$ mm	$\Delta x = M/Z$ mm	Inclination correction $\Delta Z/Z$
IMGC	1.5	-1.7	0.5	0.004	0.012	1×10^{-7}
NPL	-3.4	-1.9	0.4	0.003	-0.004	1×10^{-7}
TNO	-9.8	6.8	0.5	-0.012	0.048	1×10^{-6}

standard deviation, obviously, takes account of casual factors as well as of cause-effect factors (e.g., rotational effect) that are at all events detectable and consequently the estimate of S_x and of S_y was over-dimensioned; with Z = 80 kN, for instance, their values were less than/or equal to 3,5 N whereas the maximum deviation resulted R≤ 4 N for all the three machines. The figure supplies as well the values of side component X obtained after rotation, β, of the main machine frame in the direction of the greatest machine flexibility, so as to annull the X side component (β= X/Z=1.5x10^{-4}).

Fig. 2b gives the values of the twisting moment (N) vs axial load and with the dynamometer at four angular positions.

Table 1 gives the values of the ratio of side components (X, Y) and bending moments (L, M) to axial load (Z), of the inclination correction ΔZ/Z, and of the twisting moment (N) at Z = 80 kN for the 3 standard deadweight machines.

4.2. The NPL 500 kN standard deadweight machine

Fig. 3a shows the values of side components X and Y vs axial load Z, with the dynamometer positioned at 0° to the machine reference axes, in conditions of normal machine operation. They show, additionally, the component values obtained with the adjustable crossbeam in different positions and with different weight-piece combinations.

Fig. 3b shows the values of side components vs axial load and gives average ($\bar{X}$, $\bar{Y}$) and standard deviation (S_x, S_y) values at each load level calculated for several angular positions.

Table 2 gives the mean values of side components and twisting moment at four angular positions with the ball both eccentric and centered and with constant load Z = 40 kN for the NPL and TNO machines.

TABLE 2. Side components and twisting moment at Z = 40 kN with ball centered and eccentric (NPL and TNO machines)

		Ball centered	Ball eccentric (± 1 mm)
NPL	$\bar{X}$ (N)	-10.02 ± 0.07	-9.94 ± 0.7
	$\bar{Y}$ (N)	- 6.66 ± 0.056	-6.47 ± 0.06
	$\bar{N}$ (N m)	0.76 ± 0.003	0.77 ± 0.06
TNO	$\bar{X}$ (N)	-60.5 ± 0.44	-59.0 ± 6.3
	$\bar{Y}$ (N)	26.4 ± 0.1	26.3 ± 4.5
	$\bar{N}$ (N m)	- 0.13 ± 0.06	- 0.19 ± 0.1

4.3. The TNO 500 kN standard deadweight machine

Fig. 4a gives the values of side components vs axial load concerning the use of PB as LTS.

Fig. 4b shows the values of side components vs axial load with three weight-piece combinations (A, B, C), the dynamometer at 0°, and PB as LTS.

Combination A can be defined "normal", since it was usually employed in tests. In combination B the number of 5-kN weight pieces remains unchanged and there are added in succession the 10-kN weights of the second weight-carrier. Consequently, this combination employs six 10-kN weight pieces instead of the three of combination A. In combination C 5-kN weights are still employed to obtain the first load increment, and the 20-kN weight pieces of the third weight carrier are also employed. The use of the second and the third carriers involves application of a different number of oscillation-damping rods connecting the loading frame to the main frame.

Fig. 5a, b, c show the rotation effect on the three national standard deadweight machines.

5. RESULT ANALYSIS

5.1. Side components (X, Y)

5.1.1. Normal tests. According to the above results, the three national deadweight machines produce only small side forces (X, Y). The X to Z (applied load) ratio is about $(1.5 \pm 0.4)\times10^{-4}$ for the IMGC, $(3.4 \pm 0.3)\times10^{-4}$ for the NPL and $(9.8 \pm 0.3)\times10^{-4}$ for the TNO machines.
The Y to Z ratio is about $(1.7 \pm 0.15)\times10^{-4}$ for IMGC, $(1.9 \pm 0.2)\times10^{-4}$ for NPL and $(6.9 \pm 0.3)\times10^{-4}$ for TNO machines.

The maximum deviation of X and Y values over two successive cycles carried out with the dynamometer at the same angular position were usually lower than or equal to 2 N.

The linearity of side components X and Y vs axial load for IMGC and NPL machines indicate that these components mainly depend on the initial geometrical machine condition (inclination); in other words, the main machine structure has not undergone subsequent noticeable distortions, which would have introduced a quadratic element in side component values vs axial load.

A slight deviation from linearity was observed for component X of the TNO machine, presumably caused by oscillation-damping rods along this direction.
Good reproducibility of the values of parasitic components on the IMGC-machine measured at the beginning and at the end of this exercise indicates as well that the behaviour of the machine, dynamometer and instrumentation did not change in the course of time.

Inclination correction $\Delta Z/Z$ was calculated on the basis of the values of side components as follows

$$\Delta Z/Z = 1 - \cos\beta = 2\sin^2\beta/2 \simeq \beta^2/2$$

Since $\sin\beta = X/Z$ (β being the angle between the load action line and the dynamometer axis), one obtains

$$\Delta Z/Z = X^2/2Z^2$$

Given these side component values (Table 1), inclination correction $\Delta Z/Z$ consequently always results lower than 1×10^{-6} for the three standard machines and, therefore, lower than the sensitivity of the measurement chain; it is also lower by one or two orders of magnitude than the rated machines accuracy (1 to 2×10^{-5}).

Side components, even when they do not introduce noticeable variations in axial load Z, may cause - as is well known - variations in the output signal of the load cell used in intercomparison exercises (rotational effect). Care was therefore taken to have the machine properly arranged and settled, with the purpose of reducing such parasitic components.

The results given in Fig. 2a concerning the introduced counter-rotation $\beta = X/Z = 1.5\times10^{-4}$ of the main frame of the IMGC machine, confirm the validity of the precaution taken, and at the same time demonstrate the possibility of improving the characteristics of a standard machine by acting on the machine structure, on the basis of measurement values obtained with six-component dynamometers.

An examination of the behaviour of side components X and Y as a function of angle φ of dynamometer rotation around the machine axis, with a constant Z load (Fig. 5) indicates the existence of a rotational component in the three machines, which is slightly higher in connection with component X and which corresponds, as a rule, to the direction of greater machine flexibility. Components X and Y can therefore be expressed by an equation of the type $X = X_o + a\sin(\varphi \pm \varphi_o)$ where the part varying with angle φ can be considered the representation of the machine-dynamometer interaction (rotational effect). The low value of this parameter, a, with respect to the average X_o value, and the fact that (a) appears to be very little dependent on the absolute values of components X and Y generated by the machine are an indication of a very small machine-dynamometer interaction, as well as of high machine stability. They point out, additionally, that the values of the side components of the three machines tested mainly depend on the initial inclination, β , of the adjustable crossbeam and/or on the $\Delta\beta$ rotation of the loaded beam.

Furthermore, on the basis of the foregoing considerations and as regards the calibration of standard deadweight machines, it is pointed out that the choice of an assembled-type dynamometer is the solution that is less affected by the factors (such as support surfaces, interface conditions, load transfer system) that are responsible for machine-dynamometer interaction to a greater extent and that, consequently, are apt to widen out the fan of the values observed with the dynamometer at the same angular position and at different positions as well (rotational effect).

With one-piece dynamometers the situation is obviously different. In this case variations in the loading point and interface conditions modify the behaviour of the force flow lines through the various strain-gauge bridges and this fact can alter, even sizably, measurement results.

5.1.2. Special tests. Tests were carried out at different load application rates, from about 3 min, during which no weight-piece oscillations occurred, to about 10 s, the shortest application rate permitted by the machines.

The values obtained from measurements do not reveal significant differences for side components X and Y, in spite of the fact that the weight pieces oscillated with different amplitudes.

A number of tests were carried out at different load Z levels during which the sequence and the type of weight pieces were varied (Figs. 3a and 4b) and the positions of the adjustable crossbeam changed. Two measurement cycles were carried out for each combination (A, B, and C). The repeatability of side components resulted unchanged in the different combinations and at the various load levels, namely maximum deviations were less than or equal to 2 N for all the standard machines.

The values of X and Y measured in these different conditions (Fig. 3a) did not differ significantly for the IMGC and NPL machines. For the TNO-machine the difference is not important as regards component Y measured in the three weight-combination conditions (A, B, C of Fig. 4b) but the use of a third weight carrier caused X to decrease ($\Delta X = X_C - X_A = 62$ N), may be for the action of a different number of oscillation damping rods.

The use of a PB instead of a CK caused small variations in the values of side components X and Y. These variations were not such as to affect the overall TNO machine behaviour.

The application of a 40 kN load with $\pm$ 1 mm eccentricity (L = M = 40 N m) did not sizably alter the values of the transverse components and of the twisting moment.

It must be remarked that also the summation of vertical-channel values (axial load) did not vary with changes in the load application point of $\Delta x = \Delta y = \pm 1$ mm, with an applied axial load of 40 kN, though the signals from the individual vertical channels varied by about 600 digits.

These results confirmed that the machines tested can very satisfactorily work even with $\pm$ 1 mm offset, namely, without this causing the weight carrier to touch the main frame - thus discharging a portion of the load - and without the M and L bending moments, resulting from 1-mm eccentricity in load application, causing marked rotations of the loading crossbeam.

5.2. Twisting (N) and bending moments (L, M)

5.2.1. Twisting moment N resulted to be lower than 1 N m for IMGC and NPL machines at all load levels. For the TNO-machine the values of twisting moment N resulted initially of the order of -4.9 N m with Z = 80 kN, after adjustment of the rods used to block weight-piece oscillations, these values changed to -3.48 N m with CK as LTS; and to 0.5 N m with Z = 80 kN at all the dynamometer positions, with PB as LTS, so that the twisting moment can be neglected.

This moment was not affected by the type of weights employed and by the position of the adjustable crossbeam. For diagnostic purposes the twisting moment is in fact the most sensitive tool and a very important component for evidencing possible contact points between the loading and the main frames.

5.2.2. Bending moments (L, M). These moments indicated that the kind of interface (CK, PB) used in the tests is highly satisfactory for defining the load application point.

As a rule, bending moment L resulted smaller than $\pm$ 3 N m and M smaller than $\pm$ 2 N m at Z = 100 kN with CK as LTS; L was smaller than $\pm$ 4 N m and M smaller than $\pm$ 3 N m at Z = 80 kN with PB as LTS.

This means that the reproducibility of the load application point was better than $\Delta y = L/Z = \pm 0.03$ mm and than $\Delta x = M/Z = \pm 0.02$ mm at rated load, with the dynamometer at different angular positions and with the use of CK, and better than $\Delta y = \pm 0.05$ mm and than $\Delta x = \pm 0.03$ mm with the use of PB as load transfer system.

The repeatability of components M and L at the different load levels appears to be slightly higher when CK is used than with PB. Repeatability values observed between two successive cycles are equal to a few micrometres with CK.

The results show that the value of bending moments L and M, and consequently of the load application point (eccentricity) were not affected by:

a) the type or combination of weight pieces;
b) different positions of the adjustable loading crossbeam;
c) different load application rates.

5.3. Axial load

The mean value and the maximal deviation (in digit) during two or three test cycles were determined at several load levels and for each angular position. All the results showed that deviation remained, as a rule, within 1 or 2 digits, with about 200.000 digit full-scale signal.

Reproducibility was determined by changing the position of the dynamometer (0°, 90°, 180°, 270°) to machine axis.

The variation coefficient at Z = 80 kN was always lower than 1.2×10^{-5} for the IMGC machine, 1.1×10^{-5} for the NPL machine and 1.8×10^{-5} for the TNO machine during nine test cycles carried out with the dynamometer at four angular positions.

6. CONCLUSIONS

The tests carried out on the IMGC, NPL and TNO machines in the course of this intercomparison campaign as regards the measurement of parasitic components and the influence of several parameters on the values of such components, allow the following main conclusions to be drawn:

a) Side forces are small and repeatable for the three standard deadweight machines. Their values in any case are so small that the relevant inclination correction $\Delta Z/Z$ would be lower than 1×10^{-6} for the axial load component.
b) Component values depend mainly on the initial machine inclination (β) and on a possible rotation ($\Delta\beta$) of the adjustable crossbeam.
c) The values of bending moments L, M show that a repeatability of the load resting conditions within a few micrometres is possible when a crossknife joint is used as a load transfer system.
d) Twisting moment N is usually lower than 2 N m at all load levels. This component is a powerful diagnostic tool for evidencing possible contact points along the load trasmission line.
e) For intercomparison purposes, it is better to check every time that oscillation-damping rods, where existent, are free from contacts, and to use the same loading plan (time, weight combination, weight-carrier).

The results supply, additionally, an experimental confirmation of the possibility of using a 6-component dynamometer to detect anomalous situations in standard machines and improve, when necessary, their metrological characteristics.

ACKNOWLEDGMENT

The authors are indebted to Mr. Jenkins of the NPL and all force group staff, to Mr. Wieringa of the TNO and his staff for their cooperation during the work carried out in the United Kingdom and the Netherlands.
The fruitful discussion and encouraging leadership of prof. A. Bray director of IMGC are gratefully acknowledged.

The work described was carried out under a BCR-EEC contract.

REFERENCES

1. Debnam RC, Wieringa H: An intercomparison of force standard machines. VDI - Berichte Nr. 212, 1974.
2. Sawla A, Weiler W, Peters M, Bray A, Levi R, Vattasso M: A comparison of force standards between the IMGC and the PTB. 6th Confer. IMEKO TC3 Odessa, 1-13, 1977.
3. Jenkins RF, Debnam RC: An intercomparison of force standard machines at the NBS and the NPL. NPL Report MOM61, 1982.
4. Bray A, Ferrero C, Levi R, Marinari C: An investigation on parasitic effects on force standard machines. VDI-Berichte n. 312, 1978.
5. Ferrero C, Marinari C, Martino E: Deformazioni elastiche nella macchina campione di forza IMGC. XI Congr. AIAS, 1983.
6. Bray A: Interaction deadweight machine - Load cell. IMEKO TC3, 7th Round Table Discussion, London, 1976.
7. Dubois M, Bourateu JP, Gosset A, Priel M: Intercomparaison de trois bancs d'etalonnage dynamometrique de capacite 250-300 kN, agree par le BNM. Bulletin d'information du Bureau National de Metrologie, vol. 11, 1980.
8. Ferrero C, Marinari C, Martino E: Analysis and calibration of IMGC six-component dynamometer. BCR Technical Report, contract n° 1304/1/0/062, 1983.
9. Ferrero C, Marinari C, Martino E: A six-component dynamometer to measure parasitic load components on deadweight force standard machines. Proc. of the 10th Conf. of IMEKO TC3, Kobe, Japan, 1984.
10. Bray A: Structural properties of force standard machines. Proc. of Intern. Conf. of Stress Analysis, Haifa, 1982.
11. Peters M: Reasons for and consequences of the rotation and overlapping effect. Prof. of the 10th Conf. of IMEKO TC3, Kobe, Japan, 1984.
12. Bray A, Levi R, Vattasso M, Weiler W, Sawla A, Peters M: Comparison of force standard between IMGC and PTB. BCR Technical Report, 1976.
13. Peters M: Limits to the uncertainty achievable in force transfer, Weightech'83, 1983.

THE EFFECT OF TIME RESPONSE OF FORCE TRANSDUCERS ON THEIR TRANSFER PROPERTIES IN FORCE COMPARISONS

Jürgen Paetow
Dipl.-Ing.
HOTTINGER BALDWIN MESSTECHNIK GMBH
Darmstadt

1. INTRODUCTION

If a measuring instrument contains time-dependent portions in its characteristics, these are more or less transmitted to the object the instrument is testing. If several links of the calibration chain display such defects, then it is almost impossible to eliminate such effects by specific corrective measures at the end of the calibration chain. When extreme demands in accuracy are at stake, such deficiencies can no longer be ignored. Cause and effect of such relationships will be shown in the following along with measures to reduce the resulting errors.

2. EVERY MEASURING CHARACTERISTIC DEPENDS ON TIME

To test a measuring instrument, e.g., determining its characteristics, a measuring device is needed. This device must be much more accurate than the object measured if the measuring results are to make a reasonable statement. The argument , however, that a measuring result can be only as good as the measuring intruments used, is correct to a limited degree only. It is easy to prove, though, that "good" measuring values can seemingly be achieved only as a result of an interrelationship between measuring device and object measured. The experienced measuring engineer is well aware of this; whether he adheres to it is a different story. Often enough, a good measuring result is indeed desired.

Let us take a look at the first figure 1a: it shows the deviation diagram of a force transducer measured with a specific measuring device. The result is rather pleasing; the transducer apparently can be used for precision measure-

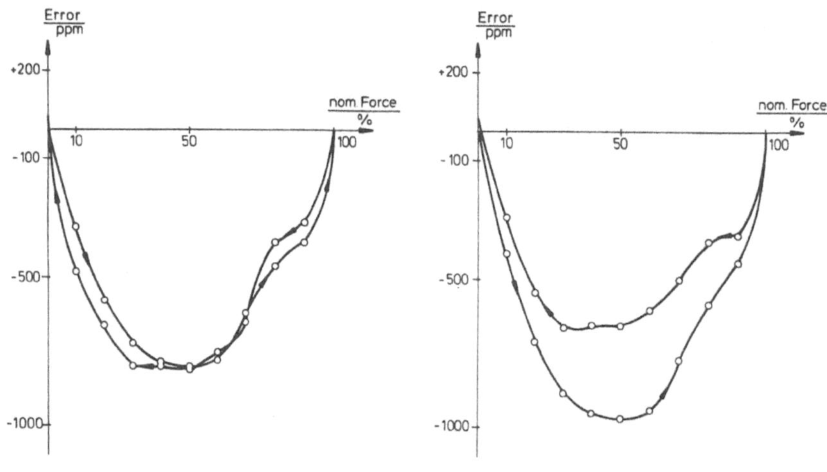

Fig. 1a. Calibration Curve resulting from Test Standard No. 1

Fig. 1b. Calibration Curve resulting from Test Standard No. 2

ments. The following figure 1b shows the deviation diagram of the very same transducer obtained in a different measuring device under otherwise identical conditions.

The results speak for themselves . It is worth mentioning that the measuring characteristics of both test setups were obtained in standard measuring setups of a higher accuracy class so that due allowance could be made for both. By repeating the measurements, it was found that the results are of a systematic nature. Obviously , it is not simply enough to use the simple measuring characteristic of the testing device as a basis.

It is true that the effect of influence variables on measurements has been thoroughly dealt with theoretically. There are numerous suggestions on how to allow for them in order to obtain a measuring result as "pure" as possible. A particulary unpleasant effect - because it is so hard to eliminate - is the time, or better: the course of time. What we refer to here are not wear or ageing, but to predominantly short-term reproducible events. These are usually ignored in measuring; at the most, they are allowed for by determining the chronological course of the measuring program. Primarily, this only reduces the uncertainty - non-repeatability - of a measurement. The systematic proportion of errors is retained if the measuring program is not adapted to the time characteristic of the test standard. This very thing explains the apparently "good" characteristic in Fig.1a. The characteristic in Fig.1b is closer to the truth. The first measuring device - also a force transducer - does have a markedly time-dependent characteristic while the second device has a very reduced response on time.

3. CAUSES OF TIME DEPENDENCIES

What exactly are such time dependencies? As we mentioned above, long-term effects will be ignored here. Short-term effects - we are looking at periods lasting seconds or minutes - in principle occur with the most measuring processes. In electrical methods, they show up, among other things, as zero drift or slope change of the amplifier. The "power-up drift" of the measuring device is mentioned here as one single example, albeit a primitive one. Mechanical processes display analog phenomena; the slope of the transducing characteristic is subject to systematic and nonsystematic chronological changes. Primarily, elastic and thermal imperfections of elements determining the measuring value are the cause. Vibration-induced characteristics are consciously ignored in our considerations.
Two time phenomena sufficiently known in the sense of the preceding considerations to the user of electromechanical transducers, e.g., force transducers, are creep and hysteresis. Creep expresses itself as a time-based change of the measuring value after the change of the measuring variable, see Fig.2. The direction of the load change plays a subordinate part in this case. This creep consequently is a "time elasticity" or an elastoplastic event in contrast to the notion of creep introduced in material technology, particularly for high-temperature materials. The hysteresis as per Fig.3 also displays a time-base which surely must be

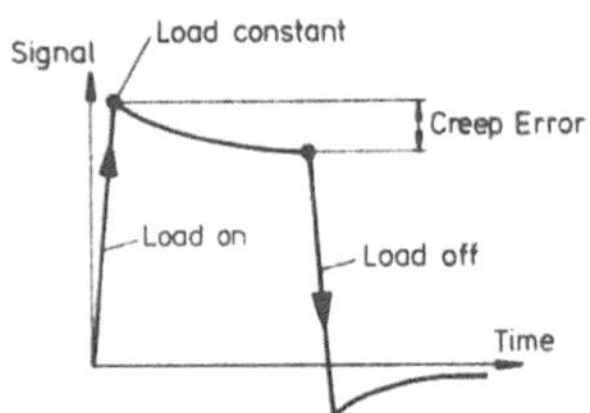

Fig. 2. Creep Error

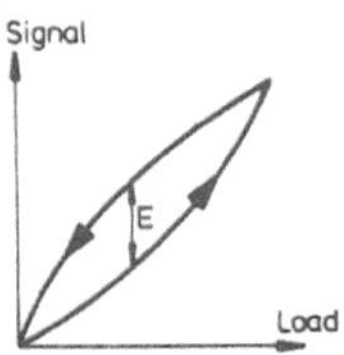

Fig. 3. Hysteresis Error

viewed in connection with creep. Its modeltype description by creep to date has not been successful. No measuring system is free of such phenomena. They play a part only for higher demands on accuracy for the most part; frequently, they set the limits of accuracy of a measuring method.

4. CONCRETE EFFECTS OF THE TIME PHENOMENA

This time phenomenon takes on a many-faceted aspect if we compare two measuring setups. Let this be shown by the example of a force comparison. The direct event consists of two force measuring devices being mutually loaded; both placed at different positions of the "accuracy hierarchy". In the most simple case, this comparison is used to "calibrate" in compliance with DIN 1319, i.e., the determination of the characteristic of the lower-ranked force measuring device. The most demanding force comparison is used to retain or to expand the force scale. In each case, the characteristic of the higher ranked system is assumed to be known, no matter where this knowledge may have been gained. The time-dependencies in the sense of the above considerations for the most part are ignored.

In the following, we intend to show systematically effect and interaction of time responses of the two measuring setups participating in this type of measurement.
For a better understanding, certain idealised simplifications form the basis for the following considerations.

The point of origin of a "hierarchical" measurement is a primary setup (primary standard, S1), its characteristic may be known sufficiently. It shall be considered absolutely linear and independent of time (Fig.4a).
This setup is used to determine the characteristic of a secondary standard (S2); to wit, once in the shortest possible time, (Fig.4b), once markedly slow, (Fig.4c). To be sure the situation (4c) does not represent the time characteristic - this would require a three-dimensional image at a minimum-but it does give the characteristic for a specific time course.

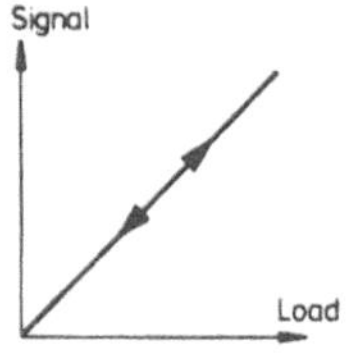

Fig. 4a. Ideal Primary Standard (S1) loaded quickly and slowly

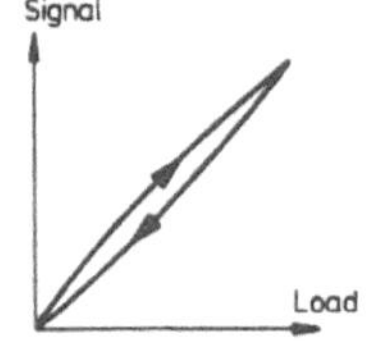

Fig. 4b. Secondary Standard (S2) loaded quickly

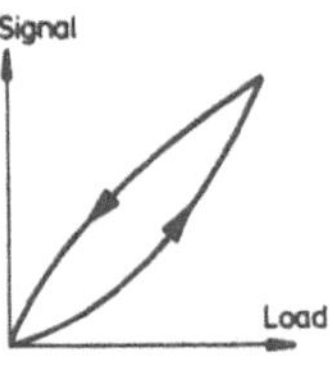

Fig. 4c. Secondary. Standard (S2) loaded slowly

In the following, let S2 be used as a test setup for various test objects (T) - transducers, for example. Let the first test object (T1) have an ideal, similarly time-independent characteristic, as if it were S1, (Fig.5a). When tested with S2, four differing characteristics may result depending on the load being applied slowly or quickly or on the characteristic (4b) or (4c) being used of S2. This is to be demonstrated by the following four figures.

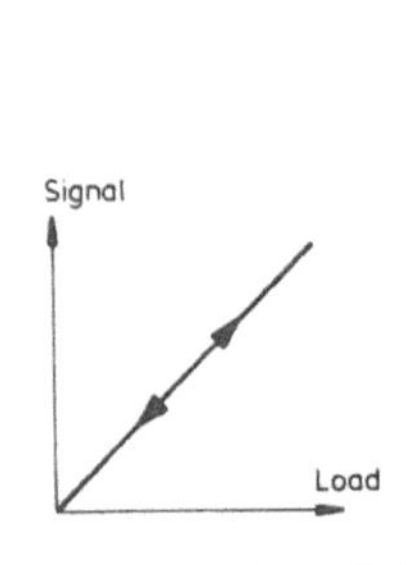

Fig. 5a. Test Object (T1) with ideal time independent characteristic

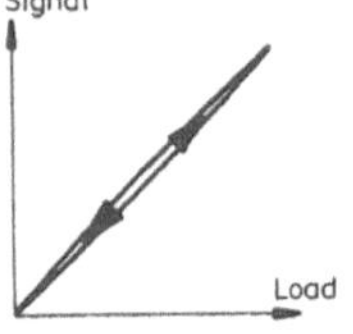

Fig. 5b. Test Object (T1), calibrated with S2, char. 4c, loaded slowly

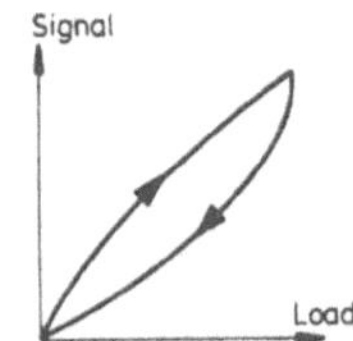

Fig. 5c. Test Object (T1), calibrated with S2, char. 4b, loaded slowly

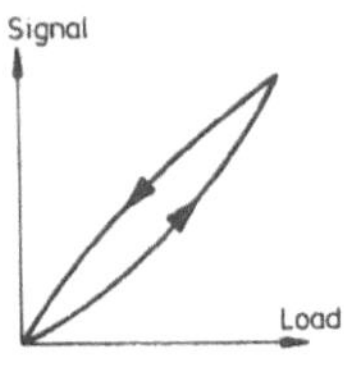

Fig. 5d. Test Object (T1), calibrated with S2, char. 4c loaded quickly

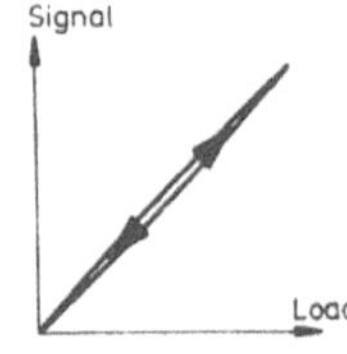

Fig. 5e. Test Object (T1), calibrated with S2, char. 4b loaded quickly

S2 characteristic 4c; slow application of loads: Fig.5b
S2 characteristic 4b; slow application of loads: Fig.5c
S2 characteristic 4c;quick application of loads: Fig.5d
S2 characteristic 4b;quick application of loads: Fig.5e

Clearly, the characteristic of the testing measuring device is transmitted to the object tested in dependence of the chronological sequence of the test operation. In case of high demands on accuracy, this fact may be ignored only with more or less ideal test conditions. It is interesting to note that the hysteresis may change its sign as shown by Fig.5c and 5d.

The situation as per Fig.5b and 5d achieves higher practical significance. In force measuring technology at least, it is true that the test setup is the slower, the higher it is ranked in the accuracy hierarchy. In theory, this means that the test speed of even the last link in the hierarchy depends on the chronological capacity of the primary standard if there are any time deficiencies. There is only one solution for this problem: a highly reduced time dependency of the characteristic, at least of the standard used in the production of measuring devices.
In practical applications, the object tested (T2) will also display time dependencies. Its possible characteristics are shown in Fig.6a. For the test instrument (S2), only the characteristic calculated as per Fig.4c may be used. For T2, this produces characteristics as shown in Fig.6b and 6c. Seemingly, the object tested loses its time base, the more so the more the time characteristic of test device and test object agree. In such cases, efforts to determine the hysteresis of a transducer as a function of the duration of a load cycle may become senseless. Even though the results are obvious, let us mention the situation in which the test device S2 does not show any time-based proportions in its characteristic, i.e., as per Fig.4a. Then and only then can the characteristic of the test object be correctly figured in its time base.

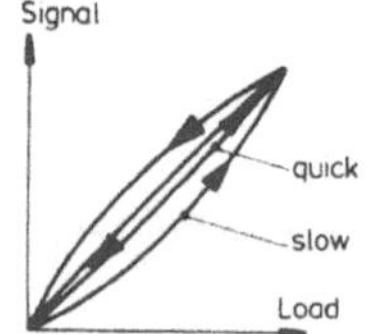

Fig. 6a. Test Object (T2) with time dependend characteristic

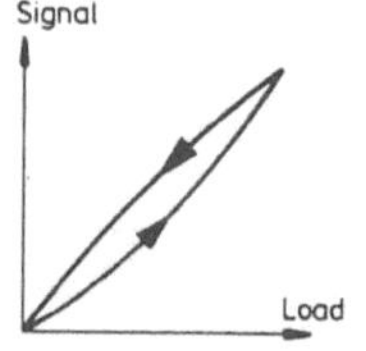

Fig. 6b. Test Object (T2), calibrated with S2, char. 4c, loaded quickly

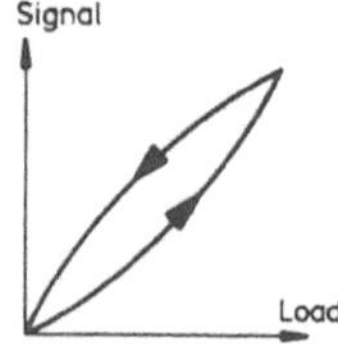

Fig. 6c. Test Object (T2), calibrated with S2, char. 4c, loaded slowly

The question to what extent time based errors are transmitted from step to step in the hierarchy of test devices may be answered with reference to the above considerations.

In summary, the following consequences arise for measuring applications:
If a measuring device shows time-based proportions in its characteristic, these are transmitted more or less but still within reason to the objects to be tested by the device. If several links of a calibration chain (hierarchy) display such deficiencies, their effects cannot be clearly shown and for the most part cannot be eliminated by specific corrective measures at the end of the calibration chain. Then universally true statements on sign and size of such errors resulting at the end can no longer be made.

5. POSSIBLE MEASURES

We need not simply accept these errors. A prerequisite for improvement first of all is the development or absorption of a specific "feeling for time" in measuring techniques. At the start of calibrating a pretentions measuring device, we must first ask the question if it may be subject to the physical principle of short-term effects (such as hysteresis, creep). If this cannot be clearly negated, at least two calibrations, markedly differing in their chronological operations, should be performed. If there are differences in quantities which cannot be ignored, we should warn of a universal application of this measuring device. Methods to determine these differences to date have not been defined; there are no quantitative statements on what is "permissible". Probably, this is neither practicable nor does it make sense. It is much more practical at this stage of the state of the art to adapt the use of the measuring device to its time responses , in other words, to specify the measuring sequence more accurately.

This dilemma, however, brings us to a suggestion for all manufacturers and users, particularly in the field of force measuring engineering: simple standard tests are lacking which would describe quantitatively and practically the time responses for measuring devices. Perhaps the creep test known in transducer technology will give us a point from which to start. To date, however, not even the interrelationship between creep and hysteresis seems to be quantitatively recorded.

This raises the question for the practical significance of these "time errors". To date, no quantitative statements have become known on this subject. If we concentrate on creep errors, usable figures are available. Their size under adverse circumstances may exert its full effect on the measuring result. For transducers, for example, it may amount to 0.1%. Transducers of topmost quality display less than one tenth of this.

The following Fig.7 shows a force transducer especially developed for force transfer which, in like form, is also used in testing machines. It is equipped with a very effective creep compensation which in practice combines two

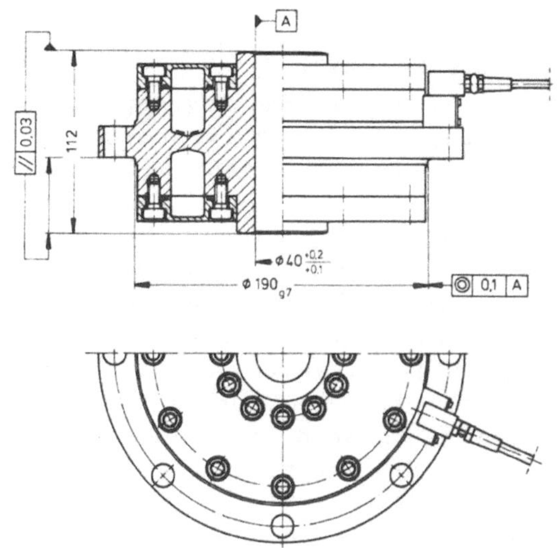

Fig.7 Force Transducer Z12

opposing creep characteristics. This is demonstrated in Fig.8 which shows that creep reductions 1:20 are technically feasible. Unfortunately, we do not yet have any figures relating to what extent this ratio is transmitted to the "time characteristic" of the transducer. However, it should be considerably better than shown in Fig.9. This shows two chronologically varying measuring characteristics of a transducer which was in use a few years ago with the PTB, its creep error in 20 min. amounting to 0.009 %.

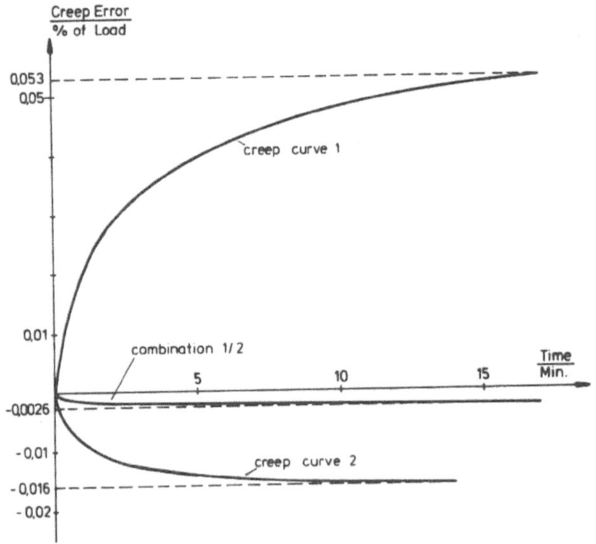

Fig. 8. Super Creep Compensation

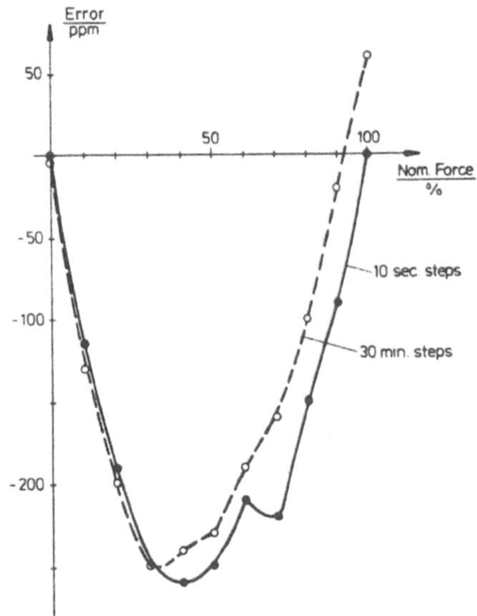

Fig. 9. Calibr. Curves
at different Loading Speeds

CALIBRATION AND VERIFICATION OF MULTICOMPONENT DYNAMOMETERS IN THE MEGANEWTON RANGE

G. BARBATO, A. BRAY, A. GERMAK, R. LEVI

Istituto di Metrologia "G. Colonnetti" and Politecnico - Torino - Italy

1. INTRODUCTION

Nowadays multicomponent dynamometers are required with capacities in the meganewton range, for a number of purposes ranging from improvement of existing force standards to routine checks on material testing machines.

Owing to size constraints, elastic element configuration is almost invariably limited to single-billet style; advantage is taken of current strain gage technique for first order compensation right at bridge level of the influence of the main error originating sources. Transducer construction is therefore well within the reach of most laboratories having the required proficiency in strain gage work; commercially available instrumentation is more than adequate for excitation, signal conditioning and recording as required. Data processing is no problem either thanks to the ubiquitous microprocessor.

The catch however is that for measurement exploitation purposes the relationship between applied load components and output signals must be known within a level of uncertainty consistent with the requirements of the problem at hand. As these may not infrequently entail evaluation of minute components in view of their compensation or elimination as far as possible, accurate assessment of some second order terms may then be necessary.

At medium to low force/moment capacities, that is within the range of large wind tunnel testing work, say up to the order of 10^5 N and 10^4 N·m, the problem is only one of time, and cost. Facilities and procedures are available capable of yielding routinely all reasonable requirements at state of the art level. The picture however is a completely different one when the limits of these facilities are exceeded, and the outlay of the huge sums of money required to make a large capacity installation is out of question.

In these conditions one finds that but a subset of the component combinations required for regular second order model fitting and validation may be applied, and educated guesses only instead of hard facts are available as basis for estimation of some effects. Theoretical analysis and accumulated experience are drawn upon to offset at least to some extent the lack of specific experimental evidence.

2. CALIBRATION SETUPS

For model fitting purposes to either calibration or exploitation sets of equations, that is giving (in the linear case) output vector $\{U\}$ in terms of applied loads $\{F\}$:

$$\{U\} = [C] \{F\}$$

or F in terms of $\{U\}$:

$$\{F\} = [E] \{U\}$$

besides covering each single component range most two by two component combinations at least at capacity value are to be applied for interaction term evaluation. This requirement entails use of a full six component calibration facility, allowing application to the unit under test of any component combination within a specified volume on both force and moment spaces XYZ and LMN, as shown in Fig. 1a.

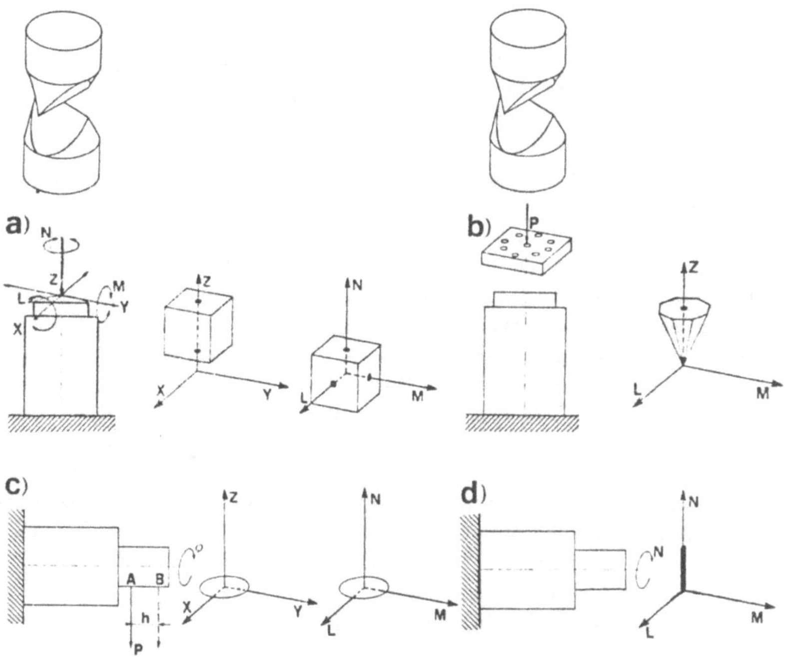

Fig.1 Different types of multicomponent calibration setups, with the relevant calibration volume: a) Full six-component setup; b) Z & L,M setup; c) X,Y & L,M setup; d) N setup.

Lacking such a facility, one must put up with alternative setups /1/, which can provide however only a subset of the information which only complete six component calibration may give access to. The arrangements adopted in the case at hand are shown in Fig. 1 b,c,d. along with the corresponding sample spaces, which make apparent the inherent limitations. Combination of these methods together, that is multiple loading points plus tilting of unit's axis, and rotation along the same axis, extend somehow the sample space, leading up to almost complete exploration, but for the inherent limitation concerning ZN, YM and XL combinations /2,3/. As deadweight load application is almost mandatory, these methods can be exploited up to the capacity of the largest deadweight machine available, using rather simple rigs. The setups of Fig.1 b,c,d, used for the evaluation of elements of the (linear) exploitation matrix are listed in Table 1. It may be appreciated that some estimates are definitely more reliable than other ones, as in some instances only cross-checking is available.

TABLE 1. Calibration setups used for estimation of terms of linear exploitation matrix $[E]$, see Fig. 1

$$\begin{bmatrix} b & b & c & b & b & d \\ b & b & c & b & b & d \\ b & b & c & c & b & d \\ b & b & c & b & c & d \\ b & b & c & b & b & d \\ b & b & c & b & b & d \end{bmatrix}$$

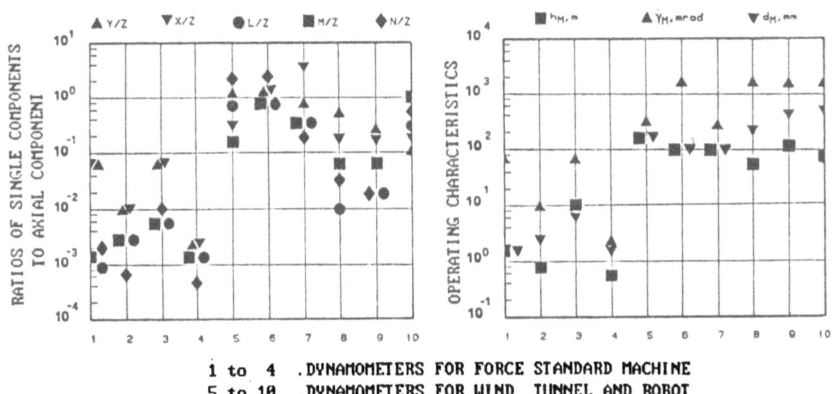

Fig. 2. Plot of component ratios for different types of multicomponent dynamometers.

Location of the line of action of load with a small uncertainty is mandatory, if even moderately accurate results are aimed at; this spells out the inadequacy in most case of ball and socket joints, unless extraordinary care is paid in their manufacture and exploitation. Knife edges and/or elastic hinges may prove necessary if the line of action of load is to be located within say a tenth of a millimeter or less; hydrostatic and air bearings may also provide an answer (the latter at lower load level) although they are best limited to permanent applications, on account of the complications entailed.

Now a peculiarity of units used for control of force standard machines is to be pointed out, namely the required capability of evaluation of small parasitic components in presence of axial loads several order of magnitude larger. This is the main difference between these units and multicomponent transducers developed for other purposes, as shows a perusal of Fig. 2, related to ten different units. Ratios of components to the axial one, and three main performance evaluating parameters, namely maximum axial load offset d_M, inclination angle γ_M and torque arm h_M are considered /4,5/. As the deadweight machine used for calibration purposes cannot be assumed to be free from parasitic components, considerable precautions are required in order to obtain unbiased estimates of parameters.

For the unit at hand, tests were performed with the following step:

1) axial load, generating Z component;
2) load offset from but parallel to axis generating Z and L and/or M components;
3) load applied orthogonal to axis, generating L and/or M, X and/or Y components;
4) torque applied on torquemeter calibration stand, generating component N.

The first four sets of tests were performed on a 100 kN deadweight machine, relying upon systematic rotation of unit and replication of tests for detection and cancellation of spurious effects due to setup.

A 2 MN six component dynamometer, developed at IMGC mainly for

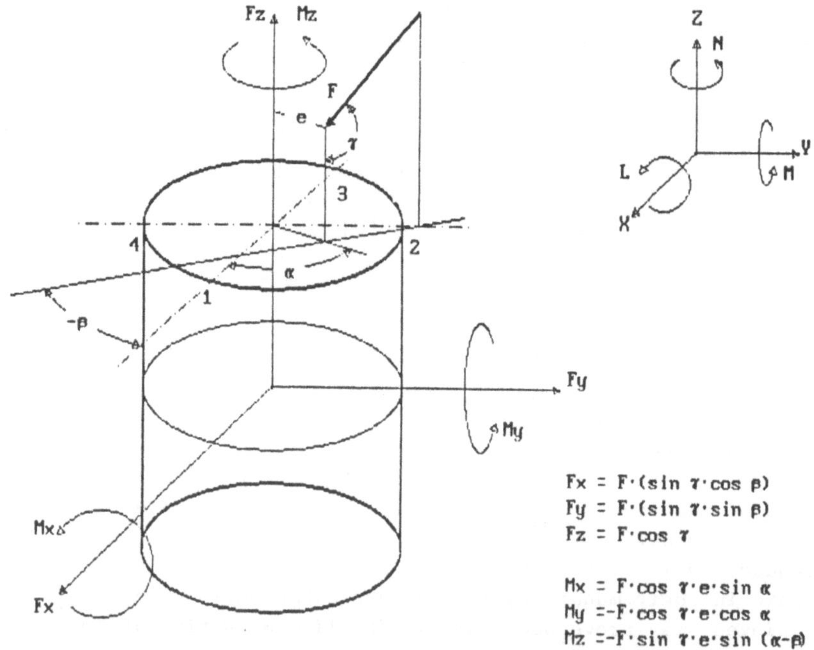

Fig. 3 Unit's reference systems. A) X, Y, Z, L, M, & N stationary system. B) Fx, Fy, Fz, Mx, My & Mz dynamometer system.

verification of testing machines /6/, was considered in these tests. Owing to peculiar requirements, size had to be kept down to a minimum; the unit is therefore made up by a short, single billet elastic element, 100 mm dia by 200 mm high. Foil resistance strain gages cemented in axial and circumferential direction at mid-height are wired up with the usual arrangements to measure axial load and bending moments, while "chevron" gages evaluate side loads and torque through shear strain.

With a design load of 2 MN for the axial load, capacity is 20 kN for side loads, 4.4 kN·m for bending moments, and 1.5 kN·m for torque, according to service requirements. Accordingly, only axial load induced deformations are liable to originate non linearities justifying evaluation of second and higher order terms in calibration and exploitation equations.

Unit's reference system is shown in Fig. 3; for clarity sake, component rotation Fx, Fy, Fz and Mx, My, Mz was selected, reserving the usual X, Y, Z and L, M, N designation to components referred to a stationary reference system, such as e.g. that pertaining to a deadweight machine. Unit's output channels are U1 to U6; however U1 and U2 are calculated from U_{T1} and U_{T2} corresponding to side loads T1 and T2, lying on the Fx, Fy plane and forming angles of 292.5 and 22.5 degrees with Fx axis.

A HBM DK38 S6 indicator with scanner, interfaced with an Olivetti M24 Personal Computer catered for data acquisition and recording for further evaluation.

3. REMARKS ON DATA REDUCTION AND MODEL FITTING

For tests performed with the setups listed above, systematic replication of loading schedules at eight angular settings of the dynamometer were resorted to; faulty conditions could thus be spotted quickly and corrections made when necessary. When a regular "rotation effect" was observed, Fourier analysis yielded three main pieces of information for a given loading combination, namely constant term, corresponding to main effect, first order term, corresponding to direction dependent sensitivity, and phase angle, whose discrepancies from expected values provided useful checks towards such parasitic effects as e.g. rig distortion and strain gage position errors. Exploitation of these three information for each output channel helped a lot to offset some of the inherent shortcomings of the system, by allowing separation of otherwise tangled contributions /6/.

An observation is in order concerning some departures between expected and observed results. Owing to the compact size of the unit (h/d = 2), strain gage signal is affected by stress distribution at the interfaces, as departure from De St. Venant conditions is a marked one /7/. Superimposing an additional stress pattern by tightening screwed nose pieces or pull rods, as required for some setups, may not only affect zeroes but interact with sensitivity as well, especially at low load levels. Such effects may be detected, and under favourable conditions their magnitude may be assessed; however, routine quantitative evaluation and further correction are out of question at the present state of art in all but a few cases.

Let us consider now in detail some test results. Side loads were applied with several moment arm values, in order to allow for separate evaluation of bending moment effect, the latter to be further checked against the effect of a shift of axial load. U1 and U2 channel outputs were closely approximated by a regression equation of the form $U = A \sin(\alpha - \varphi) + C + 2s$ with a constant term C of the order of 0.1% of the amplitude, and random (residual) term s within 1%. Phase angle exhibits systematic deviations up to some degrees from the nominal value; however, upon closer examination these deviation are found to be proportional to moment arm, and correction for moment effect brings the discrepancy to fractional degree level, which is acceptable.

U4 and U5 (moment) channel outputs exhibit a better fit to the sine law model, as constant term is within 0.05% of amplitude, random term is of the order of 0.5%, and difference between actual and nominal phase angles is within a few tenths of a degree.

Tests with eccentric load lead to much the same effects of moments, but for some minor discrepancies accounted for by an overall uncertainty of line of action of vertical load within less than 0.1 mm. Fourier analysis leads to estimation of a phase difference between moment effects obtained with eccentric load tests, and side load tests, of about one degree on U4 and U5 channels.

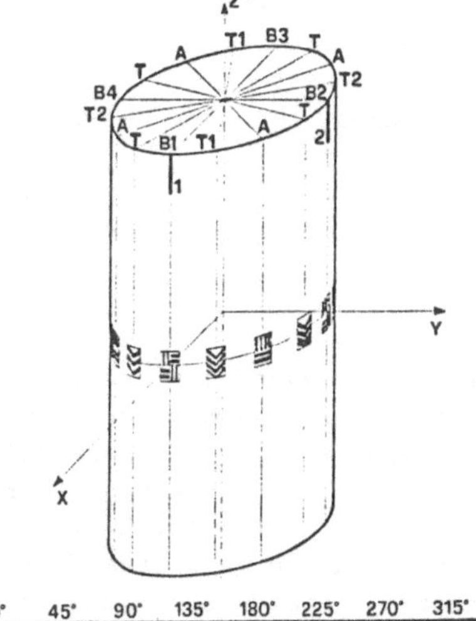

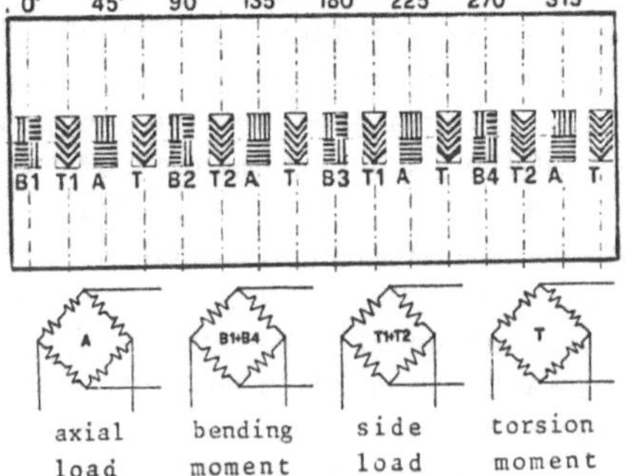

Fig. 4. Strain gage location and bridge wiring.

Axial load tests up to capacity value, performed against a PTB - calibrated 3 MN load cell on a 10 MN hydraulic press, lead to a third order polynomial equation; as no knife edges were used beyond 100 kN, localization of line of action of load could not be made within an uncertainty consistent with estimation of effect of moments. Rotation was therefore relied upon mainly in order to average out setup defects, as opposed to dynamometer's departures from ideal performances /8/.

Torque tests were performed with lever and deadweights; cardan suspension and splined shaft were used in order to prevent unwanted effects to creep in. The main results of tests are summarized in the exploitation coefficient table, which enables computation of applied force and moment components in terms of output signals.

4. ERROR SOURCES IN SINGLE-BILLET UNITS

Given the strain gage location and bridge wiring arrangement of Fig. 4, pertaining to the unit under test, the effects of such departures from nominal situation as angular and linear positioning errors γ and δ, and scatter in gage factor k, may be estimated from a theoretical point of view, according to lines developed for cross-sensitivity evaluation and compensation /9/.

Taking e.g. the case of torque bridge, evaluating Mz through output U6, individual gage outputs due to moment Mx are given in Fig. 5 according to Mohr's circle and strain distribution, in terms of nominal bending strain ϵ_{Mx} at the corresponding location. Individual gage factors then enter bridge equation to affect output.

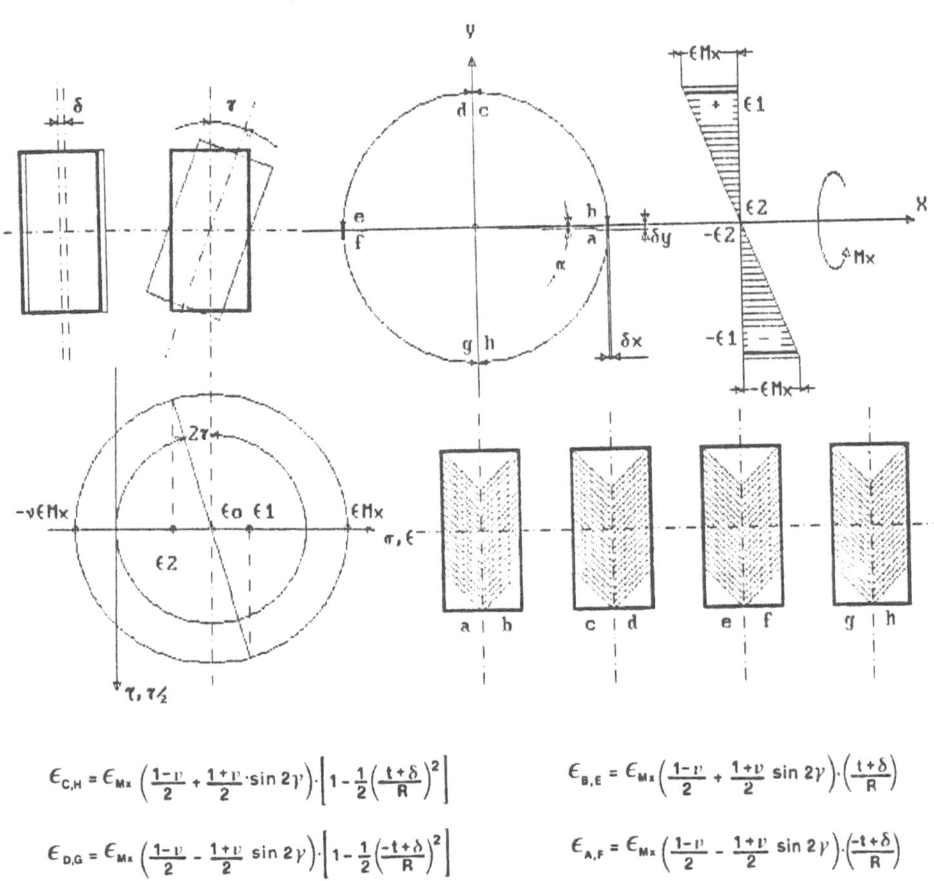

Fig. 5. Strain distribution and Mohr's circles for calculation of the effects of γ and δ on torque output U6 due to Mx.

Algebra of normal functions may then be resorted to in order to estimate output average and variance as functions of individual error components/10/. Rather unwieldy equations are obtained in all but the simplest cases, and recourse to Montecarlo methods is found to be more expeditious as a rule. Owing to the form of gage output equations obtained, normal distribution of parameters ν, δ and k does not imply normality of distribution of dependent variable. Departures are however rather moderate, and appropriate empirical distributions are available offering a better fit if need be /11/. Main results are given in Table 2, where standard errors of force and moment estimates are given (in kN/(mV/V), and kN·m/(mV/V)) as produced by a standard error of gage factor of 0.5%, and of linear and angular positioning errors of 0.1 mm and 0.01 rad respectively.

Mean values obtained by computer simulation are found to agree substantially with experimental values, given in Table 3. The existence of sizable departures from the expected zero value of several non diagonal terms is explained by the data of Table 2, but for the case pertaining to sensitivity to bending moments Mx and My respectively of side force channels U1 and U2. In these instances the unwanted sensitivity is explained by the distance between strain gage grids and neutral axis, which could only be averaged out in the bridge at the expense of torque sensitivity.

TABLE 2. Estimates obtained by Montecarlo simulation of standard errors, of the exploitation coefficients for evaluations of forces and moments, these being given in kN, N·m, and outputs in mV/V.

2	1	4	0	3	2
1	2	4	3	0	2
9	9	3	1	1	13
1	209	56	21	11	216
209	1	56	11	21	216
27	27	155	56	56	45

TABLE 3. Experimental values of the exploitation coefficients for evaluation of forces and moments, these being given in kN, N·m, and outputs in mV/V.

495	0	-7.3	-6.5	0.4	0.3
0	499	6.0	1.6	7.2	0.5
0	0	1250	0	0.4	15.5
0	138	-48.5	7150	3.4	60.3
-6.6	0	29.5	20.6	7150	386
0	0	99	0	-43.4	15100

5. DISCUSSION

Owing to the lack of comprehensive six-component deadweight calibration facilities, verification of multicomponent dynamometers in the meganewton range turns out to be a rather demanding exercise, whose results are invariably of a lower quality than what can be achieved at lower load order of magnitude.

Several setups are required to make up even a barely adequate number of loading combination, and the combined effect of changes in interface conditions and positioning errors does not fail to affect results.

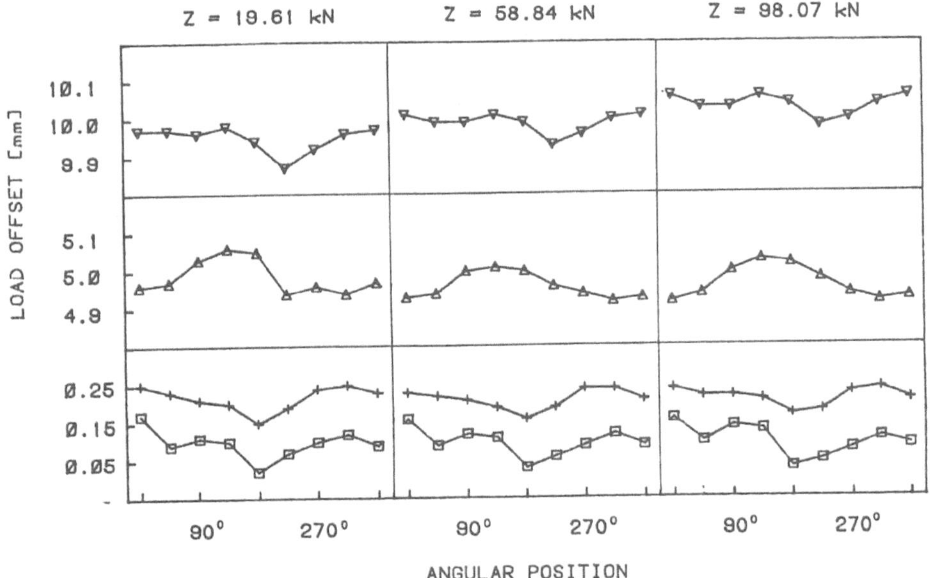

Fig. 6. Load offset measured in different loading conditions:
10 mm nominal offset (▽) and 0 mm nominal transverse offset(+);
5 mm nominal offset (△) and 0 mm nominal transverse offset(□).

The need of resorting to different setups produces however also positive results, when some overlapping of loading combination can be designed among the different arrangements.

Redundancy may then be exploited to provide validation of results, and estimate of deviation between planned and actually applied loading combinations. An example is given concerning the analysis of test results obtained from eccentric load application, axial and bending moment sensitivities being derived from axial tests to capacity load and separate side load testing, with several moment arm levels and angular orientations. Fig. 6 shows the main results; the expected sine law oscillation of moments is observed on unit's reference system, while on a stationary reference system moments variation is consistent with ± one tenth of a millimeter scatter in moment arm, inclusive of knife edge centering and fixture positioning errors. Similarly, axial loading with knife edges yields typical deviation of measured side loads from the expected zero value of the order of 20 N for axial loads up to 100 kN, corresponding to an overall inclination error of 0.2 mrad.

On the basis of the foregoing results, it can be expected that this 2-MN dynamometer, originally developed to check materials testing machines, can yield useful indications when checking primary high-capacity force standards, if proper account is taken of end effects.

Summing up, by comparing estimates of effects obtained with different methods, by collecting and scrutinizing carefully every bit of evidence obtained experimentally, and by checking them against simple theoretical models, information loading to reasonable sets of calibration and exploitation coefficients may be gathered, thus enabling meaningful analysis of parasitic components also on large capacity machines.

REFERENCES

1. Bray A: The role of stress analysis in the design of force-standard transducers. The William M Murray Lecture, 1980. Experimental Mechanics, v. 21, n. 1, p. 1,20, 1981
2. Dubois M: Private communication, 1971
3. Levi R: Multicomponent calibration of machine tool dynamometers. Trans. ASME, B, v. 94, p. 1067, 1972
4. Barbato G, Bray A, Levi R: Multicomponent dynamometers for control of parasitic components on force standard machines. Proceedings of the International Conference on Experimental Mechanics, p. 784, Science Press, Beijing, China, 1985
5. Barbato G, Bray A, Germak A, Levi R: Criteri di progettazione ed analisi delle caratteristiche metrologiche di dinamometri a più componenti. Atti XIII Convegno AIAS, p. 255, Bergamo, 1985
6. Barbato G, Germak A, Li Z Q, Trevissoi F: Determinazione delle caratteristiche metrologiche di un dinamometro multicomponenti da 2 MN per calcestruzzo. Rapporto Tecnico IMGC R198, Torino, 1984
7. Bray A: The influence of contact stresses on the characterization of a load cell. VDI Berichte n. 176, p. 43, 1972
8. Dubois M, Bourateau JP, Gosset A, Priel M: Intercomparaison des trois bancs d'etalonnage dynamométrique de capacité 250-300 kN, agreés par le BNM, Bulletin BNM, n. 41, 1980
9. Levi R: La sensibilità trasversale dei dinamometri per la misura delle forze di taglio, Atti e Rass. Tecn. Soc. Ing. Torino, v. 19, n. 2, 1965.
10. Haugen EB: Probabilistic Approaches to Design, Ch. 3, J. Wiley, New York, 1968.
11. Hahn GJ, Shapiro SS: Statistical Models in Engineering, J. Wiley, New York 1967.

INVESTIGATION OF THE CALIBRATION SIGNAL OF FORCE MEASURING DEVICES

G. RAMM AND M. PETERS

PHYSIKALISCH-TECHNISCHE BUNDESANSTALT, FRG

ABSTRACT

In international comparison measurements of force standard machines, the uncertainty of measurement is limited in essence by the uncertainty of measurement of the transfer standards, i.e. force transducer and measuring instrument. Separate investigations are necessary in order to allow the individual influences to be distinguished. The problems encountered in connection with force transducers as, for example, rotation and overlapping effect have been reported on repeatedly /1,2/.

For calibration of the measuring instrument, the transducer is replaced by a so-called bridge standard which allows defined voltage ratios to be realized and applied to the measuring instrument. Bridge standards comprise either resistance dividers (suitable for alternating and direct current) or transformers (suitable only for alternating current), the voltage ratio 2 mV/V being the calibration signal most frequently used. In the present paper a special measuring device for the calibration of bridge standards will be presented with which voltage ratios of about 2 mV/V can be determined with a relative uncertainty of $5 \cdot 10^{-6}$. The measuring frequency is 225 Hz, input voltages of 5 V to 10 V are possible.

1. INTRODUCTION

Fig. 1a shows a force measuring device consisting of the transducer FT and the measuring instrument MI. For the force transducer the ratio of input voltage U_{ad} to output voltage U_{bc} is proportional to the force F introduced:

$$U_{bc}/U_{ad} = K_1 \cdot F \qquad (1)$$

The measuring instrument MI arranged behind supplies the input voltage to the force transducer and amplifies its output voltage. In a digital/analog converter DAC the indicated value I_v is formed which is proportional to the voltage ratio U_{bc}/U_{ad}:

$$I_v = K_2 \cdot U_{bc}/U_{ad} \qquad (2)$$

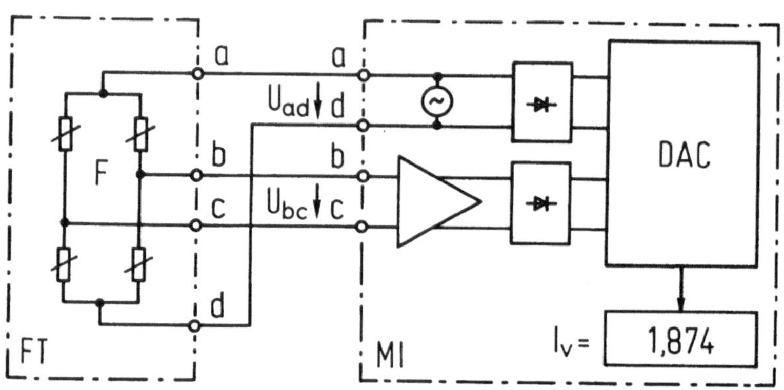

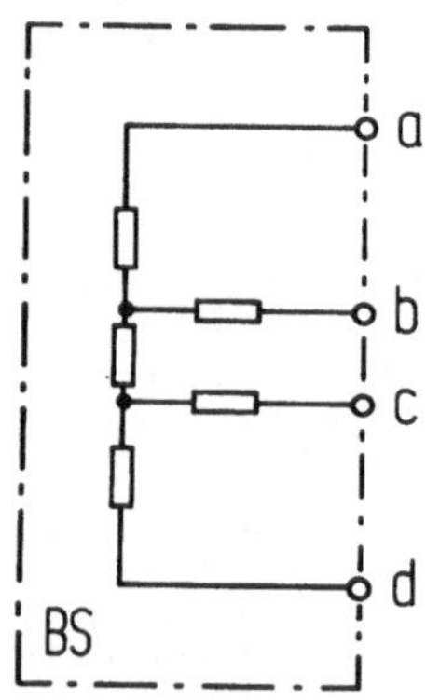

FIGURE 1. a) Force measuring device consisting of force transducer FT and measuring instrument MI. I_v is the indicated value.

b) Bridge standard BS for the calibration of the measuring instrument.

With (1), the following relationship between force and indicated values is obtained for the force measuring device:

$$I_v = K_1 \cdot K_2 \cdot F \qquad (3)$$

For electrical calibration of the measuring instrument the force transducer is uncoupled and a bridge standard BS (Fig. 1b) is substituted. Bridge standards allow exactly defined ratios of output voltage U_{bc} to input voltage U_{ad} to be adjusted. With the signal emitted the transfer factor K_2 of the measuring instrument is calibrated. The voltage ratio

$$U_{bc}/U_{ad} = D_k = 0{,}002 \mathrel{\hat{=}} 2 \text{ mV/V}$$

is the calibration signal most frequently used. In the following, a method of measurement for the investigation of this calibration signal is described.

2. PRINCIPLE OF THE METHOD OF MEASUREMENT

Inductive voltage dividers (IVD) allow very exact AC voltage ratios to be realized largely independently of temperature and with long-term stability /3,4,5/. Fig. 2 shows the principle of an adjustable IVD. For the ratio of ouput voltage U_o to input voltage U_i, the following equation is obtained:

$$U_o/U_i = D \pm u \tag{4}$$

D is the transfer ratio; values between zero and one ($0 \leq D \leq 1$) can be adjusted. As precise IVDs contain seven or eight decades, the resolution in the transfer ratio is 10^{-7} or 10^{-8}. u is the uncertainty of the arbitrary transfer ratio D. Values of down to $u = 1 \cdot 10^{-7}$ (all uncertainties stated in this paper correspond to a confidence level of 95%) can be obtained at a measuring frequency of 225 Hz. For inductive voltage dividers, input and output voltage are related to a common terminal which is usually connected to ground. Contrary to this, bridge standards are powered so that the input voltage is balanced to ground and deliver a balanced-to-ground output voltage. The circuit shown in Fig. 3 allows the voltage ratio

$$U_{bc}/U_{ad} = D_k \tag{5}$$

of a bridge standard powered so that the input voltage is balanced to ground to be compared with the transfer ratio D of an IVD.

The transformer Tr whose primary winding lies at the generator G and whose secondary winding is grounded in the middle balances the input voltage U_{ad}. The inputs of the IVD and of the bridge standard are connected in parallel. When switch S is in its position "b", the transfer ratio D is adjusted - with regard to the null indicator NI - until the output voltage of the divider is equal to the voltage U_{bd}.

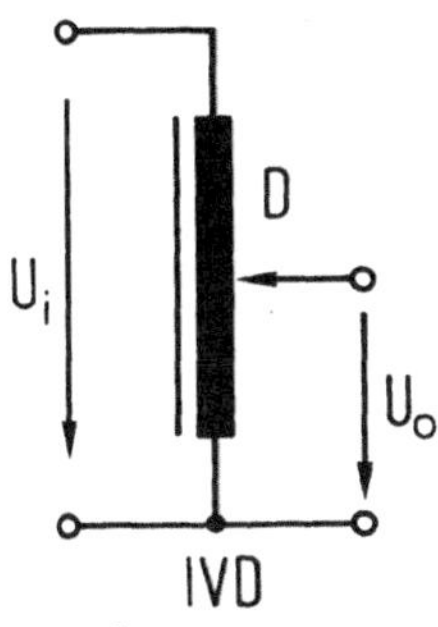

FIGURE 2. Inductive voltage divider IVD with adjustable transfer ratio D.

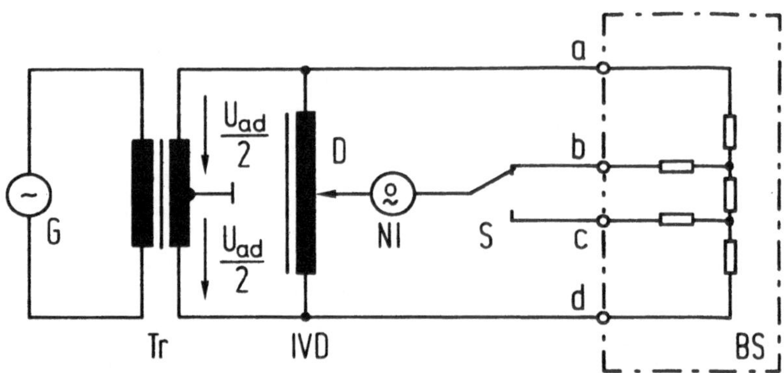

FIGURE 3. Principle of the method of measurement for the calibration of bridge standards (G=generator, Tr= transformer, NI = null indicator, S = switch)

The following then holds:

$$U_{bd}/U_{ad} = D_b \pm u_b \qquad (6)$$

A second balance with switch S in its position "c" yields

$$U_{cd}/U_{ad} = D_c \pm u_c \qquad (7)$$

The difference between (6) and (7) is the voltage ratio according to (5) searched:

$$D_k = U_{bd}/U_{ad} - U_{cd}/U_{ad} = U_{bc}/U_{ad} \qquad (8)$$

or

$$D_k = D_b - D_c \pm u_b \pm u_c = D_x \pm u_x \qquad (9)$$

The overall uncertainty u_x is determined as the root mean square of the individual components u_b and u_c to be

$$u_x = \sqrt{u_b^2 + u_c^2} \qquad (10)$$

With $u_b = u_c = u$, the following results from (9) and (10):

$$D_k = D_x \pm 1{,}4 \cdot u \qquad (11)$$

When a calibration signal of about 2 mV/V is investigated, with $D_x = 0{,}002$ and $u = 1 \cdot 10^{-7}$, this method of measurement yields

$$D_k = 0{,}002 \pm 1{,}4 \cdot 10^{-7} = 2 \cdot (1 \pm 7 \cdot 10^{-5})\,\mathrm{mV/V}.$$

For various applications, the relative uncertainty of $7 \cdot 10^{-5}$ of this method which can be attained for the determination of calibration signals of about 2 mV/V is too large.

3. SPECIAL MEASURING DEVICE FOR CALIBRATION SIGNALS OF ABOUT 2 mV/V

A special measuring device was developed with a view to reduce the uncertainty of measurement. The principle is shown in Fig. 4. Other than in Fig. 2, an additional inductive voltage divider IVD 2 is used. It comprises two fixed taps balanced towards the middle, with the transfer ratios D_{21} and D_{22}. Only a small portion of the input voltage U_{ad} of the bridge standard is thus still applied to the adjustable IVD 1 arranged behind, by means of which a very high resolution becomes possible. Again two balancing operations are carried out with switch S in its positions "b" and "c". Without taking the uncertainties into account, the following equations are obtained:

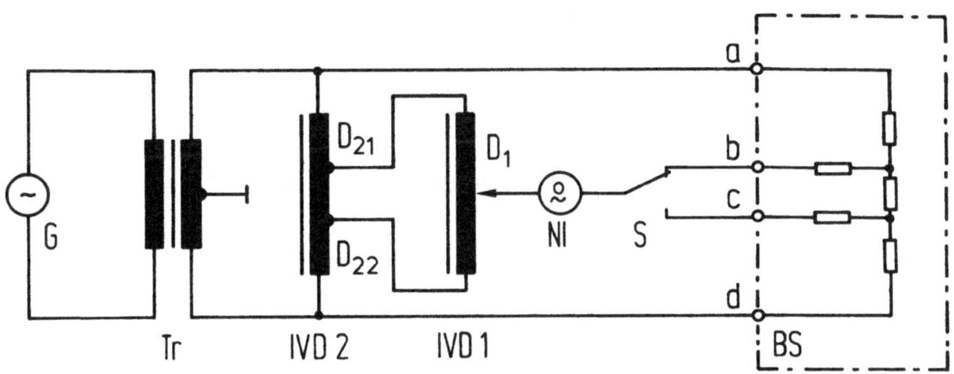

FIGURE 4. Principle of the special measuring device for calibration signals of about 2 mV/V. IVD 2 is an inductive voltage divider with fixed taps balanced towards the middle.

$$U_{bd}/U_{ad} = D_{22} + (D_{21} - D_{22}) \cdot D_{1b} \quad (12)$$

$$U_{cd}/U_{ad} = D_{22} + (D_{21} - D_{22}) \cdot D_{1c} \quad (13)$$

The difference is

$$U_{bc}/U_{ad} = D_x = (D_{21} - D_{22}) \cdot (D_{1b} - D_{1c}) \quad (14)$$

The transfer ratios D_{21}, D_{22}, D_{1b} and D_{1c} are subject to the uncertainties $u_{21} = u_{22} = u_{1b} = u_{1c} = u$.

When products of small quantities are neglected, with $(D_{21} - D_{22}) = D_2$ and $(D_{1b} - D_{1c}) = D_1$,

$$D_k = D_x \pm u_x = D_1 \cdot D_2 \pm 1{,}4 \cdot u \cdot \sqrt{D_1^2 + D_2^2} \qquad (15)$$

can be calculated.

Again $D_x = D_1 \cdot D_2 = 0{,}002$ is assumed corresponding to the calibration signal 2 mV/V to be investigated. This condition can be fulfilled with various combinations for D_1 and D_2 as, for example, with $D_1 = 0{,}1$ and $D_2 = 0{,}02$. The choice of D_1 and D_2 has an influence on the uncertainty u_x. Compared with equation (11) is the uncertainty in any case reduced by the factor $\sqrt{D_1^2+D_2^2}$. The smallest uncertainty is obtained with $D_1 = D_2 = \sqrt{D_x} = 0{,}04472$ ... . This value can be approximated with the transfer ratio D_1 of the adjustable inductive voltage divider IVD 1 but it cannot be realized for the fixed transfer ratio D_2 of IVD 2. Therefore $D_2 = 0{,}04$ has been selected; for D_1, a value of 0,05 is then obtained. When $u = 1 \cdot 10^{-7}$ is used again, the following equation is obtained from equation (15):

$$D_k = D_x \pm u_x = 0{,}002 \pm 1 \cdot 10^{-8} = 2 \cdot (1 \pm 5 \cdot 10^{-6})\,\text{mV/V} \qquad (16)$$

The result shows that this special measuring device allows voltage ratios of about 2 mV/V to be determined with a relative uncertainty of only $5 \cdot 10^{-6}$. In contrast to the circuit in Fig. 3, the uncertainty can be reduced by the factor 14. In practice, however, this small relative uncertainty can be attained only when the inductive voltage dividers are designed as two-stage transformers /4/ with magnetizing winding. Moreover, all lines must be thoroughly twisted and shielded and the bridge standard must be connected via a six-wire circuit /6/.

With the measuring device not only voltage ratios of about 2 mV/V can be measured: the overall range of measurement rather extends from 0 mV/V to ± 40 mV/V, and the corresponding uncertainties can be calculated with the aid of equation (15). For $D_2 = 0{,}04$ and $u = 1 \cdot 10^{-7}$, the results for several voltage ratios within the range $0{,}0005 \leq D_x \leq 0{,}04$ are listed in the following table. In all cases, the resolution of the

D_x	0,0005	0,0010	0,0020	0,0040	0,0100	0,0400
u_x in 10^{-6} mV/V	6	7	9	15	35	140
u_x/D_x in 10^{-6}	12	7	5	4	4	4

measuring device is $0{,}4 \cdot 10^{-6}$ mV/V provided that an eight-decade inductive voltage divider is used as IVD 1. To allow this high resolution to be fully utilized, the null indicator must be equipped with a low-noise preamplifier /7/.

4. SUMMARY

The measuring device presented allows calibration signals of about 2 mV/V emitted by bridge standards to be determined with a relative uncertainty of $5 \cdot 10^{-6}$. The measuring frequency is 225 Hz; input voltages of 5 V to 10 V are possible. This presupposes the use of two inductive voltage dividers connected in series, which allow the resolution to be substantially increased as compared with conventional methods and the uncertainty to be reduced at the same time. With the aid of the bridge standards calibrated by this principle, the measuring instruments necessary for force measurement can be calibrated substantially more accurately than has been hitherto possible. For international comparison measurements of forces, this opens up the possibility of attaining in future relative uncertainties of measurement of less than $2 \cdot 10^{-5}$.

5. REFERENCES

1. Peters M.: Reasons for, and consequences of, the rotation and overlapping effect. Proc. R.T. Disc. 10th IMEKO TC-3, Japan, 1-5, 1984.
2. Peters M.: Limits to the uncertainty achievable in force transfer. Proc. Weightech 83, London, 1-9, 1983.
3. Hill J. J., Miller A.P.: A seven-decade Adjustable-Ratio Inductively-Coupled Voltage-Divider with 0,1 ppm Accuracy. Proc. IEE, Vol. 109 B, pp. 157-162, March 1962.
4. Deacon T.A., Hill J.J.: Two-stage Inductive Voltage Dividers. Proc. IEE, Vol. 115, pp. 888-892, June 1968.
5. Ramm G., Vollmert R., Bachmair H.: Microprocessor-Controlled Binary Inductive Voltage Dividers. IEEE Trans. Instrum. Meas. IM-34, No. 2, pp. 335-337, 1985.
6. Kreuzer M.: Kalibrieren des Digitalen Präzisions-Meßgerätes DMP 39 mit einem speziellen Brückennormal. Messtechnische Briefe 17, No. 3, S. 67-73, 1981.
7. Ramm G.: Rauscharme Wechselspannungsverstärker für den nV-Bereich. Elektronik, No. 5, S. 82-86, 1983.

ECCENTRIC LOAD SENSITIVITY OF FORCE SENSORS

R. A. MITCHELL, R. L. SEIFARTH, C. P. REEVE

NATIONAL BUREAU OF STANDARDS, GAITHERSBURG, MARYLAND 20899 USA

1. INTRODUCTION

Strain gage force sensors are ordinarily designed with a symmetry of form and gage location that makes them relatively insensitive to eccentric load. In general, however, there is a residual eccentric load sensitivity due to imperfections in machining, assembly, gage location, etc. that may be significant. Eccentric load sensitivity is a particularly important characteristic of force sensors that are used as transfer standards either to intercompare force standard machines or to calibrate testing machines or other force sensors. Since load alignment in these processes is not perfect, there are unknown errors due to differences in alignment of the different setups that are compared. Eccentric load sensitivity is an important factor contributing to these errors.

A relatively simple and inexpensive test has been developed at the United States National Bureau of Standards (NBS) to measure the eccentric load sensitivity of universal (tension and compression) force sensors loaded in compression. The same test can also be used to calibrate the transverse bending moment bridges of multi-axis force-moment sensors. SSome results of both applications are given here. The NBS approach was suggested in part by the work of Levi (Ref. 1 and 2) who used eccentric loading fixtures to study force sensors and machine-tool dynamometers.

The measured eccentric load sensitivity could be used in several ways. It could serve as an objective index of performance in either the development, production control, or procurement of force sensors. It could be a basis for estimating the uncertainty of a particular force measurement due to either a controlled or an assumed degree of load eccentricity. Perhaps its most important application will be in the selection of force sensors for the extremely precise measurements needed to intercompare force standard machines. An important example of the latter application is the determination of the effective lever ratio of a deadweight machine that has either mechanical or hydraulic load multiplication.

2. TEST DESCRIPTION

Five commercially produced force sensors used to evaluate and demonstrate the test method are shown in Fig.1. Sensors I, II, III, and IV each have two axial force sensing bridges, although one bridge of sensor IV was found to be too unstable to use. Sensor V has one axial bridge and two transverse bending bridges. Installed in each sensor is the threaded loading fixture and spherical ball through which eccentric loads were applied. The stainless steel threaded loading fixtures, removed from the force sensors, are shown in Fig. 2. Two eccentric conical ball seats are visible in fixture V and one eccentric seat is visible in each of the other four fixtures. The other end of each fixture has a concentric ball seat. The machined flats are used to rotate the fixture in the force sensor. Table 1 further describes the force sensors, threaded fixtures, and loading balls.

TABLE 1. Description of force sensors and loading fixtures.

Force Sensor	Serial Number	Sensor Cap. klbf (kN)	Sensor Ht. mm	Sensor Dia. mm	Thread(UNF) Dia. in (mm)	Thread(UNF) Pitch in^{-1} (mm^{-1})	Fixture Ht. mm	Fixture Ecc. mm	Ball Dia. mm
I	83874	5 (22.2)	150	90	1 (25.4)	14 (0.55)	28.6	2.08	11.1
II	34655	2.5 (11.1)	117	86	0.5 (12.7)	20 (0.79)	14.2	1.04	10.0
III	84835	50 (222.4)	290	166	2 (50.8)	12 (0.47)	57.2	3.93	15.9
IV	45629	25 (111.2)	215	123	1.5 (38.1)	12 (0.47)	42.8	2.98	12.7
V	19724	25 (111.2)	89	155	1.25 (31.8)	12 (0.47)	35.6	2.51 6.99	12.7

Three NBS deadweight machines were used to apply vertical test loads. The estimated uncertainty of the vertical forces applied by the machines is 0.002 percent. A test load of only 40 percent of nominal capacity was applied to force sensor V to avoid excessive plastic deformation of the two closely spaced ball seats in that fixture. Each of the other four sensors was tested at nominal capacity load.

2.1. Test setup

The force sensor is placed on the compression platen of the deadweight machine with the cable connector box oriented toward the front of the machine (see Fig. 3). The threaded loading fixture is screwed into the sensor until the threads are approximately fully engaged. The fixture is not allowed to touch the bottom of the threaded hole and therefore the entire load is transmitted through the threads. The fixture is then rotated until its axis, as defined by the eccentric ball seat location, coincides with the axis of the sensor, as defined by the cable connector. This is the zero degree orientation of the fixture. The sensor is then

Figure 1. Force sensors with loading fixtures and loading balls installed, shown in the order (left to right) III, V, IV, II, I.

Figure 2. Loading fixtures, shown in the order (left to right) III, II, IV, I, V.

translated on the machine platen, without rotation, until the ball seat is located in line with the vertical loading axis of the machine.

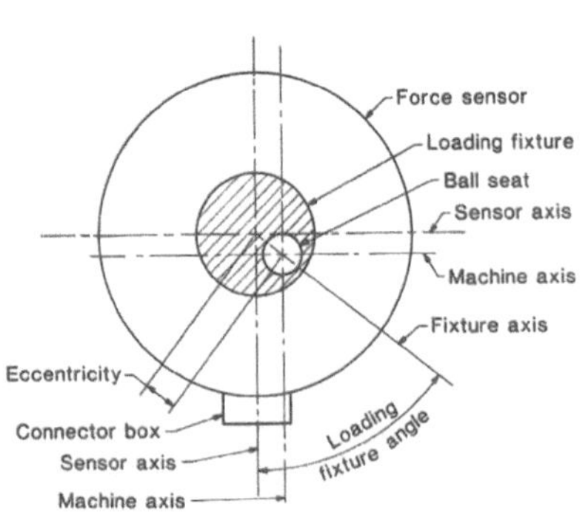

Figure 3. Geometry of test setup, as viewed from above the force sensor.

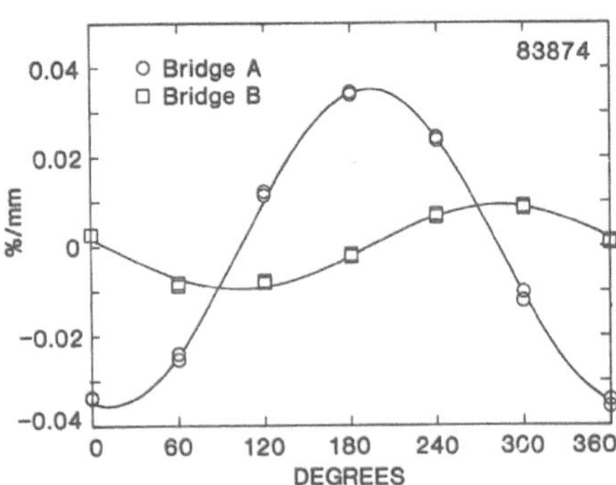

Figure 4. Eccentric load sensitivity of bridge A and bridge B of force sensor I.

2.2. Loading procedure

With the sensor and loading fixture in the initial setup position, three or more preload cycles are applied until the sensor output is stable. Two replicate test loads are then applied. The threaded loading fixture is next rotated 60 deg. counterclockwise and the sensor is translated, without rotation, until the ball seat is again located in line with the vertical loading axis of the machine (see Fig. 3). Two replicate test loads are then applied at this 60 deg. orientation. The process is repeated in 60 deg. increments to 360 deg. The 360 deg. orientation of the loading fixture coincides with the zero deg. initial orientation; however, one less thread is engaged.

3. STATISTICAL MODEL

The test results were analyzed in terms of a sinusoidal model which appeared to be a good representation of the effects of the underlying physical processes. The model assumes that the response of a load cell under eccentric loading consists of a constant term C, a sinusoidal term of period 2π with unknown phase ϕ and amplitude A, and a small error term. Measurements are made at n equally spaced angles $\theta_1, \theta_2, \ldots, \theta_n$. Let m be the number of replicate measurements at each θ_i ($m \geq 1$). The model can then be written

$$y_{ij} = C + A \sin(\theta_i - \phi) + \varepsilon_{ij}$$

for $i = 1, 2, \ldots, n$ and $j = 1, 2, \ldots, m$ where y_{ij} is the response and the ε_{ij} are independent error values from a distribution with mean zero and variance σ^2.

The above model, which is nonlinear in the unknowns, can be written in the linear form

$$y_{ij} = C + a \sin\theta_i + b \cos\theta_i + \varepsilon_{ij}$$

where a = A cosϕ and b = -A sinϕ. Estimates of the unknowns are then computed by linear least squares. When the θ_i are equally spaced these estimates may be expressed in the simplified form

$$\hat{C} = \frac{1}{n}\sum_{i=1}^{n} \bar{y}_i \quad , \hat{a} = \frac{2}{n}\sum_{i=1}^{n} \bar{y}_i \sin\theta_i \quad , \text{ and } \hat{b} = \frac{2}{n}\sum_{i=1}^{n} \bar{y}_i \cos\theta_i$$

where $\bar{y}_i = \frac{1}{m}\sum_{j=1}^{m} y_{ij}$ for i = 1, 2, ..., n. The least square estimates of A and ϕ are then

$$\hat{A} = (\hat{a}^2 + \hat{b}^2)^{1/2} \quad \text{and} \quad \hat{\phi} = \arctan(-\hat{b}/\hat{a})$$

where arctan is the four quadrant inverse tangent function.

4. TEST RESULTS

Results of eccentric-load tests of the axial force sensing bridges of the five sensors are given in Figs. 4 through 9. Bridge output, in percent of mean output per millimeter of load eccentricity, is indicated by the vertical scale. Angular orientation of the threaded loading fixture (see Fig. 3) is indicated by the horizontal scale.

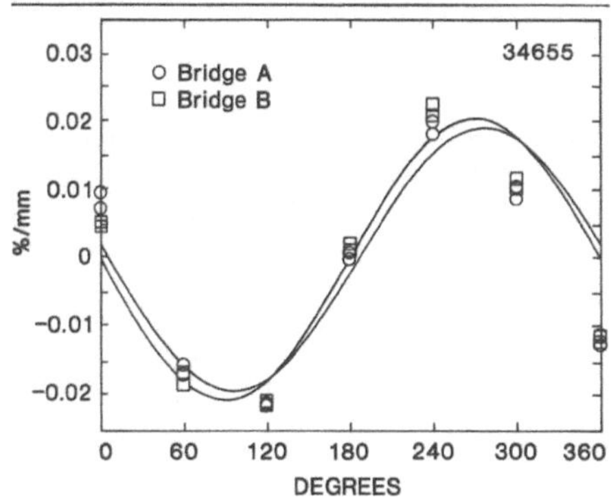

Figure 5. Eccentric load sensitivity of Bridge A and bridge B of force sensor II.

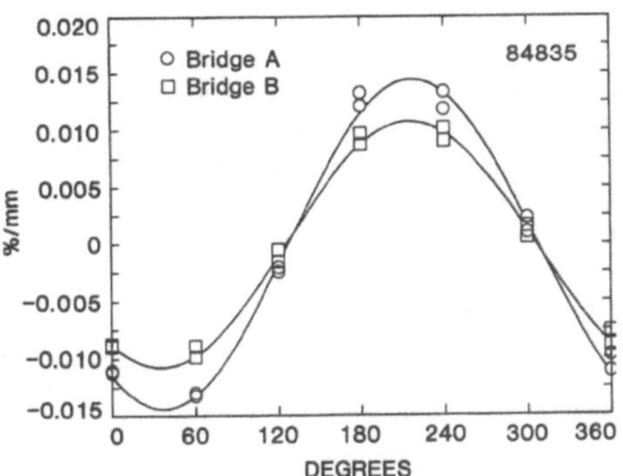

Figure 6. Eccentric load sensitivity of bridge A and bridge B of Force sensor III.

Results of eccentric-load tests of the transverse bending moment bridges of sensor V are given in Fig. 10. Bridge output, in millivolts per volt, is plotted versus the angular orientation of the threaded loading fixture.

The plotted sine curves were fitted to the replicate test data for only the six angular orientations from zero to 300 deg. The test data for 360 deg. are plotted for comparison even through they had no influence on the curve fit. Table 2 gives the eccentric load sensitivities, bending moment sensitivities, and phase angles obtained by fitting the sinusoidal model to the test data. The tabulated eccentric load sensitivity is the amplitude of the plotted sine curve. The tabulated bending moment sensitivity is the plotted curve amplitude divided by the applied bending moment (the product of the test load and the load eccentricity). The tabulated phase angle is the angle at which the sine curve has zero value and positive slope.

TABLE 2. Test results.

Force Sensor	Serial Number	Test Load klbf (kN)	Load Mach. klbf (kN)	Bridge	Ecc. mm	Ecc. Load Sens. %/mm	Bend. Mom. Sens. mV/V / N•m	Phase Angle deg.
I	83874	5	6.1	A	2.08	0.0353		102
		(22.2)	(27)	B		0.0093		191
			25.3	A		0.0375		100
			(113)	B		0.0088		192
II	34655	2.5	6.1	A	1.04	0.0191		186
		(11.1)	(27)	B		0.0204		180
III	84835	50	112	A	3.93	0.0143		126
		(222.4)	(498)	B		0.0107		125
IV	45629	25	25.3	B	2.98	0.0523		170
		(111.2)	(113)					
			112			0.0521		169
			(498)					
V	19724	10	25.3	Z	2.51	0.0153		30
		(44.5)	(113)					
				X	2.51		0.00148	255
				Y			0.00149	347
				X	6.99		0.00162	256
				Y			0.00160	346

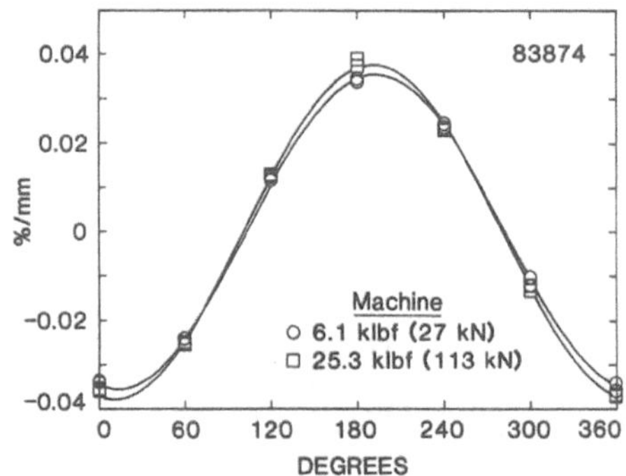

Figure 7. Eccentric load sensitivity of bridge A of force sensor I as measured in the 6.1 klbf (27kN) machine and as measured in the klbf (113 kN) machine.

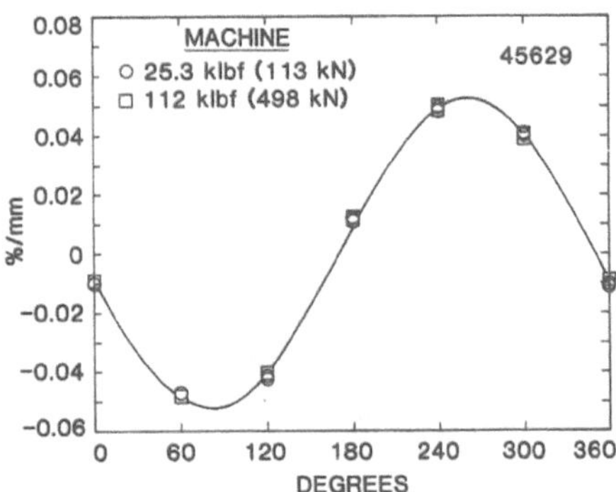

Figure 8. Eccentric load sensitivity of bridge B of force sensor IV as measured in the 25.3 klbf (113 kN) machine and as measured in 25.3 the 112 klbf (498 kN) machine.

4.1. Dual-bridge sensors

Both axial bridges of force sensors I, II, and III were tested. Figure 4 shows that bridge A of sensor I is about four times more sensitive to load eccentricity than is bridge B and that the two sensitivities have a phase difference of about 89 deg. Figure 5 shows a difference in eccentric load sensitivity and phase of the two bridges of sensor II of only about 7 percent and 6 deg., respectively. Figure 6 shows a difference in eccentric load sensitivity and phase of the two bridges of sensor III of about 34 percent and 2 deg., respectively. The actual configurations of the strain gage bridge circuits of these three force sensors are unknown to the authors. However, one might expect phase differences to correlate with differences in angular orientation of the bridge circuits.

4.2. Test in different machines

Force sensors I and IV were each tested in two different deadweight machines. Figure 7 shows a difference in eccentric load sensitivity of bridge A of sensor I, as determined in the 6.1 klbf (27 kN) and 25.3 klbf (113 kN) machines, of about 6 percent. Bridge B of sensor I indicated a difference of about 6 percent in the opposite direction (see Table 2). Figure 8 shows that the fitted curves for sensor IV, tested in the 25.3 klbf (113 kN) and 112 klbf (498 kN) machines, approximately coincide. These results for two force sensors, each tested in two different machines, suggest that the test method is essentially independent of the characteristics of the three deadweight machines.

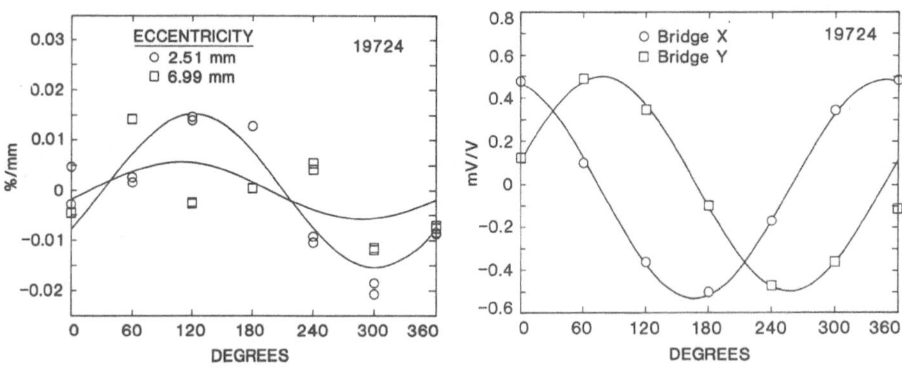

Figure 9. Eccentric load sensitivity of axial bridge Z for force sensor V as measured with a 2.51 mm eccentricity and as measured with a 6.99 mm eccentricity.

Figure 10. Bending moment sensitivity of transverse bending bridges X and Y of force sensor V as measured with a 6.99 mm eccentricity.

4.3. Test with different eccentricities

Axial bridge Z of force sensor V was tested using two different load eccentricities and the results are plotted in Fig. 9. The larger amplitude sine curve has been fitted to the data for an eccentricity of 2.51 mm; the smaller amplitude curve has been fitted to the 6.99-mm eccentricity data. The computed least squares fit to the 2.51-mm data appears to be good enough to justify its use as an approximate measure of eccentric load sensitivity. However, the computed least squares fit to the 6.99-mm data indicates that the sinusoidal model is inadequate and the results of this computation are not included in Table 2. The internal construction of force sensor V is in the form of multiple shear webs radiating out from a central hub to an outer rim. The strain gages of the force sensing bridge are mounted on the radial shear webs. Perhaps a model with two sinusoidal terms, one of period 360 deg. and one of a shorter period such as 90 deg., fitted to eccentric load measurements made at 15 deg. intervals, would better describe the response of such a complex force sensor.

4.4. Bending moment sensitivity

Transverse bending bridges X and Y of sensor V were tested using two different load eccentricities. The results for the larger eccentricity are plotted in Fig. 10 and the computed bending moment sensitivities and phase angles are given in Table 2. The bending moment sensitivities indicated on the name plate of sensor V for bridges X and Y are respectively 0.001641 and 0.001631 mV/V/N•m. These manufacturer's values differ from the results obtained with a 6.99 mm eccentricity by only 1.3 and 1.9 percent respectively for bridges X and Y.

5. CONCLUSION

A relatively simple and inexpensive test to determine the eccentric load sensitivity of force sensors has been developed. The use of the same test to calibrate the transverse bending moment bridges of a three-axis force-moment sensor has also been demonstrated. Most of the test data were well represented by a sinusoidal statistical model. If greater measurement precision is needed, it could probably be obtained by the following:

1. More precise control of the angular orientation of the loading fixture.
2. Measurements at a greater number of equally spaced angular orientations.
3. More replicate measurements at each angular orientation.
4. Measurements at different load levels and at different eccentricities to characterize the higher order effects of these variables.

Useful test results could probably be obtained with less precise equipment than was used here. A loading fixture even simpler than those shown in Fig. 2 was used in a preliminary test of sensor I. That fixture was made by sawing a 30 mm long piece from threaded bar stock and machining a ball seat with the conical point of a drill bit. The results obtained with that relatively crude fixture were consistent with those reported here. A complete deadweight machine would not necessarily be required to apply the test load. A test could be performed with a single test weight (not necessarily precisely known), a level and stiff compression platen, and a symmetrical loading yoke.

6. ACKNOWLEDGEMENT

The computer program for data analysis and plotting was developed by J. A. Hermoza. E. G. Erber provided valuable consultation in the precise measurement of load eccentricity.

7. REFERENCES

1. Levi, R.: Multicomponent Calibration of Machine-Tool Dynamometers, Journal of Engineering for Industry, Trans. ASME, Nov. 1972, pp. 1067-1072.
2. Levi, R.: Performance Evaluation of Strain-Gage Dynamometers, Proc. Round Table Discussion on Force Transfer to Force Measuring Devices, 6th IMEKO Congress, Dresden, June, 1973, pp. 61-90.

WAYS FOR THE REMOVAL OF THE MAIN DISADVANTAGES OF FORCE TRANSDUCERS WITH SEMICONDUCTOR STRAIN GAGES

J.LUKAS

Aeronautical research and test institute, Beranových 130, 199 05 Prague, Czechoslovakia

1. INTRODUCTION

The up-to-date electronic components allow to design such devices that are capable of very precisely measuring and processing the electric signals even in the microvolt range. This facility makes very easy the design of equipments for measurement of mechanical quantities, the input part of which is consituted by transducers containing the metallic strain gauges. The accuracy attained by using these equipments is admirably high. The violent development of electronics has influenced even an application of semiconductor strain gauges, the high sensitivity of which is no longer so much sought-for feature as it was before.

However, there are still areas in application of transducers where a high level of the primary output signal considerably facilitates solution of technical problems, and there are also the manufacturers, by whom the semiconductor strain gauges are being applied with a certain tradition, as it has become even in several establishments in Czechoslovakia.

Production of precise transducers of mechanical quantities with semiconductor strain gauges represents a whole complex of problems. A number of them is common to those encountered with widely applied transducers with metallic strain gauges however, the different behaviour of semiconductor strain gauges results in some more difficult problems such as the non-linearity of the output signal and its dependence on temperature. These last are the two problems that will be discussed in this paper.

2. ELECTRIC CIRCUIT FOR COMPENSATION OF THE OUTPUT SIGNAL FOR NON-LINEARITY AND TEMPERATURE EFFECTS

The main causes of a non-linear waveform of the output signal in the transducers followed are the intrinsic non-linearity of dependence $\Delta R/R = f / \varepsilon /$ of semiconductor strain gauges, the non-linearity of the output signal of the Wheatstone bridge and a scatter of parameters of semiconductor strain gauges - Ref.1. Especially for the series production is advantageous such a connection that allows to reach a precisely defined modification of the calibration dependence. Such a modification can be obtained, for example, by using the electric circuit depicted in Figure 1.

2.1. Description

Semiconductor strain gauges $A_{1,2}$ and $K_{1,2}$ bonded to the sensing element according to Fig.2 are connected in the Wheatstone bridge depicted in Fig.1. The bridge is supplied via a voltage divider which is formed by temperature dependent resistor R_4, by strain gauge A_3, the resistance of which depends also on load F to be measured, and by the other adjustable resistors. The change of resistor A_3 with the load applied allows the control of bridge

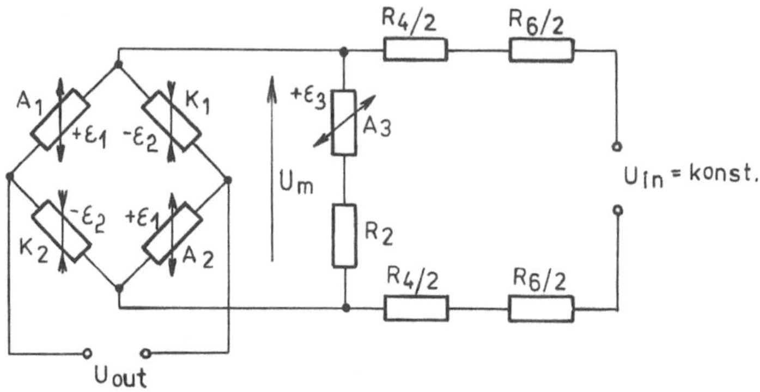

$A_{1,2,3}$; $K_{1,2}$ ... semiconductor strain gages

R_2 .... resistance for adjustment of nonlinearity

R_4 .... resistance for compensation of sensitivity change with temperature

R_6 .... resistance for adjustment of full scale

FIGURE 1 Compensation circuit of transducer

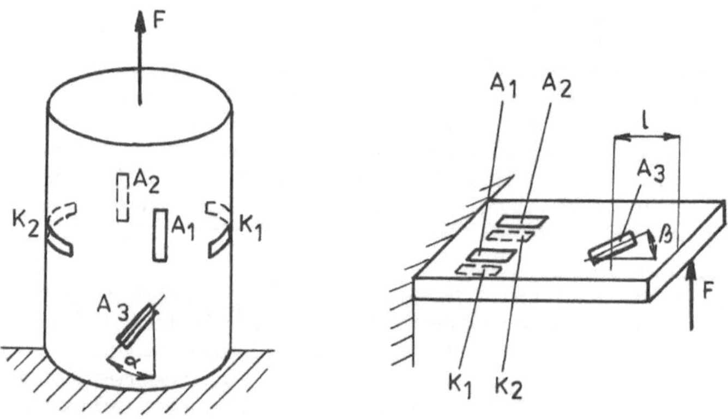

FIGURE 2 Location of strain gauges on sensing element

supply voltage U_m, and thus even the compensation of the transducer output signal for non-linearity. The control efficiency can be affected by changing parameters α , β and $\underline{1}$ according to Fig.2, by varying the value of resistor R_2 or by changing the behaviour of strain gauge A_3 which can be made of both the P- or N-type semiconductor material. It can be mathematically proved that the circuit depicted in Fig.1 allows decreasing the effect of temperature on the output signal even if temperature-dependent resistor R_4 is applied. However, in some cases this decrease is not sufficient and so it is necessary to use resistor R_4 with a negative temperature coefficient resistance realized, for example, by the parallel connection of a thermistor and an adjustable temperature-independent resistor. More preferable is application of such a semiconductor element A_3 which can compensate both the non-linearity and the temperature dependence of the output signal. The parameters of such an element are determined below.

2.2. Calcutation of resistance for non-linearity compensation

The dependence of the output signal for many types of transducers with statically balanced strain gauges connected in the Wheatstone bridge can be described by the following equation

$$U_{out}(F) = A.F - B.F^2 \qquad /1/$$

where A and B are constants. The non-linearity of this dependence can very well be compensated if the product of dependence of output signal U_{out} on bridge supply voltage U_m, and of dependence of supply foltage U_m on supply voltage U_{in} of the whole circuit constitutes a linear function. However, that is uneasy to be used for calculation of individual values of the circuit elements. Easier is to utilize the condition of equality of the percent drop of voltage U_{out}, caused by the quadratic term, between zero and the rated load, in accordance with the following equation

$$\frac{\Delta U_m}{U_m} = \frac{B.F_n^2}{U_n} = E \qquad /2/$$

where F_n is the rated load and U_n is the rated output. The applicability of condition /2/ can be judged even according to the scatter of values Δ U_m/U_m, determined for various values of the circuit resistors, at the same non-linearity of the calibration curve.

TABLE 1

Combination	M	A								
		1	2	3	4	5	6	7	8	9
$R_4 + R_6$ /Ω/	0	20			60			120		
R_2 /Ω/	∞	20	60	120	20	60	120	20	60	120
U_{in} /V/	6	8			11			17		

Table 1 lists various combinations of resistor values used in the circuit depicted in Fig.1, for which increments $\Delta U_m/U_m$ have been calculated and the non-linearity has been measured. In the above tests has been used the bending-stressed sensing element with the same $|\varepsilon|$ for strain gauges $A_{1,2,3}$ and $K_{1,2}$ which have been made of silicon with specific conductance $\varrho = 0.02\ \Omega$ cm. The combination designated by M corresponds to the system without any compensating strain gauge.

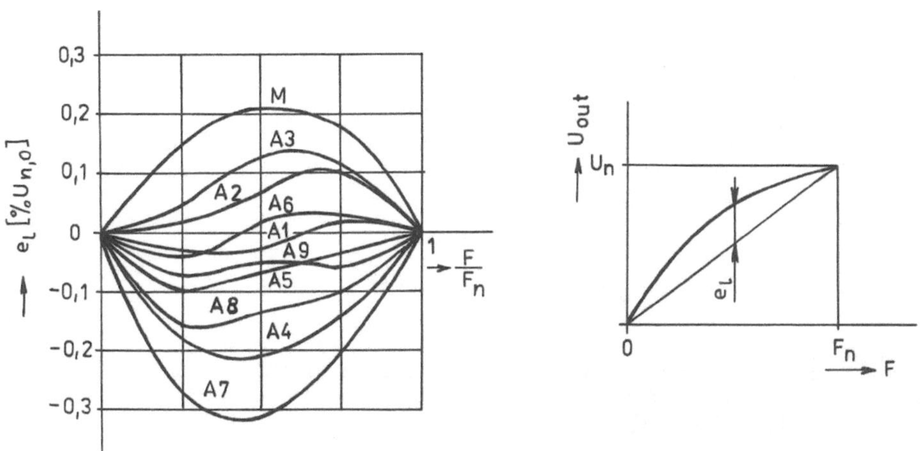

FIGURE 3 Non-linearity of calibration curve for various combination of resistors

Fig.3 illustrates the non-linearity for various combinations of resistor values according to Table 1 and the method of its evaluation. The graphic expression of the scatter to be examined is depicted in Fig.4. Each circuit combination is plotted by one point, the coordinates of which are the relative increment of the supply voltage and the non-linearity easured at the point for $F_n/2$. Interconnected are the points of those combinations where only resistor R_2 is changed. The maximum scatter is 0.17%. In practice, the scatter will be even lower because the probability of application of configurations A6 and A9 is very low. When respecting condition /2/, for determined of the value of resistor R_2 can be derived the following equation:

$$R_2^2 + R_2\left[(2A_{3N,0} + \Delta A_{3N}) + \frac{R_0(R_4+R_6)}{R_0+R_4+R_6}\right] + A_{3N,0}(A_{3N,0} + \Delta A_{3N}) + \qquad /3/$$

$$+ \frac{R_0(R_4+R_6)}{R_0+R_4+R_6}\left(A_{3N,0} - \frac{1}{E}\cdot \Delta A_{3N}\right) = 0$$

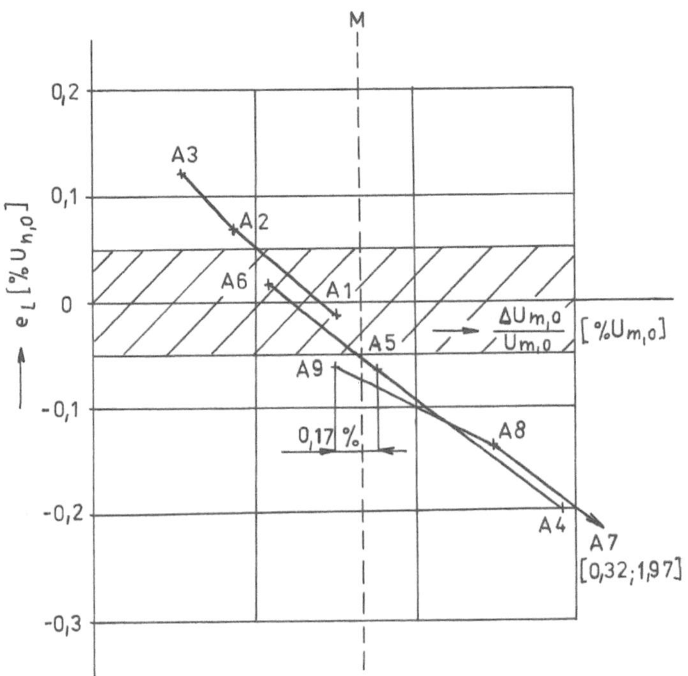

FIGURE 4 Scatter of changes in supply voltage for various combinations of resistor values

where $A_{3N,O}$ is the resistance of bonded strain gauge at temperature T_o, ΔA_{3N} is the change in resistance of strain gauge A_3 at the rated load, and R_o is the input resistance of strain gauges $A_{1,2}$ and $K_{1,2}$ in the Wheatstone bridge connection at temperature T_o.

2.3. Calculation of resistance for compensation of temperature effect on output signal

If there are used strain gauges $A_{1,2,3}$ and $K_{1,2}$ of the same type, interconnected according to Fig.1, the temperature effect on the rated output will be rather lower but will not be rejected completely. To achieve this, another temperature-dependent resistor R_4 should be added to the supply branch as mentioned in the introduction of Chapter 2. The condition of equality of the rated output at temperatures T_o and T_1 is given by the following equations:

$$\frac{U_{n,T1} - U_{n,o}}{U_{n,o}} = \frac{U_{m,o} - U_{m,T1}}{U_{m,T1}} \qquad /4/$$

and/or

$$\frac{U_{n,T1} - U_{n,o}}{U_{n,T1}} = \frac{U_{m,o} - U_{m,T1}}{U_{m,o}} \qquad /5/$$

where $U_{n,T1}$ and $U_{n,o}$ are rated outputs at temperatures T_1 and T_o, and $U_{m,T1}$ and $U_{m,o}$ are voltages U_m at temperatures T_1 and T_o. Into the calculation should be involved even resistor R_6 intended for adjustment of amount of the rated output and given by equation

$$R_6 = \frac{(A_{3N,o} + R_2)\, R_o}{A_{3N,o} + R_2 + R_o} \cdot \frac{a \cdot U_{n,o} - U_{o,n}}{U_{o,n}} - R_4 \qquad /6/$$

$$a = \frac{U_{in}}{U_{in1}} \qquad /7/$$

where $U_{o,n}$ is a desired rated output, U_{in1} is the bridge supply voltage at which are measured voltages $U_{n,o}$ and $U_{n,T1}$ without resistors R_4 and R_6. If, for example, the parallel connection of thermistor R_7 and temperature-independent resistor R_8 is used in place of resistor R_4, the value of resistor R_4 at temperature T_o can be obtained from the formula

$$R^2_{4,0} = R_{7,0} \cdot R_0 \, \frac{D(1+\alpha_R \Delta T)(a \cdot U_{n,T1} - U_{o,n}) - \frac{C}{C + R_0}(a U_{n,o} - U_{o,n})}{U_{o,n} \cdot \alpha_{R7} \Delta T \left[D + R_0 (1+\alpha_R \cdot \Delta T) \right]} \cdot \qquad /8/$$

$$\cdot \frac{\left[D + R_0 (1 + \alpha_R \cdot \Delta T) \right]}{1}$$

$$C = A_{3N,0} + R_2 \qquad /9/$$

$$D = A_{3N,0}(1 + \alpha_{A3} \Delta T) + R_2 \qquad /10/$$

where $R_{7,O}$ is the value of resistor R_7 at temperature T_O; α_R, α_{R7} and α_{A3} are temperature coefficients of resistors R_O, R_7 and of bonded strain gauge A_3; and ΔT is the difference of temperatures T_O and T_1. The value of adjustable resistor R_8 can be calculated from the known equations for the parallel connection.

2.4. Determination of parameters of elements for full compensation of temperature effect on output signal

The application of semiconductor strain gauges of the same type for a basic mechanical-to-electrical conversion of non-electrical quantity /strain gauges $A_{1,2}$ and $K_{1,2}$/ and for compensation of undesired effects /strain gauge A_3/, provides primarily any technological advantages. However, there is a disadvantage consisting in a necessary application of an additional semiconductor element connected in place of resistor R_4. More advantageous appears to use such a semiconductor element, introduced in place of resistor A_3, that would provide a capability of both the compensation for non-linearity and primarily the compensation for temperature effect on the output signal, as perfect as possible. There are two alternatives that can be applied to solve this problem:

A/ To find such an α_{A3} that allows obtaining $\alpha_{R4} = 0$ for a selected value of resistor A_3, or

B/ to find such a resistor A_3 that provides $\alpha_{R4} = 0$ for a selected value of α_{A3}.

For analysis of the two alternatives will be used designation $R_1 = R_4 + R_6$.

2.4.1. Alternative A. In dependence on temperature, strain gauges $A_{1,2,3}$ and K_1, K_2 will change only in resistance and sensitivity. When meeting the conditions given by equations /4/ and /5/, the following formula can be derived for the searched-for coefficient

$$\alpha_{A3} = \frac{R_1}{A_{3N,0}\Delta T\left\{\frac{U_{n,T1}}{U_{n,0}}\left[1+\frac{R_1}{R_0}\left(1+\frac{R_0}{F}\right)\right]-\left[1+\frac{R_1}{R_0(1+\alpha_R\Delta T)}\right]\right\}} - G \qquad /11/$$

where for $R_2 = 0$

$$F = A_{3N,O} \quad \text{and} \quad G = \frac{1}{\Delta T}$$

and for $R_2 \neq 0$

$$F = A_{3N,O} + R_2 \quad \text{and} \quad G = \frac{A_{3N,0} + R_2}{A_{3N,0}\cdot\Delta T}$$

From equation /11/ can be determined the temperature coefficient of resistance of the bonded strain gauge, which can be expressed also as

$$\alpha_{A3} = \frac{\Delta A_{3N,0}}{A_{3N,0}\cdot\Delta T} \qquad /12/$$

where $\Delta A_{3N,O}$ is a temperature change of resistance of strain gauge A_3. For application in practice, it is necessary to define such a specific conductance of silicon that would match equation /11/. The total strain of a bonded semiconductor strain gauges at temperature change ΔT and at zero mechanical load is given by equation

$$\varepsilon_C = (\alpha_M - \alpha_{Si})\Delta T + \varepsilon_{N,0} \qquad /13/$$

where α_M and α_{si} are coefficients of thermal expansion for material of the sensing element and for silicon, respectively, and $\varepsilon_{N,O}$ is the bias produced by bonding. By applying the basic strain equation for semiconductor strain gauges the relative resistance change caused by change in temperature is given by equation

$$\frac{\Delta A_{3N,0}}{A_{3N,0}} = \frac{A_{30,0}}{A_{3N,0}} \int_{\varepsilon_{N,0}}^{\varepsilon_C} (C_{1,T} + C_{2,T}\,\varepsilon)\, d\varepsilon \qquad /14/$$

where $A_{30,O}$ is resistance of bonded strain gauge A_3 at temperature T_O, and $C_{1,T}$ and $C_{2,T}$ are coefficients of the strain equation, given by the selected type of the strain gauge at temperature T.
The temperature-dependent coefficients of the strain equation can be expressed by the equations which relatively well meet the real conditions

$$C_{1,T} = C_{1,0}\left[1 + \alpha_{kT}(T - T_0)\right] \qquad /15/$$

$$C_{2,T} = C_{2,0}\left(\frac{T_0}{T}\right)^2 \qquad /16/$$

where $C_{1,O}$ and $C_{2,O}$ are coefficients of strain equation at temperature T_O, and $\alpha_{k,T}$ is coefficient given by the specific conductance of silicon. For the strain gauges made of P-type silicon with its longitudinal axis polarized in direction /111/ of the crystallographic orientation, very frequently applied in technical practice, values $\alpha_{k,T}$ are presented in Fig.5. On the same diagram is plotted curve α_{A3} allowing to determine the specific conductance of silicon of the compensation element for a value of the coefficient calculated from equation /11/. Curve α_{A3} on Fig.5 has been determined from equations /12/ and /16/ for the folliwing values

$A_{30,0} = 120\,\Omega$; $\varepsilon_{N,0} = -1.10^{-6}$; $\Delta T = 30\,°C$, $\alpha_M = 12.10^{-6}$ 1/°C

and $\alpha_{Si} = 2{,}4 \cdot 10^{-6}$ 1/°C .

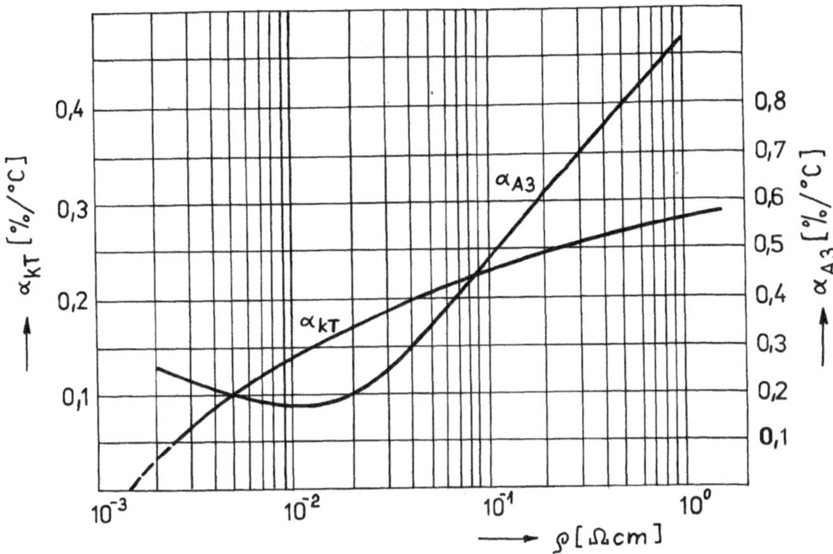

FIGURE 5 Dependence of coefficient α_{A3} on specific conductance of silicon

2.4.2. Alternative B: The basic condition for determination of such a resistor A_3 that would fully compensate the temperature changes of output signal at the selected type of semiconductor material consists again in satisfying equations /4 and /5/. By applying these equations, the explored value of resistor is for $R_2 = 0$ given by

$$A_{3N,0} = \frac{R_1\left(\frac{U_{n,T1}}{U_{n,0}} - \frac{1}{1 + \alpha_{A3}\,\Delta T}\right)}{1 + \frac{R_1}{R_0}\,(1 + \alpha_R \cdot \Delta T) - \frac{U_{n,T1}}{U_{n,0}}\left(1 + \frac{R_1}{R_0}\right)} \qquad /17/$$

For $R_2 \neq 0$, the above equation for $A_{3N,0}$ is too extensive and, for this reason, is not presented.

The following will exemplify that alternative B cannot be used for any specific conductance of silicon which would be identical for all strain gauges applied in the transducer. For example, for $\rho = 0.02\,\Omega$ cm, $R_0 = 100\,\Omega$, $R_1 = 50\,\Omega$ and $R_2 = 0$ can be obtained $A_{3N,0} = 28\,\Omega$.

For such a value, the linearizing effect of resistor A_3 is very low. However, this confirms that the desired coefficient α_{A3} for compensation of temperature effect on output signal is lower when resistor $A_{3N,O}$ is decreased.

3. CONCLUSIONS

The described internal electrical circuit of transducers with semiconductor strain gauges provides capability of applying the presented mathematical relations to attain a non-linearity better than 0.05 % and a temperature effect on the output signal better than 0.005 % /°C. The described circuit has been experimentally verified and has been with success applied in the series production of transducers. Its disadvantage consists in certain increasing the supply voltage which, however, causes no serious problems in practice.

REFERENCES

1. Dean M.: Semiconductor and Conventional Strain Gauges, Academic Press, New York, 1962.
2. Lukas J.: Electrical Circuit for Compensation of Non-Linearity on Output Signal, Research Report of Aeronautical Research and Test Institute, Prague

IMPROVING CREEP PERFORMANCE OF THE STRAIN GAGE BASED LOAD CELL

MAARTEN SPOOR

PRECISION FORCE, INC., CAMBRIDGE, MASSACHUSETTS, USA

OIML Metrological Regulations for Load Cell IR-60 require a 4-hour creep test in which the total change shall not exceed 1.5 times the absolute value of the maximum permissible error. This test has to be performed at room temperature, in addition to the lower and higher temperature ranges specified for the device which are usually from -10°C to 40°C. Because of the lack of loading equipment, it is physically impossible for a load cell manufacturer to test each and every load cell for this characteristic. As a result, shorter test time periods have to be used and that necessitates the development of some method of prediction in order to extrapolate this shorter test time period data over longer periods of time.

The creep time function for metals has been well studied, and many empirical equations can be found in the literature. Kennedy (Ref. 1) cites many functions and their sources. Functions which do not include the effects of temperature but only time are listed below.

Simple functions

Function	Source
$Cr = at/(1+bt)$	Freudenthal

Logarithmic functions

Function	Source
$Cr = a+b \log t$	Phillips, Boas and Schmid, Smith, Chevenard, Laurent and Eudier
$Cr = a(\log [1+bt])^{2/3}$	Mott and Nabarro
$Cr = a \log t+bt+c$	Weaver

Exponential functions

Function	Source
$Cr = a+bt-c \exp(-dt)$	McVetty
$Cr = at+b[1-\exp(-ct)]$	Soderberg
$Cr = a[1-\exp(-bt)]+c[1-\exp(-dt)]$	McHenry

Power functions

Function	Source
$Cr = a+bt^n$	Swift and Tyndall, and Cottrell and Aytekin
$Cr = a(1+bt^{1/3}) \exp kt$	Andrade

Power series

Function	Source
$Cr = at^m+bt^n+ct^p+ \ldots$	Graham

Logarithmic and power functions combined

Function	Source
$Cr = a \log t+bt^n+ct$	Wyatt

An early and thorough analysis by Rohrbach and Czaika (Ref. 2) uses a model of springs and dampers to describe the strain gage structure and derives an exponential curve to fit the creep. This curve is similar to the discharge or charge equation of a resistor-capacitor network (Refs. 3 and 4). The curves asymptotically approach a straight line after which the creep must stop.

Dorsey (Ref. 5) suggests that the shape of the curve is hyperbolic or quadratic in nature. Whatever curve is used for describing the creep curve, it is important that some curve fitting be done. Just to measure one point at 30 minutes or one hour after the load has been applied or removed does not use all the points available to describe the creep curve more accurately. I have used the equation

$$Cr = a+b \ln(t)$$

extensively (an equation also suggested by Jordan and Freed [Ref. 6]). This logarithmic function cannot have a zero time reading since the logarithm of zero is not defined. The curve implies that the creep never stops but keeps going, although at a substantially reduced linear rate after some time has elapsed. This curve becomes a straight line when plotted on semilog paper. When care is taken to produce properly measured data, the coefficient of correlation for this logarithmic curve is generally found to be above +.95 for positive creep and below -.95 for negative creep. The values +1 and -1 of course are the perfect correlation. However, a problem arises when the creep is extremely small and the slope becomes zero. In that case the correlation coefficient is meaningless and some judgment is required.

Since creep is time dependent, it will affect such load cell characteristics as linearity and hysteresis. Depending on the speed of loading, a positive nonlinearity coupled with a negative creep at load will tend to make the linearity more positive and hysteresis more negative. Hysteresis, which is generally positive, meaning that the decreasing load curve has a higher output than the increasing load curve, will tend to become more negative as a result of excessive negative creep as clearly shown by Yamaguchi et al. (Ref. 7). The full scale output is affected also by the direction of the creep.

These effects become extremely important when comparing the load cell performance of the same load cell on different testing machines. The loading time, as indicated by Mitchell and Baker (Ref. 8), on the 4,448,200 Newton (1,000,000 lbf) dead weight tester at NBS, Gaithersburg, Maryland, USA, requires 331 seconds for applying the weights and 529 seconds for removing the weights. Figure 5 of their report shows a surprising creep curve for a 111,210 N (25,000 lbf) tension load cell. This curve could not possibly be fitted using a simple logarithmic curve, and it would indeed be interesting to find the cause of this aberrant behavior. The same load cell tested on a hydraulic machine, instead of a dead weight tester, using a standard reference load cell will provide substantially different data owing to the fact that hydraulic systems can be loaded and unloaded in a matter of seconds.

For higher capacity load cells above 100 kN the author usually uses hydraulic loading and records the creep recovery at zero load. The data is taken using a programable controller and printer to eliminate human errors in judgment and timing. To illustrate the time effect, a load cell, which had been demonstrated to be far from perfect in its creep performance, was tested for linearity and hysteresis using readings taken

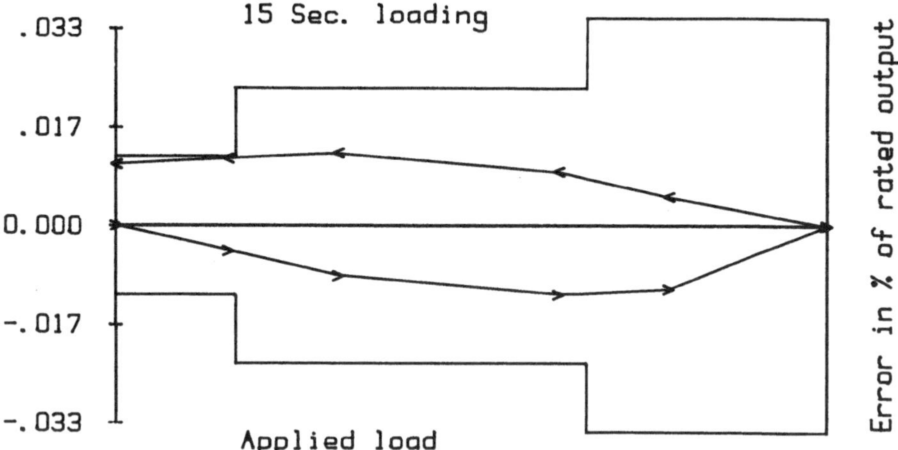

Figure 1. Load cell performance 15 sec. between loads.

15 seconds after the application of each weight (Figure 1) and comparing that data with readings taken 30 minutes after the application of each weight (Figure 2).

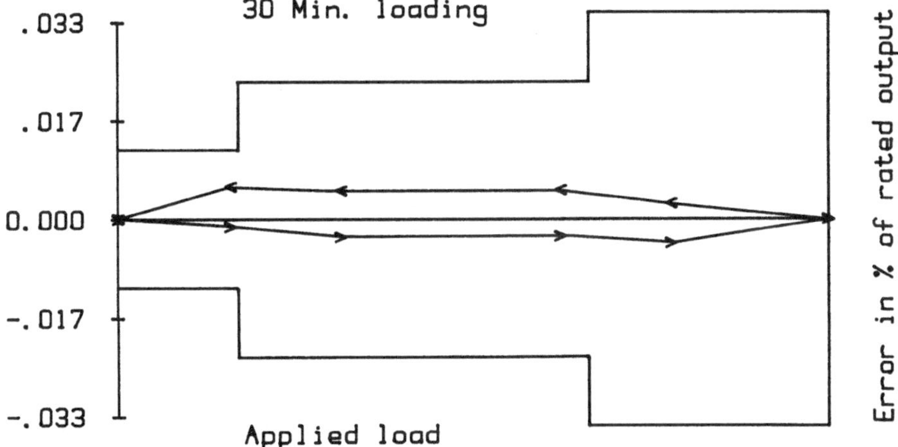

Figure 2. Load cell performance 30 min. between loads.

Figure 2 shows that after creep correction with the appropriate strain gage, this is a good load cell. Using the creep curve established for this load cell (Figure 3), it is possible to reconstruct Figure 2 by superimposing the creep curve on Figure 1. The theoretical construction assumes the creep to be proportional to the applied load, and this is roughly the case. Because of this proportionality creep can be expressed as a percentage of the applied load instead of the rated output. We could therefore test at lower loads than the rated capacity of the load cell. Testing at full capacity reduces noise caused by warmup, ambient temperature changes and instrument drift.

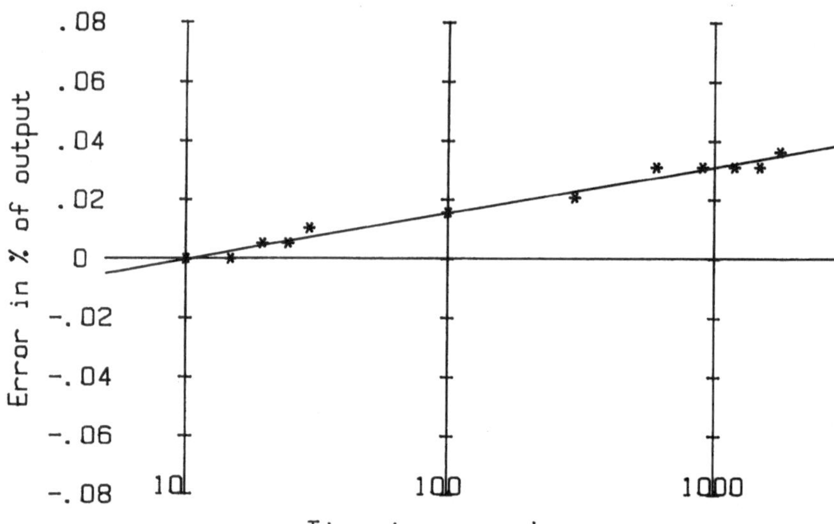

Figure 3. Creep plotted logarithmically.

NOTE: Cr = -0.016+.0068 ln(t)
R SQUARED = 0.978
Cr(30 min) = 0.0351 (Extrapolated)

The time when readings are taken to characterize a load cell is critical when creep is present. In an intercomparison of load cells, currently being undertaken by Australia, the Federal Republic of Germany, The Netherlands, the United Kingdom and the United States, I think the time sequence of the load application is a very critical factor. However, the testing methods as explained in OIML IR 60, which is used as a guide for the comparison test, are not very specific in respect to this timing sequence.

CAUSES OF CREEP

Nearly every manufacturer of load cells has had the experience of a particular load cell, which has performed very well for years, suddenly showing a detrimental creep performance. It is for this reason that a closer investigation of the creep phenomenon is appropriate in order to identify those influence factors which contribute to creep. It is possible to group these factors into two categories. One group is those that can be allocated to the deflection of the metal spring body which is influenced by the cement, strain gage and moisture proofing materials. The second group of influence factors is related to the transfer of the strain in the spring element through the cement and backing to the metalic foil and the control over the dimensional characteristics of the gage during its manufacture.

SPRING ELEMENT INFLUENCE FACTORS

The strain yield, the stress yield divided by the elastic modulus, would appear to be a good indication of a proper spring element material.

A list of frequently used spring materials in their order of strain yield are:

Material	Strain Yield x 10^{-6}
Beryllium Copper (1.9% Be)	10,800
Titanium (Ti-6Al-4V)	10,000
A6, H11, L6 Tool Steels	8,000
4340 Tool Steel	7,000
17-4 and 15-5 PH St.St.	6,500
Aluminum (2024-T351)	4,700

TABLE 1

From this list it appears that BeCu in either wrought or cast form is the best transducer material. However, 2024-T351 aluminum is an excellent spring material, although it can take less overload without producing zero shifts. This phenomenon is clearly shown by Ort (Ref. 9). The arguments between the pro aluminum people (Refs. 10 and 11) and the beryllium copper supporters (Refs. 9 and 13) are still active. It is true that aluminum is easily available in many rectangular and square shapes and that the machining of aluminum can be done extremely fast. All major transducer manufacturers in the USA sell aluminum transducers. The author, selecting BeCu for a small bending beam (Ref. 13), had serious problems after the spring bodies were fine-blanked (a Swiss developed stamping technique) and subsequently precipitation hardened at 345°C (650°F). Performance concerning hysteresis and creep was disappointing indeed. Only after a solution anneal around 800° (1470°F) and a subsequent precipitation hardening did the material become an excellent transducer material. Residual stresses due to fabrication must be reduced by annealing and subsequent heat treatment preferably using a process which uses air cooling since rapid oil or water quenching often reintroduces residual surface stresses. The search for stable materials was not initiated by transducer manufacturers, but by Massachusetts Institute of Technology (MIT) and its Draper Laboratories in the early 1950s. Their investigations were conducted to determine the "precision elastic limit" (PEL) as well as dimensional stability for parts used in gryroscopes and later on in inertial navigation devices.

A lot of information is available from those companies which were involved in actual manufacturing of the gyro components. Schetky (Ref. 14) shows the advantage of thermal cycling between -100°F to 200°F to stabilize precision parts. Weinstein (Ref. 15) investigated the microperformance of materials in order to reduce friction in the bands used for the Rolamite design. The stability of aluminum and hot pressed beryllium was investigated by Jennings et al. (Ref. 16). Honeywell (Ref. 17) reported that a hard coat anodize, (a chromic-sulphuric acid treatment) could increase the "Microscopic Yield Strain" (MYS) of aluminum substantially. The author investigated this reported decrease in element creep and found the anodize to be helpful in reducing positive element creep. The test performed by Honeywell used the Tuckerman optical strain gage. It is important to note that in many of these material creep tests performed in the literature, strain gages were used. Because of the strain gage's inherent creep performance, these studies become suspect. The first person who measured metal creep with the load cell in mind was, to the best of my knowledge, Bergqvist (Refs. 18-20) of the Aeronautical Research Institute of Sweden. His highly precise methods of measuring deflection provided creep curves which were similar to those measured with strain gages.

Despite all the literature available concerning the elastic and plastic deformation of metals, it is difficult to obtain an understanding of how creep takes place in metals, particularly considering the variety of materials and the empirical and theoretical equations used. My personal explanation, with which I am comfortable but which I cannot substantiate, is that the residual stresses which are statistically distributed are critical to a proper transducer material. We can perceive of an alloy material as consisting of grains with boundaries, each grain having a different residual stress associated with it. If one would plot the distribution of the strain levels at which yield occurs,

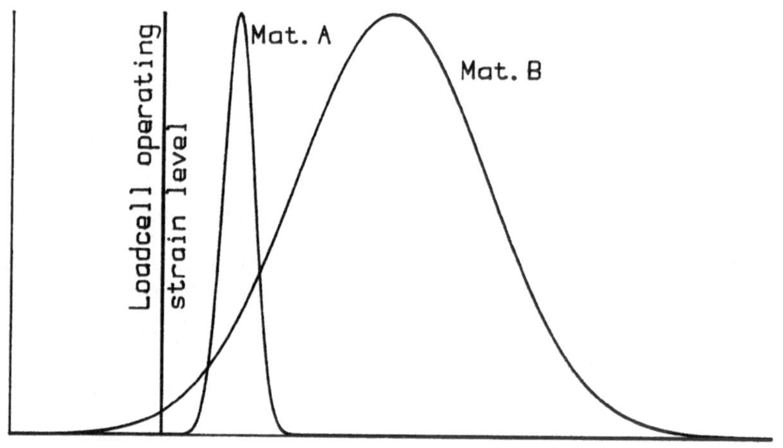

Figure 4. Comparison of two metals.

the curve might look somewhat like Figure 4. When a material is strained, the grain with the highest residual stress will start to shear first as a result of dislocations moving through its crystal structure. Materials with high strength and a small grain structure produced by cold work, such as cold drawn piano wire (material B in Figure 4), have a wide standard deviation in the distribution of their residual stresses. These materials, although they have a high ultimate strength, are very poor transducer materials. Tool steels heat treated to full hardness behave similarly and are better transducer material when used at a reduced hardness.

Another item which influences the free deflection of the spring element is the reinforcing effects of the strain gage and cement on the deflection of the load cell. The many materials used for strain gage backings such as phenolics, polyimids, epoxies reinforced with glass or other fillers do have substantially worse creep performances as compared to metals. Early calculations by Stein (Ref. 21) on the reinforcement of coatings on beams were done to establish the reinforcing effect and correction values for stress analysis. Stein's calculations go into the change of the neutral axis as well as into the change of the elastic modulus of the composite. These reinforcing effects become critical in low capacity transducers. For a family of transducers machined and heat treated from the same alloy and provided with the same gages, the creep

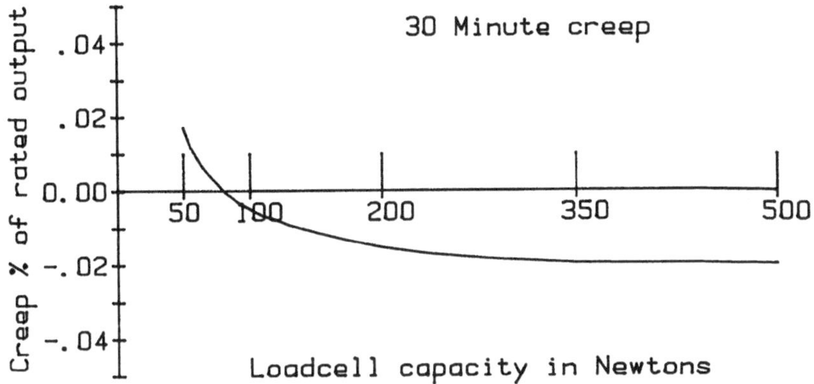

Figure 5. Creep as a function of load cell capacity.

as a function of capacity is shown in Figure 5. It is not surprising that the change in output related to temperature versus the capacity of the load cell has a nearly identical curve.

STRAIN GAGE INFLUENCE FACTORS

The second group of influence factors can be related to transmission of the strain from the spring element to the strain gage alloy. During this process yielding of the cement and backing takes place. Stein (Ref. 21) makes several references to early investigations.

A simplified solution for the strain transfer from a spring element with a constant strain applied to the strain gage was made by James H. Williams in June, 1975 (Ref. 22). The model shown in Figure 6 depicts a strain gage foil layer, the gage backing material combined with the

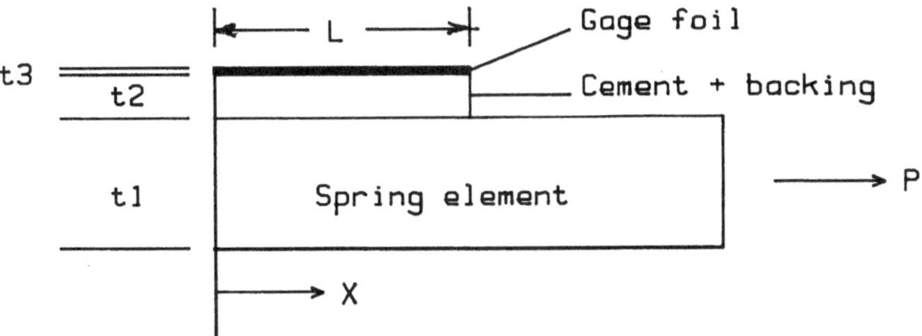

Figure 6. Strain gage model.

cement, and the metal spring material. The ratio R(x), the strain in the gage divided by the strain in the spring material, can be expressed as

$$R(x) = 1-(\cosh Zx/\cosh ZL)$$

where $Z = \sqrt{G_2/E_3t_3t_2+G_2/E_1t_1t_2}$

The shear stress in the intermediate layer as a function of the distance is

$$S(x) = (G_2e_1/t_2Z)x(\sinh Zx)/\cosh ZL)$$

Subscripts 1, 2 and 3, refer to the spring element, adhesive and backing layer, and the strain gage, respectively.

E = elastic modules
G = shear modulus
t = thickness
L = half gage length
e_1 = strain in spring element

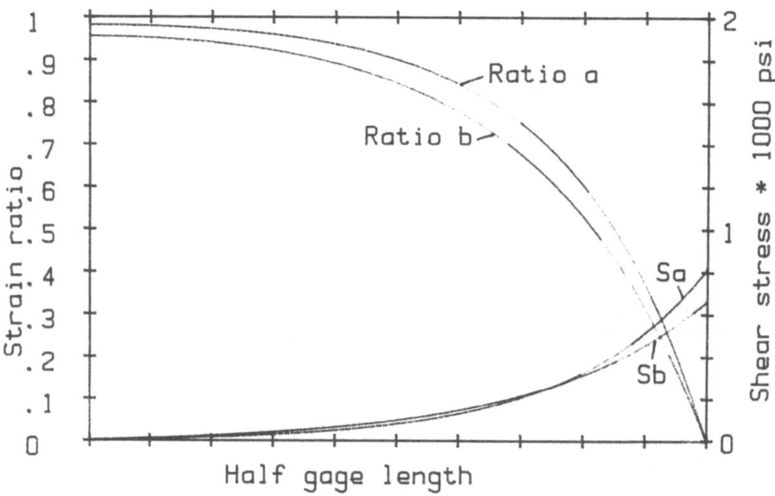

Figure 7. Strain ratio and shear stress for two backing materials of different thickness.

Figure 7 shows a plot of the strain ratio and the shear stress in the intermediate layer versus the gage length. In the same plot the b curves, an increase of 50 percent in the intermediate layer (t2), show a decrease in shear stress and a decrease in the amount of strain transmitted to the strain gage. The length of the end loops of the strain gage are designed in such a way that the local yielding of the intermediate layer reduces the strain in the strain gage so as to offset the positive spring element creep. Longer gages will show less of these end effects since a smaller percentage of the gage is exposed to the yielding of the end of the intermediate layer. It is important that the end tabs are located in a strain field so that end tab adjustments can have an influence on the gage creep performance. Since gages are made in different sizes with different widths of strands, it becomes the end tab to strand width ratio which becomes critical for the creep adjustment. It is important that this ratio be the same for all gages used in each lot of transducers.

In the manufacturing of the strain gage, variations in the end tab to strand width ratio occur because of the very nature of the manufacturing process itself. In this process the gages are etched on sheets using a photoresist process. The bulk etching is stopped before the final target resistance of 350 Ohms is reached. Post etching by hand removes strain gage material from the top surface. In the bulk etching process, increased etching will attack the strain gage foil under the photoresist. The resistance at which the bulk etching is stopped will affect the end tab to strand ratio. If, for example, at 310 Ohms the strand has a unit width of 1 and an end tab length of 10 units, subsequent bulk etching to 340 Ohms will change this geometric ratio. The strand will have a width of 0.9 and the end tab length will be reduced to 9.9 units. The end tab to strand ratio, which at 310 Ohms was 10, has changed at 340 Ohms to approximately 11.

Where some load cell manufacturers design transducers which have a constant strain under the gage, I prefer designs which have the end tab at approximately 75 percent of the peak strain in the gage. The lower strain under the end tab will reduce the end tab to strand ratio. The lower the ratio, the more consistent that ratio can be maintained.

It has been my experience that the variations in the gage backing thickness and end tab to strand ratio influence the distribution of the scatter found in creep data. Although the gage manufacturer tends to point the finger at the spring material characteristic, the gages' geometric variations, caused by the manufacturing process, are a very important influence factor.

CONCLUSIONS

1. Curve fitting of the creep data helps in obtaining better numbers to define the creep.
2. Creep will affect the linearity and hysteresis performance of the load cell.
3. The time required for the application of the load and the subsequent waiting period is critical if the creep is to be defined in meaningful numbers.
4. Removal of the residual stress with proper heat treat will reduce the scatter in the creep data.
5. The variations in the strain gage backing thickness and end tab to strand ratio during the manufacturing process are important influence factors.

REFERENCES

1. A.J. Kennedy. _Processes of Creep and Fatigue in Metals_. New York: John Wiley & Sons, Inc., 1963.
2. C. Rohrbach and N. Czaika. "A Model Describing the Mechanism and Most Important Properties of Strain Gauges." _Royal Aircraft Establishment_ (Farnborough). Translated by R. C. Murray. Library Translation No. 821. London: Ministry of Aviation, December, 1961.
3. F.M. Tovey. "Transducer Flexure, Material Behavior." Presented at Western Regional SESA Fall Meeting, Denver, CO, Sept., 1976.
4. Peter K. Stein. "Some Little-Known Literature and Effects in Strain-Gage-Based Force Transducers." _Recent Advances in Weighing Technology and Force Measurement_. Eds. Kozo Iizuka and Toshiro Ono. Proceedings of the 10th International Conference of the IMEKO Technical Committee TC-3 on Measurement of Force and Mass, Kobe, Japan, September 11-14, 1984. Japan: The Society of Instrument and Control Engineers, 1984.

5. James Dorsey. "Errors in Strain Gage Transducers." Proceedings IMEKO Conference on Force and Mass Measurement, September, 1980, Krakow, Poland.
6. E. H. Jordan and A.D. Freed. "Room-Temperature Post-Yield Creep." Experimental Mechanics, September, 1982, pp. 354-360.
7. Yukio Yamaguchi, Junsaku Ishino and Namiteru Hida. "Evaluation of Creep of Load Cells Utilizing Strain Gauges." Recent Advances in Weighing Technology . . . [See 4. above.]
8. R.A. Mitchell and S.M. Baker. "Characterizing the Creep Response of Load Cells." VDI-Berichte, Nr. 312, 1978.
9. Werner J. Ort. "A Fabricated Platform Load Cell." Framingham, Massachusetts: Hottinger Baldwin Measurements, Inc., 1984.
10. Al Brendel. "The Aluminium Load Cell." Measurement and Control, Vol 13, September, 1980, p. 312.
11. Alexander Yorgiadis. "Comparing Damping Properties of Steels and Aluminum Used in Load Cells." Weighing & Measurement, January/February, 1981, pp. 13-14.
12. J. Daniel LaJeune. "Beryllium Copper Load Cells Are Heart of Toledo 'Honest Weight' Electronic Scales." Weighing & Measurement, November/December, 1980, pp. 12-14.
13. Maarten Spoor. Miniature Load Beams. U.S.A. Patent 4,208,905, June 24, 1980.
14. L. McD. Schetky. "Dimensional Stability in Precise Parts." Product Engineering, September 12, 1960, pp. 67-70.
15. Warren D. Weinstein. "Microperformance of Metals." Machine Design, December 11, 1969, pp. 174-181.
16. C.G. Jennings, N.W. Jiron, A.G. Cross, and G.E. Patow. "Measurement of Inertial Guidance Material Dimensional Instability Parameters Using Resistance Strain Gages." Presented to Symposium on Precision Mechanical Property Measurements, 69th American Society for Testing and Materials Annual Meeting, Atlantic City, New Jersey, June 27, 1966. Anaheim, California: North American Aviation/Autonetics.
17. Technical Memo #426. Minneapolis, Minnesota: Honeywell, Inc., 1971.
18. Bjorn Bergqvist. "Method to Test Steel Material, Loaded in Pure Bending, for Creep at Room Temperature, at 0,74 MM/M Absolute Strain." Internal Memo PM MX-1282/8. Bromma, Sweden: The Aeronautical Research Institute of Sweden (FFA), March, 1979.
19. Bjorn Bergqvist. "Creep Tests at 0,74 MM/M Absolute Strain and Room Temperature, Performed with a Beam (No. 25) Made from Swedish Steel 2541-04." Internal Memo PM MX-1282/9. Bromma, Sweden: The Aeronautical Research Institute of Sweden (FFA), March, 1979.
20. Bjorn Bergqvist. "Creep Tests at 1,00 and 1,25 MM/M Absolute Strain and Room Temperature, Performed with a Beam (No. 25) Made from Swedish Steel 2541-04. Summary of Creep Tests, Including Those at 0,74 MM/M." Internal Memo PM MX-1282/10. Bromma, Sweden: The Aeronautical Research Institute of Sweden (FFA), March, 1979.
21. Peter K. Stein. Non-Self Generating Transducers. Chapt. 21, Vol. II of Measurement Engineering. 2nd Edition. Tempe, Arizona: Stein Engineering Services, Inc., 1962.
22. James H. Williams, Jr. "Strain Transfer in Bonded Gages." Report to BLH Electronics, Inc., Waltham, Mass., June, 1975.

TEMPERATURE EFFECT AND COMPENSATION OF STRAIN GAGE LOAD CELL

Ma Yanbing, Zhang Zupei

1. INTRODUCTION

Temperature effect of sensitivity is one of the most important properties for the strain gage load cell. Its temperature dependence has to be precisely investigated for purpose of compensation.

This paper describes the temperature-sensitivity relation, the measuring method, the results of regression analysis and shows the possibility and significance of compensation as well.

2. TEMPERATURE EFFECT

The temperature effects of strain gage load cell can be summarized as follows:
- on the zero;
- on the sensitivity.

These effects vary with the average level of the temperature and the temperature gradient of the load cell. They are due to the temperature effects of the mechanical and electrical characteristics.
There are only three parts:
- the metal element: expansion coefficient α_m
  temperature coeff of Young's modules E_t
- the strain gage : temperature coeff of resistance α_g
  temperature coeff of K factor K_t
- the bridge wiring: temperature coeff of wire α_w

The termal effects on zero and sensitivity can be compensated with resistances inserted in the bridge arms and input terminals. These compensations are very precise.

When the temperature variation is slow and homogeneous, the compensation is able to be completed. If there is a thermal gradient inside the load cell, this gradient generates internal forces which can act on the elastic element like an external force, this effect is very difficult to compensate.

- Temperature effect on zero

Owing to the change of ambient temperature, the α_m, α_g, and α_w are combined to act on the load cell, which cause the Wheatstone Bridge unbalance and zero shifting with temperature.

Therefore the compensated method is to series a resistance (which is sensitive to temperature) in certain arm of bridge, so that balance condition will be keeping throughout, such elaborate compensation will obtain high precision. In the case of strain gage made in constantan and karma, the auto-compensated gages can be used instead. As for this aspect is so much, without going into detail.

- Temperature effect on sensitivity

Output of load cell depends on the K factor and geometric and mechanic specification of elastic element. Its sensitivity varies with the temperature. All the parameters related to temperature effeçt the output, of all the factors, the E_t and K_t are the most important ones.

It is evident why Young's modulus E varies with temperature, from the mechanism, Young's modulus is an index that metal resist elastic deformation, it means the degree of difficulty or easy that the metal generate elastic deformation, therefore the more the Young's modulus, the more the stress which produces certain elastic deformation, so it expresses the rigidity of the material. From the microcosum, it indicates that the performance of the acting force among metal atoms, which depends on the nature and crystal lattice type of metal atom and it relates closely to atomic distance. For most of metal, except the alloy of special crystic structure, its crystal lattice constant increases with a rise in temperature, the atomic interact decreases, therefore the E decreases with temperature. It is known that the various elastic elements are made of structure steel, spring steel, precipitate harden stainless steel, aluminium alloy.

When the temperature rises, the E is on the decrease. The elastic strain (ϵ) can be expressed as a function of the applied stress (σ) and the Young's modulus (E) as : $\sigma = \epsilon E$.

When the force is applied on the elastic element, the deformation appeares with the temperature rising which is more than that generated with the temperature decreasing, so that the output of load cell is increased with temperature.

The variation of the K factor versus temperature can be either positive or negative, linear or non-linear, according to the foil material and the carrier of the strain gage.

Only to measure the sensitivity-temperature relation precisely can we do compensate well.

NIM developed a shear type weighing cell for producing the weighing instrument, which meets with the OIML requirement for class c, 5000e. Such shear type weighing cell is composed of a elastic element of 40CrNiMoA structure steel and a set of strain gages of constantan with polymide carrier. The output signal of weighing cell under force is increased with temperature. Does the output-temperature have linear dependence? Can we express this relationship as a monadical linear equation?

The anwer is obtained by means of direct measuring method and regression analysis.

Temperature coefficient of sensitivity can be described by the equation:

$$S_t=(S_T - S_{20}) / S_{20}(T-20) \qquad (2-1)$$

in which, S_T and S_{20} are sensitivities under same loading with temperature at T°C and 20°C respectively. S_t may be expressed in differential form as:

$$S_t=1/S_{20} \cdot dS/dt \qquad (2-2)$$

If the output has been measured at different temperature, the S_t linear equation can be calculated, the S_t will be known.

The linear equation:

$$\hat{Y}=a+bx \qquad (2-3)$$

Based on the principle of least square, testing points (x_i, y_i) are selected. i=1,2,......n. and where

$$\sum_{i=1}^{n} (Y_i-\hat{Y}_i)^2 = \sum_{i=1}^{n} (Y_i - a - bx_i)^2 \qquad (2-4)$$

to make the square summation minimum, the regression line is the most ideal one.

The limit value is obtained by differential method, when the square summation reaches to minimum, there is a regression line.

a,b may be expressed mathematically as:

$$b=l_{xy} / l_{xx} \qquad (2-5)$$

$$a=\overline{Y}-b\overline{x} \qquad (2-6)$$

Where:

$$\overline{x}=1/n \sum_{i=1}^{n} x_i \qquad (2-7)$$

$$\overline{Y}=1/n \sum_{i=1}^{n} Y_i \qquad (2-8)$$

$$l_{xx}= \sum_{i=1}^{n} (x_i - \overline{x})^2 \qquad (2-9)$$

$$l_{xy}= \sum_{i=1}^{n} (x_i - \overline{x}) (Y_i - \overline{Y}) \qquad (2-10)$$

It should be approved whether the regression line is significant or not, the regression coefficient has to be introduced.

$$\gamma = \frac{l_{xy}}{\sqrt{l_{xx}\, l_{yy}}} \qquad (2-11)$$

The more it approaches to 1, the batter the linear relation of x and Y is.

Where $$l_{yy}= \sum_{i=1}^{n} (Y_i-\overline{Y})^2 \qquad (2-12)$$

3. MEASURED DATA

The load cell is calibrated on the deadweight machine (see the photo) with precision better than 0.03 class, and it is mounted on the machine with high and low temperature furnace installed for measuring the relation of the output-temperature.
Then a lot of data could be obtained at five load steps and five temperature points.
Table I shows a set of measured data obtained from rated load and five testing temperature points. Here the output value is instead of sensitivity.

Table I

Temperature (°C)	27	35	44	60	70
Output (μV)	24205	24255	24309	24425	24499

In order to make easy calculation, the equation (2-1) can be expressed as follows:

$$S_T = S_{20} + S_{20}\, S_t\, (T-20) \qquad (3\text{-}1)$$

Here: $S_T=y$, $S_{20}=a$, $S_{20}\, S_t=b$, $T-20=x$ (3-2)
Then: (3-1) can be written as:

$$y=a+bx \qquad (3\text{-}3)$$

According to above data assume:

$$y' = S_T -24309=y-24309 \qquad (3\text{-}4)$$

The monadic regression calculated table has been given:

$\bar{x}=27.2,\quad \bar{y}'=29.6,\quad \bar{y}=24338.6$ (3-5)

$$l_{xx}=\sum_{i=1}^{n}(x_i-\bar{x})^2=1250.8 \tag{3-6}$$

$$l_{yy}=\sum_{i=1}^{n}(y_i-\bar{y})^2=58907.2 \tag{3-7}$$

$$l_{xy}=\sum_{i=1}^{n}(x_i-\bar{x})(y_i-\bar{y})=8576.4 \tag{3-8}$$

$$b=l_{xy}/l_{xx}=6.85673 \tag{3-9}$$

$$a=\bar{y}-b\bar{x}=24152.097 \tag{3-10}$$

$$S_t=b/a=2.8389\times 10^{-4}/°C \tag{3-11}$$

Table II

No	T	x=T-20	x^2	y	y^I	$y^{I\,2}$	xy^I
1	27	7	49	24205	-104	10816	-728
2	35	15	225	24255	- 54	2916	-810
3	44	24	576	24309	0	0	0
4	60	40	1600	24425	116	13456	4640
5	70	50	2500	24499	190	36100	9500
$\sum_{i=1}^{5}$		136	4950		148	63288	12602
$\sum_{i=1}^{5}/5$		27.2		24338.6	29.6		

To check the above regression line, the regresive coefficient γ should be gained.

Where: $$\gamma=l_{xy}/\sqrt{l_{xx}l_{yy}}=0.999 \tag{3-12}$$

To compare the checking table of regresive coefficient when $i=5, \gamma\geq 0.959$, the confidence is 99%.

As for other loading steps through the same calculation, the measured data are found to be quite linear.

The testing conclusions:

- The dependence of output signal of the strain gage load cell with temperature is ideal linear.
- The relation of output signal-temperature doesn't matter to the applied load, so its temperature dependence at rated load can be investigated precisely to instead the other loading steps.
- According to the measured data obtained, it has been proved that the linearity, repeatability, hysteresis of load cell range at normal case.
- Many years' experience in making load cell shows that the load cell, made up of a same batch of elastic material, strain gage and adhesive, under the same treatment and process, has a closer temperature coefficient of sensitivity. It provided a fast and precise temperature compensation in sensitivity.

4. THE COMPENSATION OF SENSITIVITY

The compensated method:

- This compensation can be completed by means of two series resistors symmetrically in both exciting terminals. When the impedances of the resistors change with temperature, it is therefore neccessary to adjust in order to compensate sensitivity that caused by temperature changing.
- The self- compensating strain gage is used for compensation of sensitivity. Which has negative temperature coefficient of K factor to martch the temperature coefficient of Young's modolus of elastic element.

At present, the first method used to be adopted. The calculation of resistor for compensation:
Based on the principle of compensation, when the resistors were connected in the exciting terminals, the acture exciting voltage at t_1 and t_2 individually is U_1 and U_2 by whcih R_c value can be calculated. The typical compensating circuit is shown in the following figure.

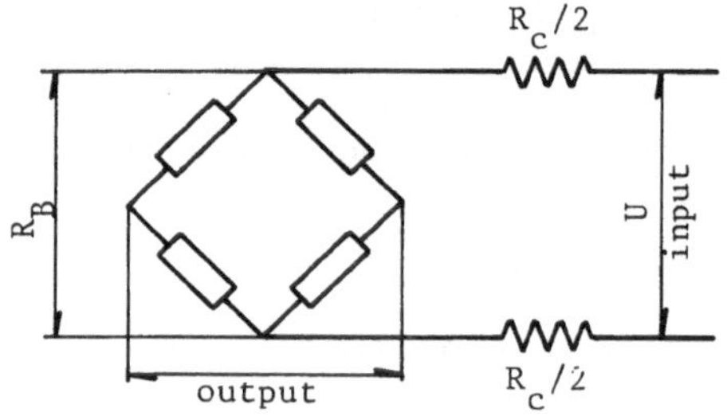

$$U_1=UR_B\ /(R_B\ +R_c) \tag{4-1}$$

$$U_2=UR'_B\ /(R'_B\ +R'_c) \tag{4-2}$$

Where R_B, R'_B ; Impedance of bridge at t_1, t_2;
R_c, R'_c ; Impedance of compensating resistor at t_1, t_2;

$$R'_B = R_B \ (1+ \alpha_g \Delta t) \qquad (4-3)$$

$$R'_c = R_c \ (1+ \alpha_c \Delta t) \qquad (4-4)$$

Here α_g : The temperature coefficient of strain gage;
α_c : Temperature coefficient of compensating resistor.

When the output had been compensated

$$S_1 U_1 = S_2 U_2 \qquad (4-5)$$

S_1, S_2: Sensitivity of load cell at t_1, t_2.

Which can be achieved by putting (4-1)-(4-4) to (4-5)

the R can be written in the form:

$$R_c = \frac{R_B \ (S_2 - S_1)}{S_1 \ \left(\frac{1+ \alpha_c \Delta t}{1+ \alpha_g \Delta t}\right) - S_2} \qquad (4-6)$$

Let $$S_t = (S_2 - S_1) \ / S_1 \Delta t \qquad (4-7)$$

turn to (4-6)

$$\because \alpha_g \ll \alpha_c \quad \therefore 1+\alpha_g \ \Delta t \cong 1$$

then: $$R_c = R_B S_t / (\alpha_c - S_t) \qquad (4-8)$$

The equation (4-8) shows: for the precision compensation of sensitivity the precision measurement of R_B, S_t, α_c are required.
According to the above principle and method, the S_t and α_c must be measured selectivily by a certain proportion for same batch of load cells which have to be found in same type and same material. If these parameters are measured precisely, the precision compensation of sensitivity can be obtained.

The compensation of shear type load cells developed in our Institute had been completed with this method and gained the higher precision.

Actual example of compensation for weighing cell with purpose of developing the weighing cell for weighing instrument class c, 5000e recommended by OIML. A compensation of its sensitivity should be made to a certain precision, so that the weighing instrument has kept the accuracy during a regulated period of calibration, the temperature range is from -20°C - 45°C even of under the rough environment with.

Through the regression analysis, we have got the coefficient of sensitivity: $S_t = 2.84 \times 10^{-4}$/°C. That means, if the temperature changes one degree, the sensitivity increases approximately 0.03% . How surprising it is. In manufacture, the compensation of sensitivity should be made precisely, otherwise, the accuracy of the weighing instrument may not be guaranteed both in calibration and in service.

The range of temperature coefficient of resistor for compensation is : 3.2–5.5 x 10^{-3}/°C. The α_c must be measured precisely (it is easy to realize) specially for the temperature range of installed local.

After having been measured, the temperature parameters of the shear weighing cell as follows:

$$S_t = 2.84 \times 10^{-4}/°C$$
$$\alpha_c = 5.3 \times 10^{-3}/°C$$
$$R_B = 900\ \Omega$$

Based on the equation (4-8)

$$R_c \cong 51\,\Omega$$

The 25.5 Ω resistor will be connected in series with the input terminals symmetrically.

The test equipments used are:

- 300kN dead weight machine
- 9601 high-low temperature furnace
- measuring temperature instrument:
  CBW-1A surface temperature sensor
  191 digital voltmeter
- 335D direct stable supply
- 1071 digital voltmeter.

Measured temperature points: -29.5°C, 21.0°C, 38.4°C, 72.0°C.

The measured weighing cell has been placed into the furnace and installed in the dead weight machine. It is important to allow sufficient time for stabilization of the temperature, at each temperature point for four hours and then applied the rated load on the weighing cell. The first loaded is pre-load, the average value of subsequent three times is as output of the weighing cell at the temperature point. The measured data are shown in tabel III.

Table III

Temperature (°C)	-29.5	21	38.4	72
Output (μV)	24116.9	24133.9	24133.2	24114.1

According to the terminilogy, temperature effect on sensitivity: the change in rated output due to a change in ambient temperature. Usually expressed as the percentage change in rated output per 10°C change in ambient temperature. After the test, the temperature effect on sensitivity is: -1.6 x 10^{-2}%F.S/10°C (from 21°C-72°C), -1.1 x 10^{-3}%F.S./10°C (at terminal temperature points -29.5,72°C).

If the temperature compensation of sensitivity completed on the dead weight machine with furnace, it would cost much and waste time. For batch process, the same method is used to compensate the effect in the same type and same procedure. According to a certain rate the parameters S_t, α_c can be measured, the R_B is known, it is easy to calculate the R_c.

It is significant to compensate precisely the temperature effect on sensitivity because the weighing cell is always in tough environment. A weighing instrument of large capacity, it's plaform is supported by several weighing cells, if we match intentionally the sign of compensation, when the temperature changes, the sign is cancelled with each other, this compensation will be better.

5. THE PROSPECT OF SELF-COMPENSATION OF SENSITIVITY

Due to the temperature coefficient of K-factor of strain gage is related to the alloy constituent, heat-treatment process, the constantan foil has the positive value, but the karma foil has the negative value. If the proper constituent of karma alloy could be adjusted and the annealing process could be controled, it is possible to produce the strain gage which is matched with elastic element, the sensitivity of load cell varies with temperature at a certain precise range. The linear relation has been verified by using direct measuring method and regression analysis. The confidence is 99%, the linearity of temperature sensitivity is quite ideal.

It is expected by further development that self-compensation strain gage can be made in matching with some elastic elements of load cell, the compensation can also be realized.

PILOT TESTS TO DETERMINE MICRO-ELASTIC EFFECTS IN LOAD CELL RECEPTOR MATERIALS

B.M.BERGQVIST

THE AERONAUTICAL RESEARCH INSTITUTE OF SWEDEN (FFA)
BOX 11021, S-161 11 BROMMA, SWEDEN

ABSTRACT

The FFA precision equipment for symmetrical beam bending and deflection measurement was used on one single beam for two purposes. First, the methods for determination of the strain-temperature relationship within a narrow interval around room temperature were investigated. High accuracy was achieved; the total errors are of the order of ± 10 ppm. Next, the methods for determination of the degree of non-linearity for bending stress-strain curves in the "elastic" region, both at small stresses on both sides of the zero and at normal working stresses, were studied. The accuracy was less good in the latter case; at present the total not compensated errors are of the order of ± 50 ppm. This, however, can be brought down to about ± 10 - 20 ppm with additional, known means.

1. INTRODUCTION

Load cell indication temperature dependence, as well as non-linearity with change in load, are disturbances that are compensated for by most manufacturers. Both the bonded resistance strain gauge applications and the load cell receptor materials contribute to these disturbances. Accurate knowledge beforehand regarding the material contributions may be an advantage.

The present paper shows how the FFA precision equipment for symmetrical beam bending and deflection measurement, described in Ref. 1 and 2, can be used to determine:

(i) beam strain ε as a function of beam temperature θ at normal working stresses and over a small temperature interval around room temperature

(ii) the deviations from a straight line of the stress-strain curve, both at small stresses on both sides of the zero and at normal working stresses

The paper presents only one test series of each kind, all performed with one single beam, being sufficient for a comprehensive description of the methods.

The novelty of the methods used for problem (i) does not rest so much with the relatively high resolution of the capacitive deflection measurement system used as with the successful methods to correct for the thermal errors.

In Ref. 3, BETHE and GERMER present an inductive extensometer having 1 nm

Wieringa, H. (ed), Mechanical Problems in Measuring Force and Mass.

resolution, for use on tensile specimens. They state, however, that the extensometer's "considerable temperature sensitivity is a severe problem". Apart from this, a tensile speciment always shows some degree of bending under tensile forces, which may cause appreciable errors in tensile strain evaluation. Conceptually, then, the present bending tests with the thermally rather insensitive capacitive transducer may be preferable.

The following definition is fundamental for the presentation of errors in this paper: "inaccuracy" is the sum of unknown but estimated systematic errors and 2σcim random errors, where 2σcim is an abbreviation for "2σ(95%) confidence interval about the mean".

2. DEFLECTION-TEMPERATURE TESTS AT NORMAL WORKING STRESSES

This kind of tests is carried out in several steps, each at constant temperature. The steps are accomplished by the aid of electrical units and coolers in the laboratory, set to the wanted power on the evening before.

Next morning, the air temperature in the insulating box surrounding the bending rig is measured by a precision thermometer, Ref. 4. The beam temperature change is measured by a diode, Ref. 4, fastened to the beam and recorded to 0.01^{o}C resolution. The diode was carefully calibrated beforehand against the thermometer, Ref. 4. When the diode has stabilized itself, three "shakedown" runs are carried out. Thereafter, some 10 repeat runs are made, lasting in all some 20 minutes. Through uniform and slow loading creep is minimized and equalized. The signals for the capacitive equipment, Ref. 4, measuring the beam deflection, are read 15 s after reaching the load.

The bending stiffness of the leaf springs which guide the position of the spherical segment feet in the legs of the measuring rig, Ref. 2, increases rapidly with the angular deviation of the segments from their neutral position. To diminish the influence of this phenomenon on the random errors, the rig is set down on the beam when it is loaded about half way to the maximum load.

Each deflection-temperature step is evaluated in a computer, showing the beam deflection in compensating scale divisions ("csd") as a function of the number of the run. The quantities used in the final evaluation of the deflection as a function of the temperature are the intercept d_f of the fitted line at the first run and the diode signal at the same time. Normally, the deflection signal decreases somewhat with the increase in number of runs, which is probably due to the creep-relaxation relationship.

Beam no. 20, made from Swedish hardened and tempered steel 2541, with hardness H_V = 440, was investigated for the actual purpose. The surface of the contact points with the measuring rig was polished with fine emery plastic cloth to an R_a-value of about 0.1 µm. At own weight, the strain was 0.17 mm/m and at maximum load 1.30 mm/m. The temperature interval was about 3.9^{o}C, from 22.0 to 25.9^{o}C. Within this range, six temperature levels were run in 17 tests to determine the beam deflection. A maximum of six tests were run each day, i.e. for one and the same temperature level. The measuring rig was dismounted and re-assembled on the beam before each test, but the beam itself was never dismounted.

For the 17 tests, the average of the 2σcim-values for the approximately 10 individual runs making up one test was ± 1.8 csd, or ± 6.5 ppm, which is a measure of the degree of repeatability of the capacitive equipment.

The test points d_f were curve-fitted in a computer to a linear function of the temperature. In Appendix A it is shown that deflection and strain in bending tests do not show a fully linear relationship. But taking into account the small change of only 2.5μm in beam deflection over the realized temperature range, it is obvious that the strain changes could be considered as being proportional to the deflection changes.

The primary value, not corrected for any error, of the strain sensitivity with temperature, i.e. the slope of the strain-temperature curve, was 238 ppm/$^{\circ}$C. After the error calculation, Appendix B, introducing the correction -14 ppm/$^{\circ}$C for the known systematic errors, the value was 224 ± ± 16 ppm/$^{\circ}$C, where 16 is the 2σcim inaccuracy. Of the value 16, the scatter in the strain-temperature curve contributes 10 ppm/$^{\circ}$C.

Experiments with the coolers, the roof lamps and the heating elements in the laboratory have shown that the temperature range can be increased to more than 7°C within the time period of one to two weeks necessary for these tests. The narrower span of 4°C in the present tests seems therefore not to have done justice to the quality of the measuring equipment. An estimate with well-known statistical formulae for an expansion with further 20 tests between 19 and 22°C yields a 2σcim value of the scatter of some ± 4 ppm/$^{\circ}$C, decreasing the total inaccuracy of the slope to ± 10 ppm/$^{\circ}$C.

The series now described, comprising 17 tests, can also be considered as having determined the random error of a single deflection value, answering the question "how reproducible is the arrangement?". The results is a 2σcim-value of ± 10.5 ppm for 17 points. This figure is almost entirely independent of the temperature interval and can be improved only by increasing the number of tests. Because the total deflection is 2.71 mm, corresponding to about 270000 csd, Ref. 2, the imprecision is ± 28 nm, or ± 0.028μm, a value that is comparable to the R_a-unevenness of a lapped metal surface but is at least three times the resolution of the capacitive system.

It is interesting to compare these facts with those for the bending rig in the test for reproducibility, Ref. 5, when the beam was provided with four full strain gauge bridges. Here, the total gauge deflection was about 1440 μV/V, and the 3σcim imprecision ± 5 ppm, corresponding to a 2σcim-value of a good ± 3 ppm. This latter means 0.4μV/V, which is about three times smaller than the resolution of the strain indicator. Thus, the reproducibility is many times better for the method of taking the reaction in the bending rig to the gravity of the weight masses, Ref. 1, than for the method to place a measuring rig on the beam.

3. DEFLECTION-LOAD TESTS FOR SMALL STRESSES ON BOTH SIDES OF THE ZERO

A bending moment variation through zero is accomplished by the aid of counterweights at various longitudinal positions and by taking into account the contributions from the weights of the measuring rig and the beam, Ref. 1. This is done without lifting the rig, thereby ensuring continuity. The weight gravity load on each ball bearing is about 75 N, causing a noticeable frictional moment at the bending arm supports, Ref. 1. This does not, however, change during such a test and thus does not cause any error in the stress-strain curve obtained.

Beam no. 20 was tested with a bending moment variation producing a strain change from -22 to +22μm/m. Ten load steps were run. The deflections were treated in a computer. From this, the relative change in tangent modulus was evaluated as a function of strain, Fig.1. The change can be represented by a second degree parabola which is approximately symmetric around zero strain. The stress--strain curve itself does have an inflexton point at the zero.

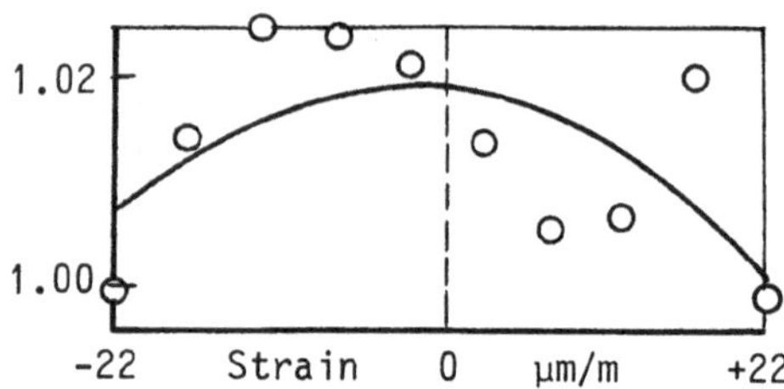

FIGURE 1. Relative tangent modulus around zero strain

In Ref. 6, describing purely compressive tests, which "do not go through zero", changes of Young´s modulus up to about 15 % for stresses less than 1 N/mm^2 were reported for several materials.

4. DEFLECTION-LOAD TESTS AT NORMAL WORKING STRESSES

This kind of test series is carried out without lifting the measuring rig. After three "shakedown" runs, the beam is loaded and unloaded in five steps, of which the lowest one is contained in the test according to Ch. 3 and gives zero moment as an average over the measuring length. This cycle is repeated twice, and every cycle takes about 7 min.

There are many methods by which a stress-strain test for non-linearity can be carried out. It should depend on the way a load cell is used in practice. In the present case, the test is carried out with as short "landings" on the various load levels as possible, in order not to increase the hysteresis unnecessarily. Through uniform and relatively slow loading, creep is minimized and equalized, and the measurement signal is read after 5 -- 10 s.

The stress is first evaluated as weight gravity bending moment divided by beam section modulus, as usual. This is not correct here, because the ball bearing frictional moments increase with the load and besides are more than proportional to it, Ref. 1. At maximum load, the calculated frictional moment is about 80 ppm of the bending moment. The stress values have been reduced accordingly. In principle, this measure of emergency is unsatisfactory, and remedies are demonstrated in Ref. 1, though they have not yet been realised. However, this drawback does not reduce the value of the conclusions that can be drawn from the tests, as will be shown soon. In Appendix B.2.1, the total errors for the stress-strain curve non-linearity are estimated to some ± 50 ppm.

The average deflection for increasing and decreasing load is evaluated, forming the stress-deflection curve. Strain is not linear with deflection and must be evaluated by a mathematical procedure, see Appendix A.

The results from the test with beam no. 20 are shown in fig. 2. It is seen that the stress-strain curve is non-linear even at 70 N/mm^2 stress, and that residual strain after unloading is many times smaller than the departure from the secant line throught 0 and 70 N/mm^2. At maximum load, the departure is 1.76μm/m. The 80 ppm correction, mentioned earlier, for the frictional moment at this load, corresponds to only 0.1 μm/m strain, which shows that this correction does in no way intrude on the conclusions now drawn regarding non-linearity.

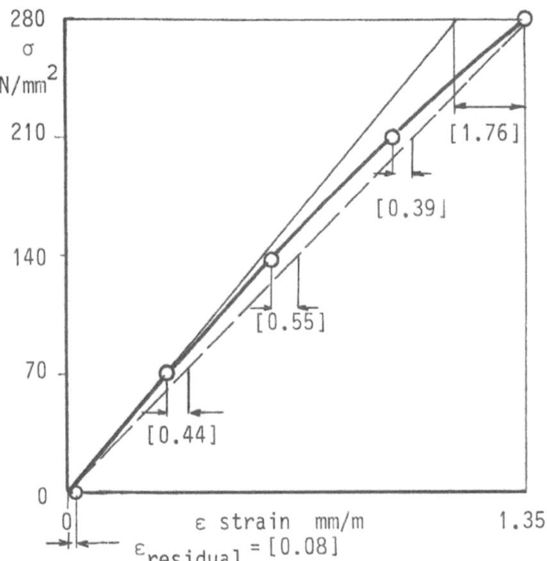

FIGURE 2. Stress - strain curve for beam no. 20. The departures in strain from straight lines are grossly exaggerated; they are shown within [ ]

5. ACKNOWLEDGEMENTS
The author expresses his gratitude to the Aeronautical Research Institute of Sweden for permission to use the test equipment installed in the precision strain laboratory. His thanks are especially due to Automatic Systems Laboratories, Ltd, the manufacturers of the capacitive measuring system, who repaired the transducer at a critical time and thus made it possible to fulfil the investigation, though at a more reduced extent than was intended from the beginning.

REFERENCES

1. Bergqvist B: A precision equipment for symmetrical beam bending. Part 1. The loading system. To be published in "Strain".

2. Bergqvist B: A precision equipment for symmetrical beam bending. Part 2. The deflection measuring system. To be published in "Strain".

3. Bethe K and Germer W: Creep and adiabatic effects in elastic materials. 8th Conference of the IMEKO Technical Committee Measurement of Force and Mass, Kraków, September 1980.

4. Bergqvist B: A Comparison between Creep on Steel in Bending and Tension. 10th Conference of the IMEKO Technical Committee on Measurement of Force and Mass, Kobe, September 1984.

5. Bergqvist B: Creep, Hysteresis and Ageing in Strain Gauge Bending Dynamometers. 8th Conference of the IMEKO Technical Committee on Measurement of Force and Mass, Kraków, September 1980.

6. Evans R H and Wood R H: Modulus of Elasticity of Materials for Small Stresses. Phil. Mag. Vol. 21, No. 138, January 1936.

APPENDIX A

METHOD OF EVALUATION OF BENDING STRAIN FROM THE DEFLECTION OF A STRAIGHT BEAM WITH CONSTANT RECTANGULAR CROSS SECTION UNDER PURE BENDING MOMENT, THE DEFLECTION MEASURED WITH A RIG STANDING ON THE BEAM

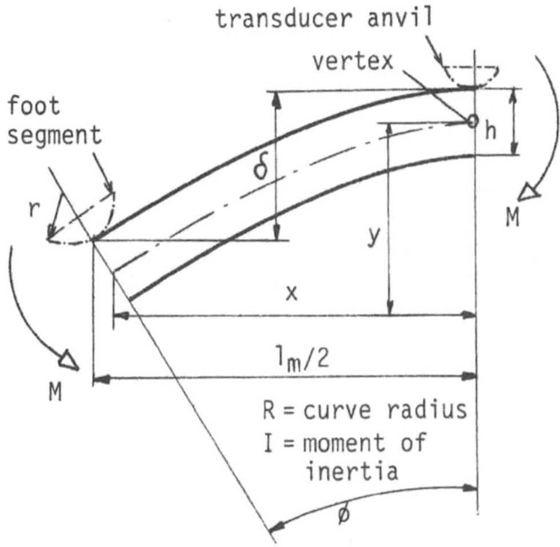

FIGURE 1A. Definitions

According to exact elasticity theory, the neutral line takes the form of a parabola, see Figure 1A:

$$y = \frac{M}{2EI} x^2 = kx^2 \tag{1}$$

If Young's modulus E in tension equals that in comprssion. Engineer's bending theory yields:

$$\frac{M}{2EI} = \frac{\varepsilon_v}{h} \tag{2}$$

where ε_v is the strain at the vertex.

Figure 1A yields the necessary geometrical relationships. The radius r = = 13.5 mm and the measuring ι_m = 375.570 mm. The sensitivity (calibration factor) for the capacitive deflection measuring equipment is

1.00023 mm/100000 csd

at 23.0°C. The unknown quantities are k (or E), x, y and ϕ. By differentiating (1), thereby stating the relation between ϕ and y, a sufficient number of equations is obtained. The evaluation procedure is straight forward and is not treated here. It is considerably simplified if the lowest strain is kept equal to zero, as was done in the load-strain tests described in Ch. 3 and 4.

ε varies along the parabola according to

$$\varepsilon = \frac{h}{2R} = \frac{h}{2}\,\frac{\frac{d^2y}{dx^2}}{\left[1 + \left(\frac{dy}{dx}\right)^2\right]^{3/2}} \approx hk\left[1 - 1.5\left(\frac{Mx}{EI}\right)^2\right] = \varepsilon_v\left[1 - 1.5\left(\frac{Mx}{EI}\right)^2\right]$$

For beam no. 20, loaded from ε = 0 to ε = 1.05 mm/m, $\varepsilon/\varepsilon_v$ = 0.9988 at the end points of the measuring length.

APPENDIX B

B.1. Identification of errors in the deflection-temperature tests

B.1.1. Inaccuracy of the sensitivity of the temperature measuring system

The thermometer readings serve as a temperature reference and a control that the results from the diode calibration which takes place under a rapidly increasing temperature, are reproduced in this static test series. However, because of the rapid increase and decrease in air temperature when opening and closing the insulating box for inspection or for dismounting and re-assembling the measuring rig, the thermometer reading does not come to rest until the end of the test. The readings obtained at the last run have shown to coincide well with the values from the diode. Using this kind of thermometer reading, it has shown that the temperature span for the whole series, in the order of 4 - 6°C, does not differ by more than about 0.5 %, i.e. 0,02 - 0,03°C, from the sum of the temperature steps obtained from the diode. These differences derive from the scatter in the calibration of the diode against the precision thermometer. The 2σim-value for this scatter was ± 0.7 % of the slope. This scatter manifests itself in the deflection-temperature tests as an unknown systematic error and has the value

$$\pm\ 0.007 \times 238 = \pm\ 1.7\ \text{ppm}/^{\circ}\text{C}$$

B.1.2. Influence of the linear thermal expansion coefficients α

In the present case, the beam deflection can be written

$$\delta = C\,\frac{\iota_b \iota_m^2}{Ebh^3}$$

where ι_m = measuring length, ι_b = length of bending arms, E = Young's modulus, b and h = beam width and thickness, respectively, C = a constant. Application for two temperature levels differing by 1°C leads to the formula

$$\delta_{\theta + 1^{o}C} = \delta_{\theta}(1 - 11.5 \times 10^{-6})$$

Here, the systematic error -11.5 was derived from the α-values for the beam material and for the BOVAC steel used in the measuring rig and the bending arms. The α-values were determined at the FFA in an insulating STYROPOL box with the method mentioned briefly in Ref. 4. The capacitive system was used here for elongation measurement, working in a MIKROKATOR stand, together with the precision thermometer and the diodes. Some further information about this would be useful. An α-test is carried out in three phases. In the first one, the transducer anvil stands directly on the MIKROKATOR table. In the second and third phases, the transducer anvil stands on the end of the quartz bar and the test bar, respectively. In this way, the expansion of the MIKROKATOR stand will be determined and the thermal drift of the transducer eliminated, whereafter the α-value of the material in question is obtained by subtraction. Both α-values were determined in only one test each, which, however, was sufficient to produce a good accuracy.

The total inaccuracy for the systematic error was ± 0.9 ppm/^{o}C, and the correction is +11.5 ppm/^{o}C.

B.1.3. Influence of the angular change between the body and legs of the measuring rig from a temperature change. According to Ref.2, the correction for this systematic error is

$$-1.0 \pm 1.0 \text{ ppm}/^{o}\text{C}$$

B.1.4. Influence of the change in the capacitive measuring system sensitivity from a temperature change. According to Ref. 2, the correction for this error is

$$-24.2 \pm 2.4 \text{ ppm}^{o}/\text{C}$$

B.1.5. Influence of the change in air lift on the weights from a temperature change. The correction for this well-known systematic error is

$$-0.5 \text{ ppm}/^{o}\text{C}$$

B.1.6. The systematic errors of the capacitive indicator. These errors were determined according to Ref. 2. The non-linearity of the potentiometer was 2 csd at most. The step errors of the inductive decades were +1 csd in most cases and +4 csd in three cases. The deflection values were corrected for these errors before the evaluation of the slope of the deflection-temperature curve. (The same was done for the deflection-load tests.)

B.1.7. Temperature drift of the capacitive system. This is of no concern, see Ref. 2.

B.1.8. Scatter in the deflection-temperature curve. This covers a lot of random errors, of which three components merit a discussion.

The first two derive from the stiffness of the leaf springs screwed to the legs of the measuring rig to guide the positions of the spherical segment feet. See Ref. 2. Calculations have shown that these errors may add up to only a small part of the observed scatter.

The third component concerns the interaction between the surface of the three contact points with the beam, and the surface itself of the beam at these three points. The surfaces are polished to about R_a = 0.1 µm. However, there are horizontal movements between the beam and the rig contact elements. At the feet, this movement is about 0.25 mm when the maximum load is reached. The anvil slides 0.022 mm, see Ref. 1. Observations of the zero load signal when the rig has been placed on the beam reveal a variation in the vertical distance between the rig and the beam from test to test of about 1 µm. To a certain extent, this is probably caused by changes in the longitudinal position of the rig. These changes, in their turn, are caused by clearances in the mounting of the rig in the stand of the framework of the bending rig, see Ref. 2. These longitudinal position changes are bound to interact with the "crests" and "troughs" in the beam-rig contact surfaces, though the effect seems to be too large for this explanation.

B.1.9. Sum of corrections and inaccuracies
The total correction is 11.5 - 1.0 -24.2 - 0.5 = -14.2 ppm/°C.

The total inaccuracy is ±(1.7 + 0.9 + 1.0 + 2.4 + 10) = ± 16 ppm/°C.

B.2 Identification of errors in the deflection-load tests at normal working stresses

These errors do not concern the slope of the stress-strain curve, i.e. the value of Young's modulus which, Ref. 1, could never be determined with a smaller error than ± 0.05 %.

Instead, the interest lies only with the error in the recorded non-linearity of the curve, where systematic error contributions such as errors in bending arm lengths, Ref. 1, are irrelevant.

The frictional moments M_f in the ball bearings supporting the inner ends of the bending arms have been calculated according to Ref. 1, in lack of present means to determine them experimentally. At zero load in the deflection load test with the beam no. 20, Ch. 4, M_f on each bending arm is 2 Nmm and at maximum load 10 Nmm. The latter is about 80 ppm of the outer bending moment for 1.35 mm/m strain. The assumed uncertainty on this figure is ± 40 ppm.

Further, the random errors from the beam contacts and from the changes in measuring rig position, see Paragraph B.1.8., also contribute.

A total uncertainty of some ± 50 ppm for the non-linearity at maximum load from these two kinds of error is estimated. By determining the ball bearing frictional moments experimentally, Ref. 1, these carried-over inaccuracy could be brought down to about ± 10 - 20 ppm.

CREEP OF LOW-RANGE-SCALES

K. BETHE und D. BAUMGARTEN, Braunschweig

The lower the measuring range of a balance, the quicker a load can (and will) be applied or removed. Thus, time dependent aftereffects, the 'creep' can be observed quite easily in spring-type scales of a few kg range, this creep being the limiting factor of useful resolution. In the case of glued-on foil strain gages, the creep of the elastic element can be compensated by a considerable degree, but high-stability deformation indicators like capacitive displacement sensors, thin film strain gages or optical interferometers quickly reveal this non-ideal behaviour of any elastic matter. It is because of this creep of dynamometric transducers, that systematic research on aftereffects occurring in elastic elements had been started several years ago 1 . These 'creep'-effects become more and more stringent, the lower the stiffness of the elastic element is, i.e. for scales in the low kg region or - e.g. - for precision navigational accelerometers. The reason for this increasing influence of creep is twofold:

1. From the applicational side: The possibility of quick and abrupt load changes.
2. Additional physical effects being due to strain gradients, which result in an increase of the absolute value of the aftereffect.

The latter now shall be dealt with.

1. Measuring setup

In those earlier works mentioned above, the samples were hollow cylindrical columns, being axially loaded to induce a homogenous strain allover the measuring region (see Fig. 1.)

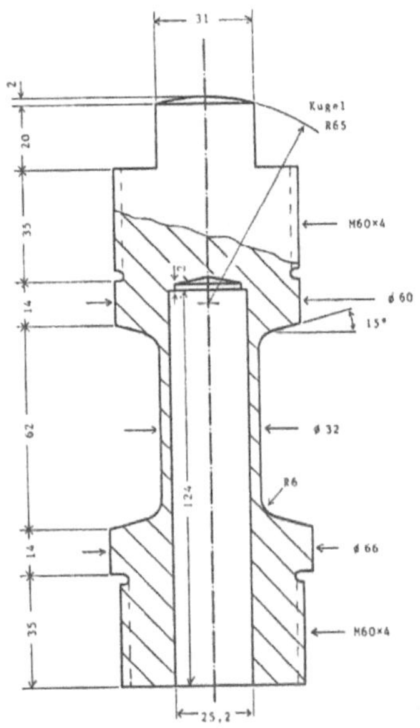

Fig. 1:

Heavy column sample (182 mm long)

In contrary to this, now straight bending beams are being used. Samples with dimensions 80 x 20 mm, the thickness t variing from 0,2 to 2 mm are supported at their ends by two hardmetal cylinders, whilest a central wedge system forces them into an elastic deformation (see Fig. 2). The deformation depth f is measured by a commercially available inductive transducer, which had been modified for this special task.

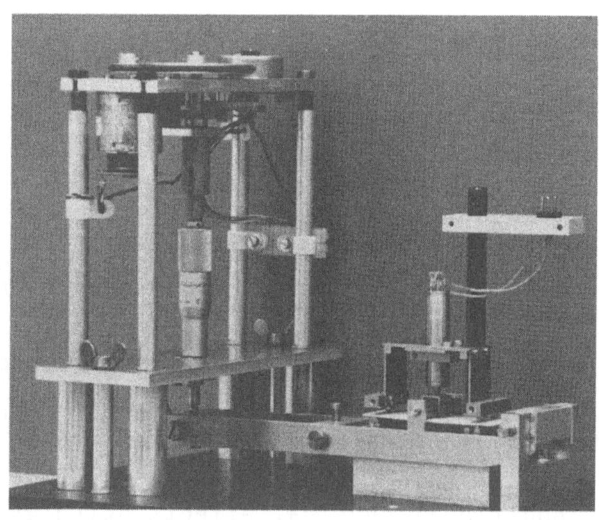

Fig. 2:

Bending beam measuring setup (center) with computer-controlled motor drive (left). For measurement, the apparatus is to be housed in a double-wall heat isolating cabinet, resting on a shock-absorbing table.

This radical change of sample geometry from a heavy homogenously strained column to a small bending beam, where positive and negative strain areas lie closely together, originally simply was caused by the increasing interest in nonmetal elastic elements. (These exotic materials are very hard to form into a large and complicated column.) But once introduced, the inhomogenous strain of the bending beam revealed a very important metrological advantage in that it renders the adiabatic temperature change effectless: This elastocaloric effect in an axially loaded column (typically 0,1 K) must be precisely determined to correct the measured ('adiabatic') creep for additional thermal expansion to get the true 'isothermal' creep. (As Fig. 3 shows, the strain field is not exactly homogenous, neither along the cylinder length nor in the radial direction; FEM analysis.) On the other hand the elastocaloric effect can be neglected in a thin bent beam, where thermal equilibrium is restored within a second by heat transport from the elongated zone (reduced temperature) to the compressed zone (where temperature is elevated), due to the large strain gradient normal to the stress direction.

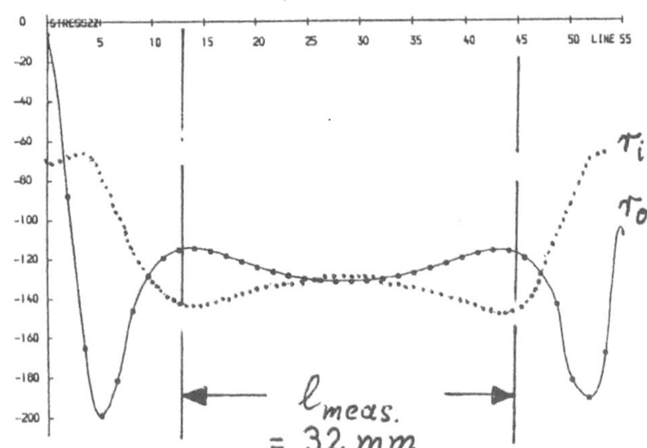

Fig. 3:
Longitudinal stress at the inner (r_i) and outer surface (r_o) of the axially loaded hollow column along its length. (FEM)

So, despite of its shaky and seismically sensitive character, the bending beam sample finally is to be preferred from the metrological point of view. Further, the bending beam is the typical elastic element for low forces due to its large complience. Unfortunately the agreement between creep data of the column (which experiments were carried on in addition) and the bending beam sample form were only fair, those gained from bending being the larger. This observation led to the discovery of the strong effect of creep enhancement by strain gradients.

2. Creep measurement results

The following creep curves are the result of a computer averaging: Each creep experiment was carried out up to 30 times under identical conditions to reduce noise and seismic interference of the individual measuring points. In Fig. 4 creep of numerous materials is compiled. The thickness of all samples was 1 mm. Some of these materials showed a considerable improvement, i.e. reduction of aftereffect, when submitted to a preaging by repeated deformation reversal. Such typical 'elastic hardening' is shown in Fig. 5. A number of further 'elastic hardening' data are compiled in Table 1, where the residual strain ε_1' (one minute after the abrupt deformation step from ε_0 to zero) is given as a characteristic figure of creep.

$$\frac{\varepsilon_1'}{\varepsilon_0} \cdot 10^3$$

	a) green sample	b) after 10^2 load reversals
CuBe2	0,1	0,04
Al_2O_3 (ceramic)	0,5	0,3 (!)
'Thermelast'	0,1	0,1
Zr	3	0,2
Bak-50	0,13	0,13
Zerodur	4	4

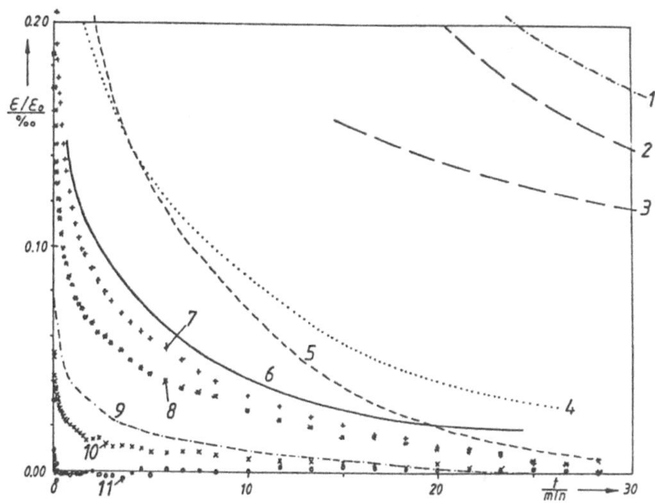

Fig. 4: Final creep of bending beams (t = 1 mm). ($\varepsilon_0 = 3 \cdot 10^{-4}$)
1. Zerodur (glass ceramic)
2. Soft glass Bk-7 (quickly cooled)
3. (slowly annealed)
4. Boron-glass 'Duran'
5. Ceramic Al_2O_3 (99%; 96%)
6. Hard glass SK-16 (annealed)
7. High-melting-point glass BaK-50
8. Fused quartz 'Homosil'; Fe-Ni-alloy 'Thermelast'
9. CuBe2
10. Silicon/Germanium
11. Sapphire (60°)

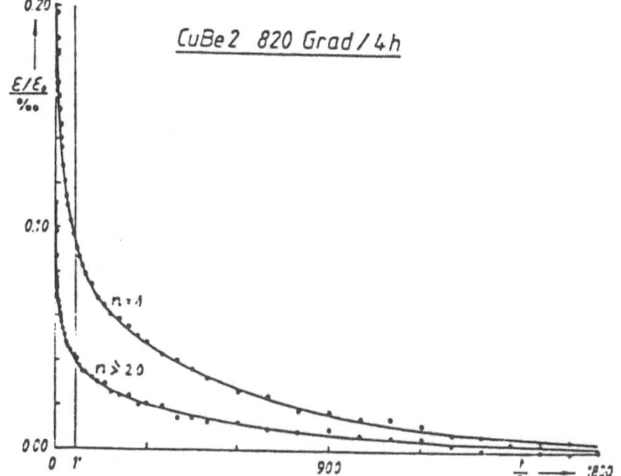

Fig. 5:

Reduction of creep by repeated load reversal. ($n \geq 20 \rightarrow$ final, 'stationary' creep).
Strain amplitude:
$\varepsilon_0 = 3 \cdot 10^{-4}$

Those creep curves depicted in Fig. 4 show the final (stationary) aftereffect ('b' in Table 1). The superior elastic material obviously is single-crystalline aluminumoxide 'sapphire' see also Fig. 6. (The crystallographic direction of cut seems to be without influence on the elastic behaviour.)

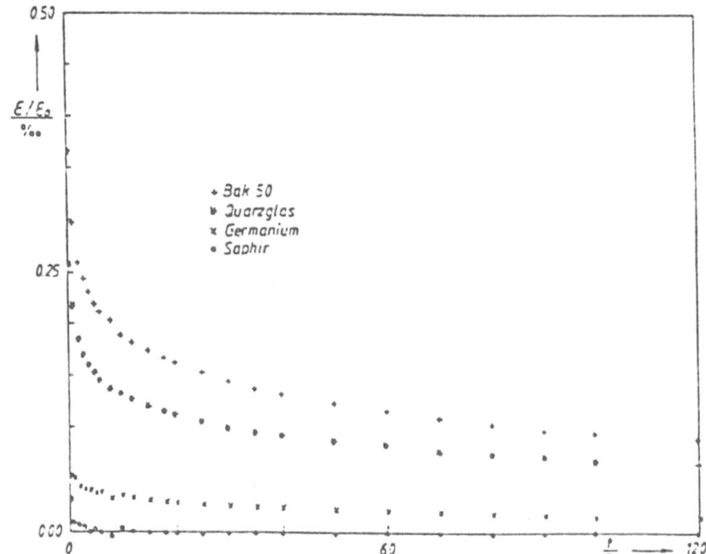

Fig. 6:
Short term creep of nonmetals

Reduction of bending beam thickness

As mentioned above, there is a considerable influence of bending beam thickness t on the absolute value of aftereffect. This enhancement of creep by strain gradients has been undoubtedly observed in fused quartz. Very recently it has been thoroughly investigated in copper-beryllium: All samples were cut from one block by low-power wire spark erosion. Initial strain ε_0 always amounted $3 \cdot 10^{-4}$. The results of this study are shown in Fig. 7, revealing an enormous increase of creep for the thin elastic elements. Finally in Fig. 8 the standard figure of creep $\varepsilon_{15}/\varepsilon_0$ is plotted versus the inverse of the beam thickness. The expected behaviour , i.e. an additional creep being proportional to the strain gradient (which is proportional to the inverse of the thickness t) approximately could be extracted from this plot. This observed exaggeration of creep by a strain gradient can be simply understood: All aftereffects in the elastic stress range are caused by diffusion, which in the case of homogenous strain is driven by the uniform lattice deformation causing non-equilibrium of atomic localisation in dimensions of the crystalline unit cell. In polycrystalline matter, on a microscopic scale strain will no longer be homogenous, giving rise to stress-induced atomic redistributions over larger distances up to grain size dimensions. In the case of a thin bending beam, the strain gradient drives a further diffusion - now transverse to the main stress direction. Thus the additional aftereffect of the bending beam must reflect the steepness of non-equilibrium (strain gradient), it further will show a large time constant, due to the range character of diffusion.

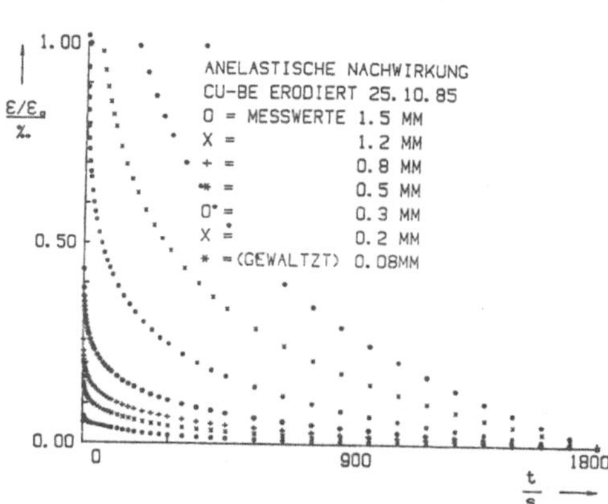

Fig. 7:

Aftereffect of bending beams of CuBe2; Parameter: thickness t

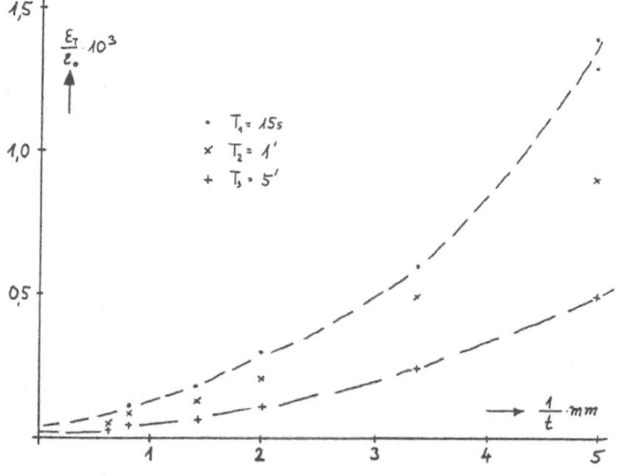

Fig. 8:

Creep figure vs. inverse of thickness of bending beams, taken at 3 different times T_n

Resumée: Thin bending beams show a strong increase of creep effects exceeding the homogenously strained aftereffects by a factor of 10, typically.

3. All-glass capacitive load cell

As seen from Fig. 4, the alkali-free optical glass Bak-50 shows an acceptable creep behaviour. A glass being very attractive from the fabricational point of view in that it allows cheap viscous forming of even complicated bodies, flat load cells for the 1 to 10 kg range were designed and tested. Fig. 9 shows a sketch of a differential capacitive cell, the two elements B and B' acting as bending beams, whilest two flat rectangular plates D and D - made from the same glass - form the static counter electrodes for the moving middle electrode of rhombic shape (E_M). (For further details see [2] .)

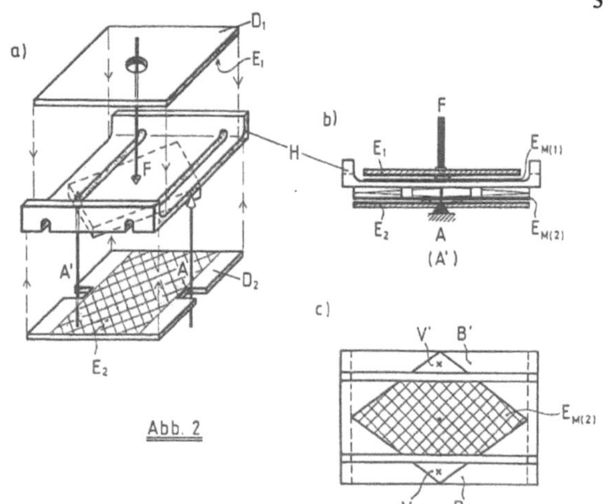

Fig. 9:

Sketch of a differential capacitance load cell made of glass

Finally Fig. 10 shows the creep of such a load cell made from hard-glass Bak-50, which indeed closely fits aftereffect data of Fig. 4. Further a load cell of same design has been fabricated from fused quartz, again yielding the creep as predicted by Fig. 4.
(It is no secret, that such time dependent aftereffects can be compensated electronically to a certain degree).

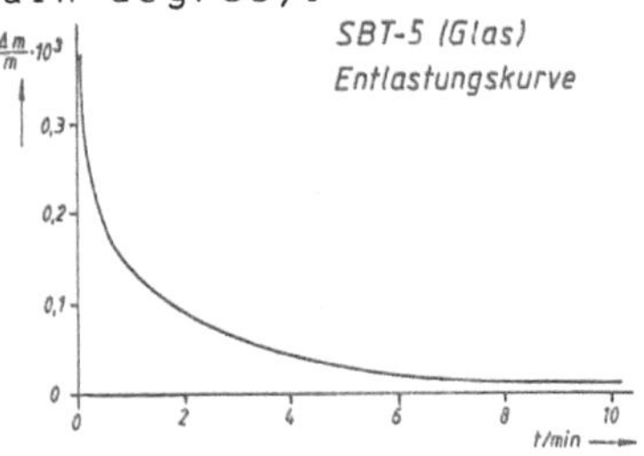

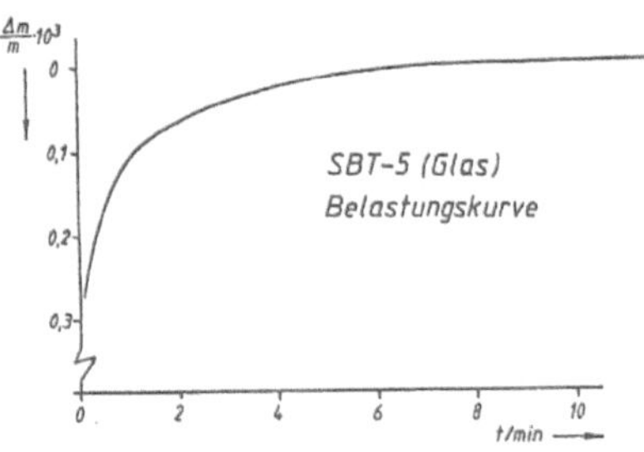

Fig. 10:

Capacitive load cell, made from hard-glass: Creep upon unloading and loading, respectively

References: Bethe, K. and Germer, W.: Creep and Adiabatic Temperature Effects in Elastic Materials, Proc. 8th Conference of the IMEKO, TC 3, Krakow 1980, p. 101

Bethe, K. and Baumgarten, D.: Möglichkeiten und Probleme nichtmetallischer Federwerkstoffe in Kraft- oder Druckaufnehmern, Proc. NTG/GMR-Fachtagung "Sensoren", Bad Nauheim 1986

INDEX